# NUCLEAR POWER'S GLOBAL EXPANSION: WEIGHING ITS COSTS AND RISKS

Henry Sokolski
Editor

December 2010

\*\*\*\*\*

Comments pertaining to this report are invited and should be forwarded to: Director, Strategic Studies Institute, U.S. Army War College, 122 Forbes Ave, Carlisle, PA 17013-5244.

\*\*\*\*\*

All Strategic Studies Institute (SSI) publications may be downloaded free of charge from the SSI website. Hard copies of this report may also be obtained free of charge by placing an order on the SSI website. The SSI website address is: *www.StrategicStudiesInstitute.army.mil*.

\*\*\*\*\*

The Strategic Studies Institute publishes a monthly e-mail newsletter to update the national security community on the research of our analysts, recent and forthcoming publications, and upcoming conferences sponsored by the Institute. Each newsletter also provides a strategic commentary by one of our research analysts. If you are interested in receiving this newsletter, please subscribe on the SSI website at *www.StrategicStudiesInstitute.army.mil/newsletter/*.

ISBN 1-58487-478-3

# CONTENTS

# FOREWORD

This volume consists of research that the Nonproliferation Policy Education Center (NPEC) commissioned in 2007 and 2008. The work was critiqued at a set of international conferences held in 2009 at the offices of the Carnegie Corporation of New York in New York City; at Chatham House in London, England; at the headquarters of Radio Liberty Europe in Prague, the Czech Republic; and at the Carnegie Endowment for International Peace in Washington, DC. Anthony Froggatt, Stephen Del Rosso, Carl Robichaud, Jeffrey Gedmin, and George Perkovich generously served as co-hosts for these events. Funding for the project came from the Carnegie Corporation of New York, Ploughshares Fund, and other charitable foundations. NPEC's project manager, Tamara Mitchell, and research intern, Kristen M. Gehringer, and the staff of the U.S. Army War College's Strategic Studies Institute (SSI) helped prepare the book manuscript; without their help, the book would not have been possible. Finally, heartfelt thanks are due to the project's authors and participants who contributed their time and ideas.

HENRY SOKOLSKI
Executive Director
The Nonproliferation Policy
Education Center

# OVERVIEW

# CHAPTER 1

## NUCLEAR POWER, ENERGY MARKETS, AND PROLIFERATION

### Henry Sokolski

## OVERVIEW

When security and arms control analysts list what has helped keep nuclear weapons technologies from spreading, energy economics is rarely, if ever, mentioned. Yet, large civilian nuclear energy programs can—and have—brought states quite a way towards developing nuclear weapons;[1] and it has been market economics, more than any other force, that has kept most states from starting or completing these programs. Since the early 1950s, every major government in the Western Hemisphere, Asia, the Middle East, and Europe has been drawn to atomic power's allure, only to have market realities prevent most of their nuclear investment plans from being fully realized.

With any luck, this past may be our future. Certainly, if nuclear power programs continue to be as difficult and expensive to complete as they have been compared to their nonnuclear alternatives, only additional government support and public spending will be able to save them. In this case, one needs to ask why governments would bother, especially in light of the security risks that would inevitably arise with nuclear power's further proliferation. On the other hand, if nuclear power evolves into the quickest and least expensive way to produce electricity while abating carbon emissions, little short of a nuclear explosion traceable to a "peaceful" nuclear facility is likely to stem this technology's further spread—no matter what its security risks might be.

Adam Smith's Invisible Hand, then, could well determine just how far civilian nuclear energy expands and how much attention its attendant security risks will receive. Certainly, if nuclear power's economics remain negative, diplomats and policymakers could leverage this point, work to limit legitimate nuclear commerce to what is economically competitive, and so gain a powerful tool to help limit nuclear proliferation. If nuclear power finally breaks from its past and becomes the cheapest of clean technologies in market competitions against its alternatives, though, it is unlikely that diplomats and policymakers will be anywhere near as able or willing to prevent insecure or hostile states from developing nuclear energy programs, even if these help them make atomic weapons.

What follows is a deeper explication of these points. The first section, "Costs," examines what the economics for nuclear power have been and are projected to be. The second, "Justifications," examines the environmental, energy security, and political reasons why nuclear power's relatively poor economic performance has been downplayed. The third section, "Concerns," explores the reasons why continuing with such downsizing is risky, and the final section "Economics: A Way Out," examines how market economic competitions could be used to help steer us towards cheaper, safer forms of energy.

## COSTS

### Nuclear Power's Past.

In the early 1950s, U.S. Atomic Energy Commission Chairman Lewis Strauss trumpeted the prospect of nuclear electricity "too cheap to meter."[2] An international competition, orchestrated under President Dwight D. Eisenhower's Atoms for Peace Program, ensued between the United States, Russia, India, Japan, and much of Western Europe to develop commercial reactors. Several reactor and nuclear fuel plants were designed and built, endless amounts of technology declassified and shared world-wide with thousands of technicians, and numerous research reactors exported in the 1950s. Yet ultimately relatively cheap and abundant oil and coal assured that only a handful of large power plants were actually built.[3]

The next drive for nuclear power came in the late 1960s just before the energy "crisis" of the early 1970s. President Richard Nixon, in announcing his "Project Independence," insisted that expanding commercial nuclear energy was crucial to reducing U.S. and allied dependence on Middle Eastern oil.[4] France, Japan, and Germany, meanwhile, expanded their nuclear power construction programs in a similar push to establish energy independence. The United States, Russia, Germany, and France also promoted nuclear power exports at the same time. Four thousand nuclear power plants were to be brought on line world-wide by the year 2000.

But, market forces, coupled with adverse nuclear power plant operating experience, pushed back. As nuclear power plant operations went awry (e.g., fuel cladding failures, cracking pipes, fires, and ultimately the Three Mile Island incident), spiraling nuclear construction costs and delays, as well as the disastrous accident at Chernobyl, killed the dream. More than half the U.S. nuclear plant orders were cancelled, and almost 90 percent of the projected plants globally — including a surprisingly large number of proposed projects in the Middle East — were never built.[5]

## Nuclear Power's Projected Future.

Today, a third wave of nuclear power promotion is underway, buoyed by international interest in reducing greenhouse gas emissions and national concerns in enhancing energy security at least as measured in terms of reliance on oil. The U.S. nuclear industry has been lobbying Congress to finance the construction of more than $100 billion in reactors with federal loan guarantees.[6] President Barack Obama has responded by proposing $36 billion dollars in new federal loan guarantees for nuclear power.[7] Other governments in Asia, the Middle East, and Latin America have renewed their plans for reactor construction as well. Even Europe is reconsidering its post-Chernobyl ambivalence toward nuclear power: Finland, France, Italy, and Eastern Europe are again either building or planning to build power reactor projects of their own. Germany and Sweden, meanwhile, are reconsidering their planned shutdown of existing reactors.

In all this, the hands of government are evident. Certainly, if nuclear power were ever truly too cheap to meter and could assure energy security, or eliminate greenhouse gas emissions economically, private investors would be clamoring to bid on nuclear power projects without governmental financial incentives. So far, though, private investors have avoided putting any of their own capital at risk. Why? They fear nuclear energy's future will echo its past. In the 1970s and 1980s, new nuclear power projects ran so far behind schedule and over budget, most of the ordered plants had to be cancelled. Even those that reached completion were financial losers for their original utility and outside investors, and the banking sector became wary.

In this regard, little has changed. In Finland, a turnkey reactor project has been executed by French manufacturer AREVA, in part as a way to demonstrate just how inexpensively and quickly new nuclear plants could be built. The project is now more than 3 years behind schedule and at least 80 percent over budget. Finland says AREVA is to blame for the cost overruns and construction delays. AREVA blames Finland and has threatened to suspend construction entirely in hopes of securing a more favorable rate of return.[8]

Meanwhile, in Canada, the government of Ontario chose to avoid this fate. It put its nuclear plans to build two large power plants on hold after receiving a $26 billion bid that was nearly four times higher than the $7 billion the government originally set aside for the project only 2 years before.[9]

In the United States, the estimated cost of two reactors that Toshiba was planning to build for NRG Energy and the city of San Antonio recently jumped from $14 billion to $17 billion. Consequently, the city board delayed its approval of $400 million in financing for the project, sued NRG, and reduced its share of the project from roughly 50 percent to less than 8 percent.[10] These estimates of the full costs to bring a new nuclear plant on line reflect a typical pattern of cost escalation, as San Antonio's experience has been replicated in many other places. Estimated construction costs (exclusive of financing) per one installed kilowatt of production capacity have jumped from a little over 1,000 dollars in 2002 to well over $7,000 in 2009. (Figure 1-1 depicts the range of rising "overnight" cost estimates over the last decade, with no interest on costs paid during construction, thus "overnight.")

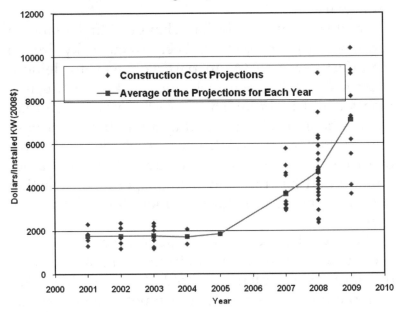

**Figure 1-1. Overnight Capital Costs Projections for New Power Reactors (2008 dollars/installed KW) — High and Rising.[11]**

To address these concerns, the U.S. nuclear industry has succeeded in getting Congress to support a growing number of subsidies, including nuclear energy-production tax credits and very large federal loan guarantees. Industry estimates indicate that proposed loan guarantees alone would save an American utility at least $13 billion over 30 years in the financing of a single modern nuclear reactor.[12] Granting these and additional government incentives, though, may not be sufficient. First, in 2003, the U.S. Congressional Budget Office (CBO) estimated that the nuclear industry would probably be forced to default on nearly 50 percent of these loans.[13] Second, in 2009, Moody's warned that barring a dramatic positive change in utility-industry balance sheets, the ratings firm would downgrade any power provider that invested in new nuclear reactor construction since these projects were "bet the farm" gambles. Moody's threat to reduce credit ratings included utilities that might secure federal loan guarantees, which Moody's described as too "conditional" to be relied on.[14]

Meanwhile, the president of America's largest fleet of nuclear power plants, who now serves as the World Nuclear Association's Vice Chairman, publicly cautioned that investing in new nuclear generating capacity would not make sense until natural gas prices rise and stay above $8 per 1,000 cubic feet (mcf) *and* until carbon prices plus taxes rise and stay above $25 a ton.[15] Yet industry officials believe that neither condition, much less both, is likely to be met any time soon. Past price history suggests why. (See Figure 1-2.)

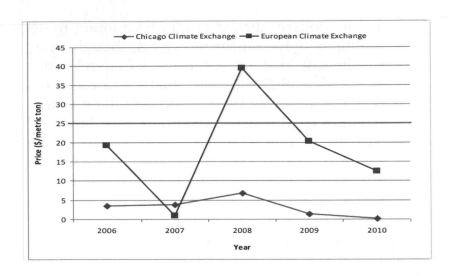

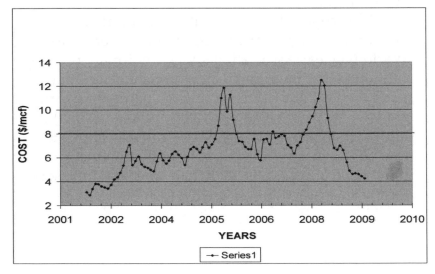

**Figure 1-2: Natural Gas and Carbon Prices —
Hardly Steady or High Enough to Underwrite
Private Nuclear Investments.[16]**

Recent developments suggest their skepticism is warranted. After the latest international conference to control carbon emissions held in December 2009 in

Copenhagen, Denmark, carbon prices in the European carbon market hit a near all-time low. U.S. natural gas prices, meanwhile, driven by reduced demand and massive increases in supplies and newly discovered reserves, have also dropped precipitously. There is good reason to believe that they are unlikely to rise significantly any time soon.[17] Conclusion: Without significant additional government financial incentives, private investments in new nuclear electricity are unlikely to be made.[18]

## JUSTIFICATIONS

### Energy Security.

Many decisionmakers in the energy sector understand this. This, in turn, has given rise to a public focus on another, less measurable but possible nuclear power benefit: Energy security. The case here, though, is also yet to be proven. In most large industrial countries, oil is only rarely used to produce electricity, but rather is being consumed at increasing rates to fuel a growing fleet of cars and trucks. This makes the link between oil imports and nuclear power quite tenuous at present. The nuclear-vice-oil argument put forth by some experts is future oriented: Some day nuclear power might supply the electricity and hydrogen to power the world's transport fleets. However, for both electric and hydrogen vehicles, much is still unknown about the costs, rate of market penetration, and even whether nuclear will prove to be the most economical way to produce the needed energy resources.

Unfortunately, few of these central issues are given serious attention in the popular news media. Instead, France, which made a massive investment in nuclear power in the 1970s and now produces about 80 percent of its electricity from nuclear energy, is held up as an energy-independence model for the United States and the world to emulate.[19] This particular nuclear example, however, has been quite costly and has not really weaned France away from its addiction to oil. France covered much of the startup and operating cost of its civilian nuclear program by initially integrating the civilian nuclear power sector with its military nuclear-weapons-production program. It also used massive amounts of cheap French government financing to pay for the program's capital construction. As a result, it is unclear how much in real francs the French program actually cost overall, or how much plant costs escalated over the life of the French program, although they clearly did.[20] What is undisputed, however, is that from the 1970s to the present, France's per-capita rate of oil consumption never declined; and that the country has needed to import increasing amounts of expensive peak-load electricity from its immediate neighbors due to the supply inflexibility of base-load nuclear electricity.[21] Despite these facts, though, the claim of French nuclear energy independence persists.

**Abating Carbon Emissions.**

Another argument nuclear power supporters frequently make is that the need to abate carbon emissions will make nuclear energy economically competitive through rising carbon prices. Once carbon is no longer cheap, nuclear proponents argue that zero-carbon emission nuclear power plants will be the clear, clean-energy victor over coal with carbon capture systems, natural gas, and renewables. Yet, by industry's own projections, new nuclear power plants have already priced themselves out of the running in any near or mid-term carbon abatement competition. Factoring industry construction cost projections, operation and decommissioning costs, and key public nuclear-specific U.S. subsidies, the total cost of abating one ton of carbon dioxide ($CO_2$) by substituting a new nuclear power plant for a modern coal-fired generator has been pegged by a leading environmental energy economist at least $120. This figure assumes fairly low capital construction costs (roughly one-half of the industry's latest high-end cost projections). If one uses industry's high-end projections, the cost for each abated ton of $CO_2$ approaches $200. This is expensive. Certainly, there currently are much cheaper and quicker ways to reduce carbon emissions (see Figure 1-3).

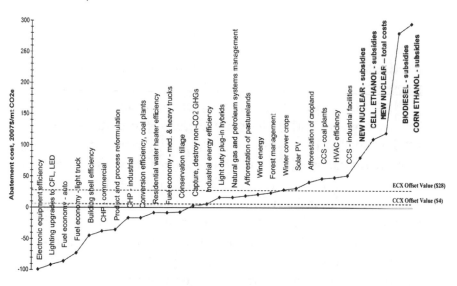

**Figure 1-3. New Nuclear Power:**
**An Expensive Way to Abate Carbon.**[22]

13

Yet another recent study completed by one of America's largest nuclear reactor operators, Exelon, confirms these points. Speaking before a May 12, 2010, Washington, DC, Resources for the Future policy leadership forum, John Rowe, Chairman and CEO of Exelon, presented analysis that clarified how expensive a new nuclear plant might be. As his central and final power point slide make clear (see Figure 1-4), carbon prices would have to rise to roughly $100 a ton of $CO_2$ before he would recommend that Exelon invest in building new power reactors. Even with federal loan guarantees, Exelon's analysis determined that carbon would have to be priced at roughly $75 per ton of $CO_2$ (which is nearly twice Exelon's projected "long-run $CO_2$ price" of $40 a ton) before it would make economic sense to build new power reactors. Before Excelon would invest in new nuclear construction, it would update its existing 19 nuclear plants, shut down its coal-fired generating stations, bring more natural gas-fired plants on line, and invest in energy efficiency programs and renewables.

Just how rapidly a nuclear approach can begin abating carbon emissions (compared to its alternatives) is also a significant issue. Certainly, if one is interested in abating carbon in the quickest, least expensive fashion, building expensive nuclear plants that take up to a decade to bring on line will have difficulty abating carbon competitively no matter how much carbon is taxed. That is why in North and South America and the Middle East, building natural gas-burning generators is currently an attractive, near-term option. Advanced gas-fired power plants can halve carbon emissions as compared to coal-fired plants, can serve as base or peak power generators, and be brought on line in 18 to 30 months rather than the 5 to 10 years needed to build large reactors. Advanced gas-fired generator

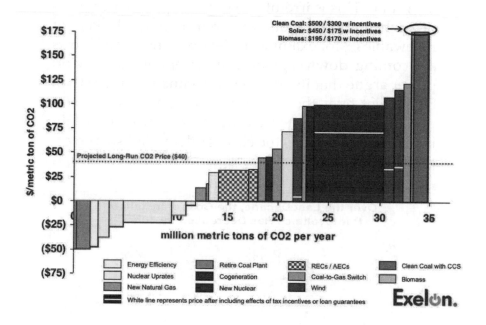

Figure 1-4. Exelon's View of Carbon Abatement
Options for 2010.[23]

construction costs, meanwhile, are a fraction of those projected for nuclear power.[24]

Where natural gas is plentiful, as it clearly already is in the Middle East and the United States, these economic facts should matter.[25] The benefits of gas become even more evident once one factors in the nuclear-specific burdens for nations with no current capacity to create proper regulatory agencies and prepare the grid for a large base-load generator.[26]

## A Future Unlike Our Past?

The counter to the foregoing argument, of course, is that fossil fuel resources are finite and will run out over time. This is irrefutable in principle, but, in practice, when and how one runs out matters. Backers of renewables,[27] for example, insist that renewables' costs are coming down significantly. Proponents of wind power argue that its costs have declined by more than 80 percent over the last 20 years.[28] The cost of solar photovoltaic-generated electricity has also been falling (see, for example, the costs of delivered solar electricity in Figure 1-5).

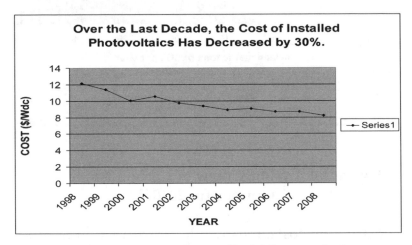

**Figure 1-5. Cost of Installed Photovoltaics**.

Many energy experts contend that significant changes would have to be made in how electricity is currently distributed and stored before intermittent generators like renewables could compete in addressing base load demand. Yet, as renewables' costs continue to decline, the incentives needed to prompt these changes are likely to increase.[29] Meanwhile, nuclear power's costs are high and rising. Finally, with new sources of oil and gas now projected to come on line, it is unclear when or how much fossil fuel prices might increase. All of this generates significant uncertainty and risk for nuclear power investors.

In the mid-term ( i.e., the next 2 decades) when nuclear advocates see this power source reemerging, a number of energy developments could easily destroy whatever value might be credited to investments made in commercial nuclear energy today. As noted, new electrical grid concepts could be employed incrementally to make the transmission of intermittent wind and solar much more practical, as could the development of practical electrical storage and of viable distributed electrical systems.[30] Economical sequestration of carbon from coal-fired plants also may emerge along with increased efficiency in the use of electricity and smart metering that could change and reduce demand patterns.

Although none of these developments are guaranteed, any one of them could have a dramatic impact on the long-term economic viability of presently investing in nuclear systems that would operate for 60 years or more after coming on line in 2020 and beyond. In fact, the uncertainties surrounding the costs for electricity generation, distribution, transmission, storage, and consumption are all very much in play for the first time in over a century. This very fluid and uncertain situation not only argues for great caution in the allocation of public funds on any energy commercialization project, but also underscores the importance of ensuring neutral markets in which multiple solutions are forced to compete against each other.

**Government Nuclear Power.**

Governments, on the other hand, view matters differently. The energy market uncertainties noted above have only encouraged them to invest more in clean energy commercialization options. In practice, this has meant they have invested most heavily in the most capital intensive options. Thus, the current carbon and energy security challenges have been addressed by Japan, South Korea, China, India, Russia, France, and the United States by initiating investments in carbon sequestration and renewables. More important, each of these governments has continued and significantly increased massive government subsidies—e.g., loan guarantees, commercial export loans, energy production credits, accident liability caps and indemnification, and construction delay insurance programs—for the construction of new, large nuclear power plants.[31]

Several factors fortify these governments' instinct to support nuclear commercialization. First, in several important cases—e.g., in France, Russia, India, South Korea, and Japan—the nuclear industry's payrolls have long been large and are essentially public: Commercial nuclear activities in these states are run through entities that are primarily government-owned. Exposing these industries to the full force of market realities could result in significant layoffs and other dislocations large enough to produce negative political results. Continuing to subsidize them, on the other hand, is politically astute.

Second and less immediate, commercial nuclear power's historical links to national security continue to make government support seem natural. Within the oldest and most significant nuclear states—the United States, the United Kingdom (UK), France, Russia, and India—government-run, dual-use reactors were long connected to electrical grids to produce nuclear weapon fuels and electricity. In the United States, this includes the Hanford dual-purpose reactor in Washington State (no longer operating), and the Tennessee Valley Authority's tritium-producing light water reactors (whose operations are about to be expanded). It

includes Russia's RMBK reactors, which made plutonium for Russia's arsenal until the 1990s; France's gas-
cooled natural uranium and breeder reactors, which
did the same plutonium chore for France through
the 1980s; India's heavy water reactors and planned
breeder reactors, which currently provide tritium and
plutonium for India's nuclear weapons program; and
Britain's Magnox plants, which provided the bulk of
the plutonium for the UK nuclear arsenal. As for the
most popular of nuclear power systems, pressurized
light-water reactors (versions of which Germany,
France, Russia, Japan, South Korea all now export and
operate), these were originally developed in the United States for nuclear submarine and naval propulsion.

This rich history of defense-related government
involvement in nuclear energy has made the new
government financial incentives to promote the construction of additional nuclear power and fuel making
plants seem normal. Yet, pushing such government
support of energy commercialization projects, both
nuclear *and* non-nuclear, actually flies in the face of
what market forces would otherwise dictate. More
important, it hides the full costs and risks associated
with each energy option. This, in turn, is undesirable
for several reasons.

## CONCERNS

### Commercial Energy Innovation.

Conventional wisdom holds that government
subsidies to commercialize technology optimize and
catalyze commercial energy modernization. In reality,
subsidy policies are politically challenging to implement. Not surprisingly, those that do make it into law

most often support the more established and powerful players in the market independent of technical merit. As such, government promulgation of energy commercialization subsidies makes it *more* difficult for winning ideas to emerge or prevail against large-scale losers, and this difficulty can increase over time. The reason why is simple: Once government officials make a significant financial commitment to a commercial project, it becomes politically difficult for them to admit it might be losing money or that it was a mistake to have supported it, even when such conclusions are economically clear. A "lock-in" effect begins to take hold: Not only will governments not terminate funding to clear losers; they may actually shore up such projects with additional funding or legal mandates to force the public to buy the project's commercial production even when cheaper alternatives clearly exist.[32]

Thus, it was evident to most that the U.S. Government's commercial synfuels and breeder reactor projects were economically untenable years before Congress finally decided to kill the projects. The delay in terminating these projects cost taxpayers billions of dollars. These projects, though, at least died. With government mandated energy commercialization programs such as corn ethanol, however, the U.S. Government has essentially mandated that the product be produced and bought by the public in increasing amounts in the face of little or no market demand. Besides costing U.S. consumers billions of dollars annually, this program is becoming institutionalized in such a manner as to make it more difficult to phase-out or terminate in the future. In France, Japan, China, Russia, Korea, and India, where the power of the government in commercial matters is stronger, this tendency is even more pronounced.

## Nuclear Safety and Off-site Damage.

With nuclear-specific energy commercialization subsidies, such as low-priced nuclear accident liability insurance, private sector incentives that would otherwise improve operational and design safety also take a hit. Under U.S. law, U.S. commercial nuclear reactor operators (about 100 in number) must secure private insurance sufficient to cover roughly the first $300 million of damages any nuclear accident might inflict on third parties off site. After any accident, the law provides that each nuclear utility should also pay up to approximately $96 million per reactor in annual installments of $15 million each (plus a bit more earmarked for legal fees) should the first-tier policy be exceeded. This requirement, however, can be delayed or waived entirely by the Secretary of Energy if, in his judgment, it would threaten the financial stability of the firm paying it. These retrospective premiums are paid as a blanket requirement: They are virtually identical for both the safest and worst run utilities.[33]

By most accounts, such pooling and the capping of liability lessens the cost of nuclear insurance significantly to the nuclear industry as a whole.[34] A key argument for such pooling and liability caps is that it is unreasonable to ask the nuclear industry to assume the full costs of insuring against nuclear accidents and nuclear terrorism because these risks are simply too large.[35] This certainly has been the logic behind the passage of the U.S. Terrorism Risk Insurance Act of 2002 and its repeated extension.[36] Yet, these acts are claimed by their backers to be only "temporary," i.e., designed to allow private insurers the time to adjust to a new risk market.

As both the CBO and the U.S. Treasury Department have argued, capping private firms' need to insure against catastrophic losses make sense only if the risks of such losses are very low and unlikely to persist. In such cases, federal subsidies for insurance "could be justified as a means of avoiding expensive and unnecessary effort to reduce losses." If, as is more likely, in the case of nuclear safety and vulnerability to terrorist attacks, the long-term risks are either long-lived or, in view of the September 11, 2001 (9/11) attacks and the aging of the existing reactor fleet, likely to increase,[37] such federal "assistance" could be "costly to the economy because it could further delay owners of assets from making adjustments to mitigate their risk and reduce potential loses."[38] Here, it is worth noting that neither General Electric nor Westinghouse has yet succeeded in producing a reactor design that can meet the Nuclear Regulatory Commission's latest requirement that the plant be able to sustain a large, direct airplane hit. Westinghouse's latest submission to meet this requirement was actually found to be wanting and was rejected because it created unintended vulnerabilities to natural disasters such as earthquakes.[39]

Unfortunately, regarding reactor safety and the continued need for insurance liability caps, the U.S. nuclear industry has been increasingly schizophrenic. Originally, in 1957 when the nuclear industry first secured legislation capping its nuclear accident liability for damages suffered by third parties, it claimed that it only needed the protection until utilities had a chance to demonstrate nuclear power's safety record, i.e., until 1967. A half century later, though, industry officials pleaded with Congress that without another 20-year extension, commercial nuclear power would die. They also insisted that they were still unwilling to export

U.S. nuclear goods to foreign states that have not yet explicitly absolved nuclear vendors from liability for damages off-site third parties might suffer in the case of an accident.[40]

The world's nuclear future, however, is supposed to be better. Industry backers of the latest reactor designs claim that their new plant machinery will be dramatically safer than that currently operating and argue that government accident insurance caps could be phased out.[41] Certainly, industry arguments against even higher coverage requirements under their Price-Anderson coverage seem implausible. The U.S. nuclear industry is already more than willing to pay for insurance to cover damages to their own nuclear assets. In fact, for a single power plant location, most nuclear utilities are buying over 10 times the amount of insurance to protect against on-site accident damage and forced outages than Price-Anderson requires them to carry against off-site property and health damages for the entire United States. At a minimum, this suggests that the insurers and utilities are able to provide substantially more than the $300 million in primary coverage for off-site accidents that they currently must purchase by law. Finally, several U.S. nuclear reactor vendors rely heavily upon taxpayer appropriations to help pay for their advanced "safer" commercial reactor designs. These "accident-resistant" reactors are precisely the ones that industry says will come on line by 2025, the date the current nuclear insurance liability limits under Price-Anderson legislation will run out.

Though U.S. nuclear liability coverage seems quite inadequate, it is regrettably even worse abroad. Within Europe—the second largest nuclear-powered region in the world—nuclear accident insurance requirements are not just inadequate, but egregiously inconsistent. Thus, nuclear accident insurance requirements that are much lower in Eastern Europe than in the European Union (EU) currently are encouraging reactor construction in states with the least stringent liability requirements and some of the weakest nuclear safety regulatory standards. Because of this worry, some experts are now arguing that the EU should adopt a nuclear insurance pooling scheme at least as tough as that in the United States. To avoid the potential problem of allowing the pool to charge too little, they argue that the pool should require higher payments than in the United States. Yet, they note that securing any *uniform* insurance requirement would be better than continuing to have none. [42]

**Proliferation.**

With commercial nuclear energy projects, especially those exported overseas, there is a major additional worry—nuclear energy's link to nuclear weapons proliferation. Here, the security risks are real, particularly in the Middle East. Israel, the United States, Iran, and Iraq have launched aerial bombing or missile strikes against reactors at Osirak in Iraq and Bushehr in Iran, even though Iraq and Iran were members of the Nuclear Nonproliferation Treaty (NPT) and the attacked reactors were under International Atomic Energy Agency (IAEA) safeguards. If one includes the 2007 Israeli strike against Syria's reactor and Iraq's failed missile attack against Dimona during the first Gulf War, there have been no fewer than 13 acts of war directed against IAEA member state reactors.

Such facts should put a security premium on efforts to subsidize the construction of such projects both here and abroad. Certainly, the more the U.S. and other advanced economies go out of their way to use government financial incentives to promote the expansion of nuclear power programs domestically or overseas, the more difficult it is likely to be to dissuade developing nations from making similar investments. This dynamic will exist even if the nuclear projects in question are clearly uncompetitive with nonnuclear alternatives. Moreover, we should be trying to discourage subsidies that substantially assist these states to move closer to developing nuclear weapons options.

Consider Iran. The United States, perhaps more than any other country, was responsible for encouraging the Shah to develop nuclear power in the 1970s. Because the United States saw the Shah as a close ally, little thought was given to the potential security implications of our sharing advanced nuclear technology with Iran. When Iran's revolutionary government began to rebuild its Bushehr power station with Russian help, though, the United States rightly became concerned about the proliferation risks of this "peaceful" program.

Presidents Bill Clinton and George W. Bush warned that Bushehr could be used as a cover for illicit nuclear weapons related activities. It also was noted that once the reactor comes on line, it produces scores of bombs' worth of weapons-usable plutonium annually, which can be diverted to make bombs.[43] The fresh fuel, meanwhile, could be used to accelerate a uranium enrichment program.[44] It was because of these facts that during the first term of the Bush administration, the State Department went to great lengths to challenge the economic viability of the Iranian nuclear program as compared to burning plentiful natural gas. President Bush also insisted publicly that no new nuclear power state needed to make nuclear fuel to enjoy the benefits of nuclear power.[45]

In its second term, however, the Bush administration decided to add significant new nuclear subsidies to promote nuclear power plant construction in the United States under the Energy Policy Act of 2005, and to encourage an expansion of nuclear fuel-making with new technologies where it was already commercially underway. It was roughly during this period that the United States also decided to "grandfather" Bushehr and offered Iran power reactor assistance if it would only suspend its nuclear fuel-making program.

With this, the United States essentially forfeited its economic critique of Iran's power program. In July of 2007, U.S. President Bush and Russian President Vladimir Putin publicly recommended that international and regional development banks make cheap loans for civilian nuclear power programs.[46] The White House also began encouraging the development of nuclear power throughout the Middle East as a way to put the lie to Iran's claim that the United States and its partners were trying to deny all Muslims the "peaceful atom."[47] The economic merits of encouraging such nuclear power proliferation, as has already been noted, are dubious. Yet, Russia, France, South Korea, the United States, China, and India are nonetheless openly competing to secure contracts in the Middle East and beyond using a variety of government supported subsidies to drive down nuclear prices.

**Economics: A Way Out**

**Linking Economics with Security and the NPT.**

For observers and officials worried about the risks of nuclear power proliferation, merely arguing for governments to be more consistent and neutral economically in their selection of different power genera-

26

tion systems might seem cynically inattentive to the substantial security dangers posed by the expansion of nuclear power. Certainly, the United States and other states have oversold how well international nuclear inspections can prevent military diversions from civilian nuclear programs. Even today, the IAEA cannot yet reliably track spent or fresh fuel for roughly two-thirds of the sites it monitors. Worse, diversions of this material, which can be used as feed for nuclear weapon fuel-making plants, could be made without the IAEA necessarily detecting them.[48] As for large fuel-making plants, the IAEA acknowledges that it cannot reliably spot hidden facilities and annually loses track of many bombs' worth of material at declared plants. With new money and authority, the IAEA could perhaps track fresh and spent fuel better; however, the laws of physics are unfriendly to the agency's ever being able to reliably detect diversions from nuclear fuel-making plants.[49]

If international nuclear inspections cannot protect us against possible nuclear proliferation, though, what can? It would help if there were more candor about the limits of what nuclear inspections can reliably detect or prevent. But just as critical is more frankness about how little economic sense most new nuclear power programs make. It is governments and their publics, after all, which determine whether or not more large civilian energy plants will be built. If government officials and the public believe backing nuclear power is a good investment, public monies will be spent to build more plants in more countries no matter how dangerous or unsafeguardable they might be.

In this regard, it is useful to note that the NPT is dedicated to sharing the "benefits" of peaceful nuclear energy. These benefits presumably must be measurably "beneficial." At the very least, what nuclear activities and materials the NPT protects as being peaceful and beneficial ought not to be clearly dangerous and unprofitable. That, after all, is why under Articles I and V, the NPT bans the transfer of civilian nuclear explosives to nonweapons states and their development by nonweapons states. It is also is why the NPT's original 1968 offer of providing nuclear explosive services has never been acted upon and is a dead letter now: Not only was it determined that it was too costly to use nuclear explosives for civil engineering projects (the cost of clean-up was off the charts), but some states (e.g., Russia and India) falsely claimed they were developing peaceful nuclear explosives when, in fact, they were conducting nuclear weapons tests.[50]

What, then, should be protected under the NPT as being "peaceful" today? Are large nuclear programs economically competitive, i.e., "beneficial" in places like the Middle East when compared to making power with readily available natural gas? What of making enriched uranium fuel for one or a small number of reactors? Would it not be far cheaper simply to buy fresh fuel from other producers? Does reprocessing make economic sense anywhere? Can nuclear fuel making be reliably safeguarded to detect military diversions in a timely fashion? Are not such activities dangerously close to bomb making? Should these activities be allowed to be expanded in nonweapons states and to new locales or, like "peaceful" nuclear explosives, are the benefits of these programs so spurious and the activities in question so close to bomb making or testing as to put them outside the bounds of NPT protection? What of large reactors, which are fueled with large amounts of fresh enriched uranium or that produce large amounts of plutonium-laden spent fuel? Should these be viewed as being safeguardable in hostile or questionable states, such as Iran or North Korea, that have a record of breaking IAEA inspection rules?

Again, getting all of the world's nations to agree on the answers to these questions will be difficult if nuclear power is truly the least expensive way to produce low or no carbon emission power. In this case, it may be impossible to prevent nuclear technology useful to making bombs from spreading world-wide. But if civilian nuclear energy projects are not economically competitive against their nonnuclear alternatives, the case against states spending extra to promote the commercial expansion of potentially dangerous commercial nuclear projects would be far stronger.

**Uncertainties.**

The only thing certain about nuclear power's future ability to compete against other commercial energy alternatives is its uncertainty. This is so for several reasons. First, 20 years out, it is uncertain how much power will be distributed off a centralized grid and how much will come from more distributed systems (e.g., local grids, cogeneration plants, storage batteries, and the like). This is important since two-thirds of the cost of electricity at the house or business outlet is unrelated to the cost of generating the electricity. Instead, it pertains to the cost of transporting the electricity over the grid and balancing and conditioning the power inputs and outputs on that grid to assure that it does not fail.

Second, it is unclear how many base load generators will be needed 10 to 20 years out since so much of the current demand for electrical generating capacity in advanced economies is driven by the need to have instantly available follow-on load capacity that frequently remains idle.[51] If our experts could figure out how to store electricity economically (and a number of schemes are now being tried out), the current premium placed on having significant reserves of additional base load follow-on capacity generators — typically supplied by large coal-fired plants, large hydro, or nuclear reactors — could be reduced significantly.

Third, there is much uncertainty with respect to carbon charges on which nuclear economics heavily depend. Will carbon be taxed and, if so, at what rate? What sectors will be grandfathered; which sectors of the economy will benefit the most from the constraints? The EU has a cap and trade system that the U.S. Congress is considering emulating. Under this system, government authorities allocate carbon allowances to different industrial concerns and sectors. Initial grants of credits follow patterns of most subsidies, with some sectors — often the most politically powerful — benefiting far more than others. "Winners" under the new system shift from economic and technical performance to political.

All of this seems an odd way to promote cost-competitive clean energy. Instead, it would make more sense simply to focus on cost comparisons for future plants that incorporated the full value of government subsidies and reflected a standardized carbon cost (e.g., a price on the carbon content of different fuels). To foster the proper use of such information, though, we will need to rely more, not less, on market mechanisms to help guide our way.

**Policy Implications.**

Our broad conclusion is that governments should spend less time trying to determine what energy technologies should be commercialized and focus instead on how market mechanisms might best be employed to make these determinations possible. This, in turn, suggests six specific steps governments might consider:

1. Encourage more complete, routine comparisons of civilian nuclear energy's costs with its nonnuclear alternatives. The starting point for any rational commercial energy investment decision is a proper evaluation of the costs of selecting one option over another. Here, as already detailed, governments have a weak track record. A couple of mandates stand out:

a. *Account for Nuclear Power's Full Costs.* One way officials could improve their performance is to take the very few economic energy assessments now required more seriously, and conduct them faithfully and conscientiously. The CBO, for example, must score the public costs of guaranteeing commercial energy loans, including those to the U.S. nuclear industry. The CBO has been asked to do this by Congress several times in the last decade. Yet, the last time the CBO made the assessment for proposed loan guarantees in 2008, it failed to give a figure for the probable rate of default on nuclear projects. The CBO's director claims that without proprietary information, the CBO has no way to make such estimates. The CBO has not attempted such projections since 2003, when it pegged the likely default rate under proposed loan guarantee legislation at the time at 50 percent.[52] The Department of Energy (DoE), meanwhile, announced that essentially it also viewed the information necessary to project the default rate to be proprietary. It would be useful for the CBO to get the information it needs to update and qualify such projections. At a minimum, the CBO should tackle this question every time it estimates what any commercial energy loan guarantees will cost. Congress, meanwhile, should demand that DoE make all of its own estimates and information relating to these projections public. Moreover, every time the CBO or DoE makes such projections their work should be reviewed in public hearings before Congress.

b. *Compare Nuclear with Nonnuclear.* Yet another way the U.S. Government could improve its commercial energy cost comparisons is by finally implementing Title V of the Nuclear Nonproliferation Act of 1978, which calls on the Executive Branch to conduct energy assessments in cooperation with, and on behalf of, key developing states. The focus of this cooperation was to be on nonnuclear, nonfossil-fueled alternative sources of energy. Yet, for these cost assessments to have any currency, they would have to be compared with the full life-cycle costs of nuclear power and traditional energy sources. This work also should be supported by the newly proposed United Nations (UN) International Renewable Energy Agency (IRENA).[53] Finally, in order for any of these efforts to produce sound cost comparisons, more accurate tallies of what government energy subsidies are worth for each energy type will be required.

c. *Increase the Number of Energy Subsidy Economists.* The number of full-time energy subsidy economists is currently measured in the scores rather than in the hundreds. Government and privately funded fellowships, full-time positions, and the like may be necessary to increase these numbers.

2. Strengthen compliance with existing international energy understandings that call for internalizing the full costs of large energy projects, and for entering them in open international bidding competition. The Global Energy Charter for Sustainable Development, which the United States and many other states support, already calls on states to internalize as many of external costs (e.g., those associated with government subsidies and quantifiable environmental costs such as the probable taxes on carbon) in the pricing of large energy projects. Meanwhile, the Energy Charter Treaty, backed by the EU, calls on states to require any large energy project or transaction to compete in open international bidding markets.[54] Since these agreements were drafted, international interest in abating carbon emissions in the quickest, cheapest fashion has increased significantly. The only way to assure "the quickest and cheapest" is to include all the relevant government subsidies in the price of competing energy sources and technologies, assign a range of projected prices to carbon, and use these figures to determine what the lowest cost energy source or technology might be in relation to a specific time line. This suggests that any follow-on to the Kyoto, Japan, understandings should require international enforcement of such energy comparisons by at least referencing the principles laid out in the Energy Charter Treaty and the Global Energy Charter for Sustainable Development. Enforcing international adherence to these principles will be challenging. A good place to start would be to work with the G-20 (Argentina, Australia, Brazil, Canada. China, France, Germany, India, Indonesia, Italy, Japan, Mexico, Russia, Saudi Arabia, South Africa, Republic of Korea, Turkey, UK, the United States, and the EU) to agree to a modest

action plan to follow up on Copenhagen that would include establishing common energy project cost accounting and international bidding rules that track these agreements. Beyond this, it would be useful to call on the G-20 to give the IAEA notice of any state decisions the G-20 believes might violate these principles by rigging assessments to favor nuclear power over cheaper alternatives. The aim here would be to encourage the IAEA to ascertain the true purpose of such economically questionable nuclear projects.[55]

3. Discourage the use of government financial incentives to promote commercial nuclear power. This recommendation was made by the Congressional Commission on the Prevention of Weapons of Mass Destruction Proliferation and Terrorism.[56] It would clearly include discouraging new/additional federal loan guarantees for nuclear fuel or power plant construction of the type now being proposed by President Obama and the nuclear industry. Although this structure should be applied against other types of energy (e.g., coal, renewables, natural gas, etc.) as well, the security risks associated with the further spread of civilian nuclear energy make it especially salient in the case of nuclear. This same prohibition should also be applied against U.S. support for developmental bank loans (i.e., subsidized loans) for commercial nuclear development and against other states' (e.g., France, Japan, Germany, Russia, China, and South Korea) use of subsidized government financing to secure civilian nuclear exports. In some cases, these foreign export loan credits are being used in the United States in conjunction with U.S. federal loan guarantees and local state tax incentives, thereby practically eliminating the risks of investing in new nuclear power plant construction. This practice should be discouraged. In

the case of every large civilian nuclear project, domestic or foreign, every effort should be made to require as much private capital at risk as possible in order to assure due diligence in these projects' execution. Even under the existing U.S. federal loan guarantee program, 20 percent of each nuclear project must be financed without federal protection. For purposes of implementing this law, this nominal figure of 20 percent should be covered entirely by private investment, not by resorting to rate hikes for ratepayers.[57]

4. Employ more market mechanisms to guide national and international nuclear fuel cycle and waste management decisions. One of the clear advantages of civilian nuclear power plants over conventional fossil-fueled plants is that nuclear power is much cheaper to fuel. Governments, however, can negate this advantage by taking steps to increase nuclear fuel cycle costs that are unrelated to the need to assure safety or international security. In this regard, states that use public money to close the fuel cycle through commercializing any form of spent fuel recycling will actually make nuclear power less competitive with its non-nuclear alternatives. We should emphasize:

a. *Managing Nuclear Waste.* Today, the lowest-cost interim solution to storing spent fuel (active for 50 to several hundred years) is dry cask storage above ground at reactor sites. Recycling spent fuel, on the other hand, is not only more expensive, but runs much greater proliferation, terrorism, and nuclear theft risks. For these reasons, President Bush in 2004, the IAEA in 2005, and the bipartisan U.S. Congressional Commission on the Prevention of Weapons of Mass Destruction Proliferation and Terrorism in 2008, all called for imposing a moratorium on commercial reprocessing.[58] This reflects economic commonsense.

Unfortunately, in many advanced states that operate nuclear power reactors, the governments own and operate the power plants. As a result, full employment, development of nuclear weapons options, and other political or military concerns often override straightforward cost benefit analysis.[59] In the United States, this tendency can be avoided by having the nuclear utilities themselves assume a significant portion of the costs of nuclear waste management and reactor site decommissioning. This would require changing U.S. law, which currently stipulates that all of the costs of final spent fuel storage are to be paid for by off-budget federal user fees paid for by the ratepayers.

b. *Making Nuclear Fuel.* As for the front end of the nuclear fuel cycle, the preexisting procurement firm nuclear fuel contracts, rather than government appropriations or loan guarantees, should dictate when and how new nuclear fuel-making facilities should be constructed or expanded. With such contracts in hand, it should be possible to secure private financing for such projects. There currently is substantial interest in creating international fuel banks to assure a reliable supply of fresh nuclear fuel and of reprocessing services to states that forswear making their own nuclear fuel. If any such banks are created, though, they should charge whatever the prevailing market price might be for the nuclear products and services they provide. The rationale for this is simple: Subsidizing the price risks creating a false demand for risky near weapons-usable fuels, such as mixed oxide (MOX) and other plutonium-based fuels. Currently, states can satisfy their demand for fresh fuel without having to resort to any international bank, and no state has a need to reprocess for any reason. Subsidizing these fuel services has been proposed as a way to induce states to eschew making their own nuclear fuels. This proposal, however, seems unsound. First, it is unclear as to whom the customers will be. India and Canada already make their own natural uranium fuels, which require no enrichment. Several others — France, Russia, Japan, Brazil, and China — enrich their own fuel, and the remaining nuclear fuel-consuming states seem content to buy their fuels from U.S. providers, Russia, URENCO, or Eurodif. Second, it is unlikely that nuclear fuel subsidies would be sufficient to block determined proliferators. After all, only a small percent of any nuclear power plant's life cycle costs are associated with its fueling requirements.[60] Again, given the dangers of propping up dangerous reprocessing activities and the dubious requirement to provide enriched fuel, the world can well afford to depend more on market mechanisms to determine when and how these services are provided.

c. *Use of Weapons Grade Uranium Fuels.* Finally, the use of nuclear weapons-usable highly enriched uranium is a nuclear fuel cycle option no longer necessary for the production of power or of medical, agricultural, or industrial isotopes. There are fewer and fewer research reactors that use highly enriched uranium (HEU), but the few that do are more than willing to pay to continue to use this fuel rather than to pay the costs of converting to low enriched uranium alternatives. Given the direct usability of HEU to make nuclear weapons, however, the elimination and blending down of these fuels are imperative to avoid nuclear proliferation and terrorism risks. In the United States, the handful of remaining HEU-fueled plants receive government funding. This should end by establishing a date certain for these few remaining reactors to be converted to use LEU-based fuels.[61]

5. Increase and further privatize nuclear insurance liability coverage to encourage best construction and operations practices. Officials within the nuclear industry frequently note that a nuclear industry accident anywhere would impact nuclear operators negatively everywhere. Yet, the potential financial and political fall-out following a major nuclear accident would be even more significant if there were a lack of adequate nuclear accident liability insurance. For this reason alone, efforts should be made to increase the minimum amounts of liability insurance coverage currently required of any civilian nuclear plant operators and to make those requirements less subject to over-ride or forgiveness by officials of the state. Here, amounts required by the international Convention on Supplementary Compensation for Nuclear Damage (CSC)[62] should be considered to be the minimum. For the EU, which is currently struggling to set a standard for its members, the coverage requirements set by CSC should be considered to be the floor from which any specific EU standard is created. It would be far preferable for the EU to adopt insurance levels that the United States currently requires under its domestic Price-Anderson legislation. The United States, meanwhile, needs to raise international nuclear insurance standards by first announcing its intention to withdraw from underwriting insurance against terrorist incidents as it currently does and instead require private insurance firms to assume this requirement as they did before 9/11. Second, Washington needs to make good on its original objective under the 1957 Price-Anderson legislation to eventually stop underwriting coverage for damages a nuclear operator might inflict on off-site third parties. Washington would do best to start this now and incrementally by announcing that beginning

in 2025, federal Price-Anderson coverage caps will no longer apply to any civilian nuclear facility operating in the United States. This announcement should be made now so that the nuclear utility and vendor industry can develop their own alternative private system of insurance to cover off-site damages. At a minimum, the requisite amounts of capital to fund such a system should be amassed well in advance of the need to bring the new insurance system into force. Under any new system, each nuclear utility, service provider, and vending firm should be free to buy as much or as little third-party liability insurance for themselves as each sees fit from private insurance firms so long as the amount was at least as much as Price-Anderson currently requires to cover any one accident (roughly $10 billion for each accident). The rates for this coverage would be set for each firm by private insurers based on each firm's safety performance, the age of the plant, and the experience of the firm's staff, etc. Of course, each nuclear firm should be free to work with other nuclear utilities and companies to create private insurance pools. Even in this case, though, rates for each firm should be set in a manner that would reward the best nuclear operators and vendors. By doing this, the government would finally be able get industry to internalize the full costs of off-site nuclear accident liability insurance. Given that some U.S. nuclear firms already believe that their products are safe enough for them to soon forgo Price Anderson subsidies and liability limits, and that the nuclear industry is insisting that its safety record has improved and will only get better, this transition over the next 15 years should go relatively smoothly.

6. Increase experimentation in the commercial distribution of energy and the generation of alternative

sources of energy through federal government-led regulatory reform. To foster energy experimentation and competition, the federal government should promote regulatory reforms that would, among other things, (1) set standard rules for selling electricity through the grid; (2) remove conflicts of interest for existing grid or pipeline operations to block new entrants; (3) ensure that regulated utilities have similar incentives to invest in efficiencies as they do in expanding generation plants and energy supplies; (4) encourage key market constraints, be they carbon limits or liability coverage, through the market pricing systems rather than through government subsidies; and (5) increase pricing visibility for power to final customers.

## ENDNOTES - CHAPTER 1

1. See, e.g., Albert Wohlstetter *et al.*, *Swords from Plowshares: The Military Potential of Civilian Nuclear Energy*, Chicago, IL: University of Chicago Press, 1979, pp. vii-32; Matthew Fuhrman, "Spreading Temptation: Proliferation and Peaceful Nuclear Cooperation Agreements," *International Security*, Summer 2009, pp. 7-41, available from *belfercenter.ksg.harvard.edu/files/IS3401_pp007-041_Fuhrmann.pdf*; and Victor Gilinsky, *et al.*, "A Fresh Examination of the Proliferation Dangers of Light Water Reactors," in Henry Sokolski, ed., *Taming the Next Set of Strategic Weapons Threats*, Carlisle, PA: Strategic Studies Institute, U.S. Army War College, 2005, available from *www.npec-web.org/node/886*.

2. Lewis L. Strauss, Chairman of the U.S. Atomic Energy Commission, Speech to the National Association of Science Writers, New York, September 16, 1954.

3. On this history, see Joseph F. Pilat, ed., *Atoms for Peace: An Analysis after Thirty Years*, Boulder, CO: Westview Press, 1985; Richard Hewlett and Jack Holl, *Atoms for Peace and War, 1953-*

*1961: Eisenhower and the Atomic Energy Commission*, Berkley, CA: University of California Press, 1989.

4. President Richard Nixon, "Special Message to the Congress Proposing Emergency Energy Legislation," November 8, 1973, available from *www.presidency.ucsb.edu/ws/index.php?pid=4035*.

5. See Yves Marignac, *Nuclear Power, the Great Illusion: Promises, Setbacks and Threats*, October 2008, p. 42, available from *www.global-chance.org/spip.php?article89*; and the Testimony of Thomas B. Cochran before the Senate Committee on Energy and Natural Resources, Subcommittee on Energy Research and Development, June 8, 1977, available from *docs.nrdc.org/nuclear/files/nuc_77060801a_23.pdf*.

6. See Simon Lomax, "Nuclear Industry 'Restart' Means More Loan Guarantees," *Bloomberg.com*, October 27, 2009, available from *www.bloomberg.com/apps/news?pid=20601072&sid=aR1MVERYEgAs*.

7. See U.S. Office of Management and Budget, "The Federal Budget Fiscal Year 2011: Creating the Clean Energy Economy of Tomorrow," The President's Budget: Fact Sheet, available from *www.whitehouse.gov/omb/factsheet_key_clean_energy/*.

8. "Financial Crisis Nips Nuclear Revival in the Bud, WNA Told," *Nucleonics Week*, September 17, 2009, available from *www.carnegieendowment.org/static/npp/pdf/NW_Sep2009_reprint.pdf*; "Analysis-Delays, hitches hamper Areva's reactor export plan," *Reuters*, December 10, 2009, available from *in.news.yahoo.com/137/20091210/371/tbs-analysis-delays-hitches-hamper-areva.html*.

9. See Tyler Hamilton, "$26B Cost Killed Nuclear Bid: Ontario Ditched Plan over High Price Tag that Would Wipe Out 20-Year Budget," *The Star*, July 14, 2009, available from *www.thestar.com/article/665644*.

10. See Rebecca Smith, "Costs Cloud Texas Nuclear Plan," *The Wall Street Journal*, December 5, 2009, available from *online.wsj.com/article/SB125997132402577475.html*; Dow Jones, "CPS Energy, NRG Energy Complete Nuclear Power Project Settlement," March 1, 2010, available from *www.nasdaq.com/aspx/stock-market-news*

*story.aspx?storyid=201003011204dowjonesdjonline000515&title=*
*cps-energynrg-energy-complete-nuclear-power-project-settlement;*
and Anton Caputo, "Nuclear Could Still Edge Out Gas," My SA
News, December 15, 2009 available from *www.mysanantonio.com/*
*news/local_news/79283092.html.*

11. This graph, which reflects some of the most recent nuclear
cost projections, is based on a chart originally generated by Mark
Cooper and spotlighted by Sharon Squassoni. See Mark Cooper,
*The Economics of Nuclear Reactors: Renaissance or Relapse?* Vermont University, Institute for Energy and the Environment, June
2009, available from *www.vermontlaw.edu/Documents/Cooper%20*
*Report%20on%20Nuclear%20Economics%20FINAL%5B1%5D.pdf;*
and Sharon Squassoni, *The U.S. Nuclear Industry: Current Status
and Prospects under the Obama Administration,* Nuclear Energy
Futures Paper No. 7, Waterloo, Ontario, Canada: The Centre for
International Governance Innovation, November 2009, available
from *www.carnegieendowment.org/files/Nuclear_Energy_7_0.pdf.*

12. See the discussion of Constellation's calculations regard-
ing its planned reactor build at Calvert Cliffs, Maryland, in Doug
Koplow, "Nuclear Power as Taxpayer Patronage: A Case Study of
Subsidies to Calvert Cliffs," Chapter 8 of this volume.

13. U.S. Congressional Budget Office, "Cost Estimate of S.14
Energy Policy Act of 2003," May 7, 2003, available from *www.cbo.
gov/ftpdocs/42xx/doc4206/s14.pdf.* The CBO optimistically assumed
that about half of the value of the projects that defaulted would
be recovered in bankruptcy, for a net loss of around 25 percent of
guaranteed principal. The DoE has tried to discredit even these
figures, claiming that the real figures will be much lower but recently said it would not publicly disclose its own calculations of
how much of an upfront loan fee to charge to cover for potential
defaults on nuclear projects. Industry officials, meanwhile, have
made it clear that if the DoE charges them much more than 1 or
2 percent of the amount borrowed to cover these risks, they will
not take the loans. See Kate Sheppard," Energy Sec Unaware that
Nuclear Loans Have 50 Percent Risk of Default," February 16,
2010, available from *motherjones.com/blue-marble/2010/02/chu-not-
aware-nuclear-default-rates,* and *Etopia News,* "DoE Spokesperson
Says that Credit Subsidy number is 'Proprietary and Will Remain
Confidential'," available from *etopianews.blogspot.com/2010/03/doe-
spokesperson-says-that-credit.html.*

14. See Moody's Global, "New Nuclear Generation: Ratings Pressure Increasing," June 2009 available from *www.nukefreetexas. org/downloads/Moodys_June_2009.pdf*.

15. See *Nucleonics Week*, "Financial Crisis Nips Nuclear."

16. Data for these charts were drawn from Chicago Climate Exchange, "Closing Prices," December 2009, available from *www. chicagoclimatex.com/market/data/summary.jsf*; European Climate Exchange, "Prices, Volume & Open Interest: EXC EUA Futures Contract," December 2009, available from *www.ecx.eu/EUA-Futures*; *www.bloomberg.com/apps/news?pid=20601109&sid=aNykpTP9hnIo*; and the United States Energy Information Administration, "U.S. Natural Gas Electric Power Price," October 30, 2009, available from *tonto.eia.doe.gov/dnav/ng/hist/n3045us3m.htm* .

17. See, e.g., Rebecca Smith and Ben Casselman, "Lower Natural-Gas Price Leaves Coal Out in Cold," *The Wall Street Journal*, June 15, 2009, available from *online.wsj.com/article/ SB124502125590313729.html*; and Edward L. Morse, "Low and Behold: Making the Most of Cheap Oil," *Foreign Affairs*, September/ October 2009, available from *www.foreignaffairs.com/articles/65242/ edward-l-morse/low-and-behold*.

18. The most recent U.S. DoE effort to skirt nuclear power's poor economic performance is to promote federal development and construction of a variety of "small modular" reactors. See *www.nuclear.energy.gov/pdfFiles/factSheets/2010_SMR_Factsheet. pdf*. The key attraction of small reactors, between 100 and 300 MWe, is that they cost less to build than the much larger commercial light water reactors that currently range between 1,000 and 1,600 MWe. A full description of this type of reactor is available from the World Nuclear Association website at *www.world-nuclear.org/info/inf33.html*. The other attraction is that they are more adaptable to small electrical grids than much larger reactors. For most developing states, though, even these "small" reactors are too large for their grid, and the unit costs (dollars/kilowatt hour produced) for this smaller reactor are far higher than for the larger reactors they are supposed to best.

19. See, e.g., Steve Kroft, "France: Vive Les Nukes: How France is Becoming the Mode 3l for Nuclear Energy Genera-

tion, *60 Minutes*, April 6, 2007 available from *www.cbsnews.com/stories/2007/04/06/60minutes/main2655782.shtml*.

20. For the most recent and thorough attempts, see Arnulf Grubler, *An Assessment of the Costs of the French Nuclear Program, and 1970-2000*, available from *www.iiasa.ac.at/Admin/PUB/Documents/IR-09-036.pdf*; and Charles Komanoff, "Cost Escalation in France's Nuclear Reactors: A Statistical Examination," January 2010, available from *www.slideshare.net/myatom/nuclear-reactor-cost-escalationin-france-komanoff*.

21. The French civilian nuclear industry and power utility system, unlike the American one, is almost entirely nationalized. As a result, France still produces incredibly opaque financial statements regarding its civilian nuclear program. What is not in dispute, however, is that because of its over-investments in base-load nuclear generators, France must export much of its production and import expensive peak load capacity, which it still lacks. For an explanation of base-load and peak load electricity, see Jay Solomon and Margret Coker, "Oil-Rich Arab State Pushes Nuclear Bid with U.S. Help," *The Wall Street Journal*, April 2, 2009, available from *online.wsj.com/article/SB123862439816779973.html*; and Dan Murphy, "Middle East Racing to Nuclear Power," November 1,2007, *The Christian Science Monitor*, *www.csmonitor.com/2007/1101/p01s03-wome.html*. See Mycle Schneider, "Nuclear Power in France: Beyond the Myth," Chapter 6 of this volume.

22. Chart generated by Doug Koplow, based on data provided by McKinsey and Company. See Doug Koplow, "Nuclear Power as Taxpayer Patronage: A Case Study of Subsidies to Calvert Cliffs Unit 3," Washington, DC: NPEC, July 2009, available from *www.npec-web.org/files/Koplow%20-%20CalvertCliffs3_0.pdf*; and McKinsey & Company, *Reducing U.S. Greenhouse Gas Emissions: How Much at What Cost?* December 2007, available from *www.mckinsey.com/clientservice/sustainability/greenhousegas.asp*.

23. See John W. Rowe, "Fixing the Carbon Problem Without Breaking the Economy," presentation before Resources for the Future Policy Leadership Forum Lunch, May 12, 2010, Washington, DC, available from *www.rff.org/Documents/Events/PLF/100512_Rowe_Exelon/100512_Rowe_Exelon_Slides.pdf*.

24. For a detailed description of natural gas fired electrical generating technologies, their cost, and performance, see International Energy Agency (OECD), Energy Technology System Analysis Program, "Gas-Fired Power," available from *www.etsap. org/E-techDS/EB/EB_E02_Gas_fired%20power_gs-gct.pdf*.

25. On the growing availability of natural gas in the Western Hemisphere, Europe, and Asia, see "An Unconventional Glut," *The Economist*, pp. 72-74, available from *www.economist.com/ business-finance/displaystory.cfm?story_id=15661889*; Ben Casselman, " U.S. Gas Fields Go from Bust to Boom, April 30, 2009; and "U.S. Natural-Gas Supplies Surge," *The Wall Street Journal*, April 30, 2009, and June 18, 2009, available from *online.wsj.com/article/ SB124104549891270585.html*, and *online.wsj.com/article/SB1245272 93718124619.html*; and Gary Schmitt, "Europe's Road to Energy Security: Unconventional Gas Could Free the EU from Dependence on Russian Gas Supplies," *The European Wall Street Journal*, March 11, 2010, available from *online.wsj.com/article/SB100014240 52748704187204575101344074618882.html*.

26. For an analysis relevant to the Middle East, see Peter Tynan and John Stephenson, "Nuclear Power in Saudi Arabia, Egypt and Turkey: How Cost Effective?"; and Wyn Bowen and James Acton, "Atoms for Peace in the Middle East: The Technical and Regulatory Requirement," Chapters 9 and 10 of this volume.

27. See Amory B. Lovins, Imran Sheikh, and Alex Markevich, " Nuclear Power: Climate Fix or Folly?" Chapter 3 of this volume.

28. See the analysis of the American Wind Association, available from *www.awea.org/faq/cost.html*.

29. For an analysis showing that renewables are already more economical than nuclear or coal base load generations, though, see Amory Lovins, "Mighty Mice," *Nuclear Engineering International*, December 21, 2004, available from *www.neimagazine.com/ story.asp?storyCode=2033302*.

30. See, e.g., Mason Willrich, "Electricity Transmission for America: Enabling a Smart Grid, End-to-End," *Energy Innovation Working Paper Series*, Cambridge, MA: Massachusetts Institute of

Technology, July 2009, available from *web.mit.edu/ipc/research/energy/pdf/EIP_09-003.pdf*; Sharon Gauin, "Bloom Fuel Cell: Individual Power in a Box," *Business Week*, February 24, 2010, available from *www.businessweek.com/idg/2010-02-24/bloom-fuel-cell-individual-power-plant-in-a-box.html*.

31. See Peter Bradford, "Taxpayer Financing for Nuclear Power: Precedents and Consequences," Chapter 5 of this volume.

32. For a detailed case study of such effects in the case of biofuel commercialization programs, see David Victor, *The Politics of Fossil Fuel Subsidies*, Geneva, Switzerland: The Global Subsidies Initiative, October 2009, available from *www.globalsubsidies.org/files/assets/politics_ffs.pdf*.

33. On this point, see the testimony of David Lochbaum before a hearing of the Subcommittee on Energy and Resources of the House Committee on Government Reform, "Next Generation of Nuclear Power," June 29, 2005, available from *ftp.resource.org/gpo.gov/hearings/109h/23408.txt*.

34. Estimates of how much Price-Anderson nuclear accident liability limits on third party damages are worth range widely between .5 and 2.5 cents per kilowatt hour. For details, see Anthony Heyes, ,"Determining the Price of Price Anderson," *Regulation*, Winter 2002–2003, pp. 26-30, available from *www.cato.org/pubs/regulation/regv25n4/v25n4-8.pdf*; and Doug Koplow, "Nuclear Power as Taxpayer Patronage," available from *www.npec-web.org/node/1125*.

35. Cf., however, Peter A. Bradford, former U.S. Nuclear Regulatory Commissioner, Testimony before the U.S. Senate Committee on Environment and Public Works Subcommittee on Nuclear Regulation, "Renewal of Price Anderson Act," January 23, 2002, available from *epw.senate.gov/107th/Bradford_01-23-02.htm*.

36. See Public Law 107-297-Nov. 26, 2002, available from *www.treas.gov/officies/enforcement/ofac/legal/statutes/pl107_297.pdf*; and The Terrorism Risk Insurance Extension Act of 2005, available from *www.cbo.gov/ftpdocs/69xx/docs6978/s467.pdf*.

37. For post 9/11 overviews of the growing number of civilian nuclear-related terrorism concerns, see Carl Behrens and Mark Holt, "Nuclear Power Plants: Vulnerability to Terrorist Attack," Report for Congress, RS21131, Washington, DC: U.S. Congressional Research Service, August 9, 2005, available from *www.fas.org/sgp/crs/terrror/RS21131.pdf*; National Research Council of the National Academies, San Luis Obispo Mothers for Peace v. Nuclear Regulator Commission, No. 03-74628, 2006 WL 151889, 9th Cir. June 2, 2006; "Safety and Security of Commercial Spent Fuel Storage," Public Report, April 6, 2005; and Henry Sokolski, "Too Speculative? Getting Serious about Nuclear Terrorism," *The New Atlantis,* Fall 2006, pp. 119-124, available from *www.thenewatlantis.com/publications/too-speculative.*

38. See U.S. Congressional Budget Office, "Federal Terrorism Reinsurance: An Update," January 2005, section three of six, "Long-term Effects," available from *www.cbo.gov/showdoc.cfm?index=6049&sequence=2#pt3*; and The U.S. Department of the Treasury, Report to Congress, *Assessment: The Terrorism Risk Insurance Act of 2002,* Washington, DC: The U.S. Department of the Treasury, Office of Economic Policy, June 30, 2005, pp. 10-12, 111-113, and 125-140. Yet another shortcoming of the current cap on nuclear accident insurance liability for third parties in the United States is the lack of commonsense differentiation between the safest and least safe and the most remotely located reactors and those located near high value urban real estate. This too discourages industry from engaging in best practices. See notes 26 and 34.

39. U.S. Nuclear Regulatory Commission, "NRC Informs Westinghouse of Safety Issues with AP1000 Shield Building," Press Release 09-173, October 15, 2009, available from *www.nrc.gov/reading-rm/doc-collections/news/2009/09-173.html.*

40. See Letter from Omer F. Brown III to Deputy Secretary of State Richard Armitage, Re: Nuclear Liability, December 18, 2003, available from *foreignaffairs.house.gov/110/sok061208.pdf.*

41. See, e.g., the testimony of David Baldwin, senior Vice President of General Atomics before a hearing of the Subcommittee on Energy and Resources of the House Committee on Government Reform, "Next Generation of Nuclear Power," June 29, 2005, available from *ftp.resource.org/gpo.gov/hearings/109h/23408.txt.*

42. See Antony Froggatt, "Third Party Insurance: The Nuclear Sector's 'Silent' Subsidy, in Europe," Chapter 13 of this volume and Simon Carroll, "European Challenges to Promoting International Pooling and Compensation for Nuclear Reactor Accidents, Washington, DC: NPEC, January 2, 2009, available from *www.npec-web.org/node/1051.*

43. On these points, see House Permanent Select Committee on Intelligence, Subcommittee on Intelligence, *Recognizing Iran as a Strategic Threat: An Intelligence Challenge for the United States*, staff report, August 23,2006, p. 11, available from *intelligence.house.gov/Media/PDFS/IranReport082206v2.pdf.*

44. Thus, when it became clear that North Korea had reneged on its promise not to attempt to enrich uranium for weapons, the Bush administration stopped construction of the two light water reactors it had promised Pyongyang because, in the words of Secretary of State Rice, North Korea could not be "trusted" with them.

45. See Remarks by the President on Weapons of Mass Destruction Proliferation, Fort Leslie J. McNair, National Defense University, February 11, 2004, available from *www.acronym.org.uk/dd/dd75/75news06.htm.*

46. White House Press Release, "Text of Declaration on Nuclear Energy and Nonproliferation Joint Actions, July 03, 2007," available from *moscow.usembassy.gov/st_07032007.html.*

47. See Solomon and Coker, "Oil-Rich Arab State Pushes Nuclear Bid with U.S. Help"; and Murphy, "Middle East Racing to Nuclear Power."

48. See "In Pursuit of the Undoable, Troubling Flaws in the World's Nuclear Safeguards," *The Economist*, August 23, 2007, available from *www.economist.com/world/international/displaystory.cfm?story_id=9687869.*

49. On these points, see Henry D. Sokolski, ed., *Falling Behind: International Scrutiny of the Peaceful Atom,* Carlisle, PA: Strategic Studies Institute, U.S. Army War College, 2008, available from *www.npec-web.org/node/1160/.*

50. On these points, see Eldon Greenberg, "The NPT and Plutonium," Washington, DC: NCI, 1993, available from *www.npec-web.org/node/854*; and Robert Zarate, "The NPT, IAEA Safeguards, and Peaceful Nuclear Energy," in *Falling Behind*, pp. 252 ff., available from *www.npec-web.org/node/1160/*.

51. Because large amounts of electricity cannot currently be stored, electrical companies must estimate how much electricity their customers will use and secure the electrical generating capacity to supply this demand. The difference between these estimates and real demand produces temporary imbalances in the electrical grid that the electrical transmission system operator must correct for by either reducing the amount of electricity being put on the grid or by bringing more electricity on to the grid. The latter is done by accessing electrical generators that are on the ready or "spinning" to supply follow-on load capacity electricity. For a more detailed slide tutorial on these points, see "Spinning Reserves, Balancing the Net," *Leonardo Energy Minute Lectures*, available from *www.slideshare.net/sustenergy/spinning-reserve*.

52. On these points, see The Congressional Budget Office, "Congressional Budget Office Cost Estimate: S. 14 Energy Policy Act of 2003," May 7, 2003, available from *www.cbo.gov/ftpdocs/42xx/doc4206/s14.pdf*; Congressional Budget Office, Director's Blog, "Department of Energy's Loan Guarantees for Nuclear Power Plants," March 4, 2010, available from *cboblog.cbo.gov/?p=478*.

53. The International Renewable Energy Agency (IREA) was created in 2009. More on its mandate is available from *www.irena.org/*.

54. More on each of these agreements is available from *www.encharter.org/* and *www.cmdc.net/echarter.html*.

55. See Henry Sokolski, Market-Based NuclearNonproliferation, Chapter 14 of this volume.

56. See The Commission on the Prevention of Weapons of Mass Destruction Proliferation and Terrorism, *The World At Risk: The Report of the Commission on the Prevention of WMD Proliferation and Terrorism*, New York: Vintage Books, December 2008, pp. 55-

56, available from *documents.scribd.com/docs/15bq1nrl9aerfu0yu9qd. pdf*.

57. On this point, see, e.g., Steven Mufson, "Nuclear Projects Face Financial Obstacles: *The Washington Post*, March 2, 2010, p. 1a, available from *www.washingtonpost.com/wp-dyn/content/article/2010/03/01/AR2010030103975.html*.

58. See *World at Risk*, p. 51; and Mohamed El Baradei, Nobel Lecture, December 10, 2005, available from *nobelprize.org/nobel_prizes/peace/laureates/2005/elbaradei-lecture-en.html*.

59. See Frank Von Hippel, "Why Reprocessing Persists in Some Countries and Not in Others: The Costs and Benefits of Reprocessing," Chapter 12 of this volume.

60. See Steve Kidd, "Nuclear Fuel: Myths and Realities , Chapter 11 of this volume.

61. For more detail on these points, see NRDC's Petition to the U.S. Nuclear Regulatory Commission For Rulemaking to Ban Future Civil Use of Highly Enriched Uranium, March 24, 2008, available from *docs.nrdc.org/nuclear/files/nuc_08032501a.pdf*.

62. See Information Circular 367, July 22, 1998, Convention on Supplementary Compensation for Nuclear Damage, available from *www.iaea.org/Publications/Documents/Infcircs/1998/infcirc567. shtml*.

# PART I

# NUCLEAR POWER'S ECONOMIC, ENVIRONMENTAL, AND POLITICAL PROSPECTS

# CHAPTER 2

# MAPPING NUCLEAR POWER'S FUTURE SPREAD

## Sharon Squassoni

Enthusiasm for nuclear energy has surged in the last few years, prompting industry leaders to talk of a nuclear renaissance. Energy security and climate change top the list of reasons that nuclear power proponents give to pursue nuclear energy. Nuclear energy has been rebranded as clean, green, and secure, and, as a result, more than 27 nations since 2005 have declared they will install nuclear power for the first time. *Nuclear Energy Outlook 2008*, published by the Organization for Economic Cooperation and Development's (OECD) Nuclear Energy Agency, suggests the world could be building 54 reactors per year in the coming decades to meet all these challenges.[1]

It is unlikely that nuclear energy will grow that much and that quickly, but it seems clear that the distribution of nuclear power across the globe is about to expand.[2] The interest in nuclear power by more than two dozen additional states is perhaps the most notable element of the much-heralded "nuclear revival." Half of these are developing countries. Some—such as Turkey, the Philippines, and Egypt—had abandoned programs in the past, while others, like Jordan and the United Arab Emirates (UAE), are considering nuclear power for the first time. If all these states follow through on their plans, the number of states with nuclear reactors could double.

Nuclear power reactors currently operate in 30 countries and Taiwan, with a total capacity of about 369 gigawatts electric (GWe) (See Figure 2-1). Three countries—the United States, France, and Japan—host more than half of global reactor capacity. Seven developing nations—Argentina, Brazil, China, India, Pakistan, South Africa, and Taiwan—have nuclear power. Figures 2 and 3 show where commercial uranium enrichment and spent fuel reprocessing plants are located. Enrichment plants now operate in 11 countries, providing 50 million separative work units (SWU); spent fuel is reprocessed in five countries. No country yet has opened a geologic waste site for final disposal of spent nuclear fuel.

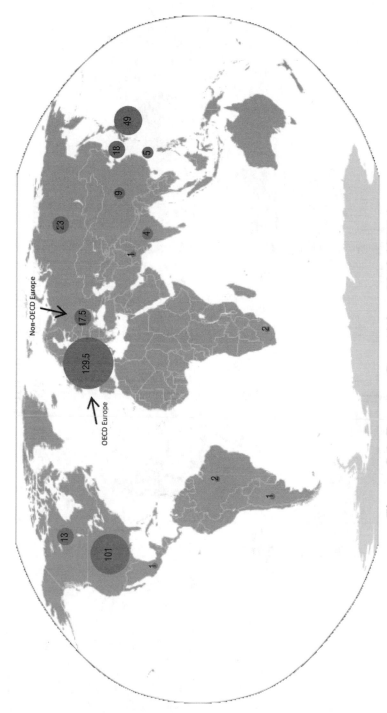

Figure 2-1. Reactor Capacities, 2010 (GWe).

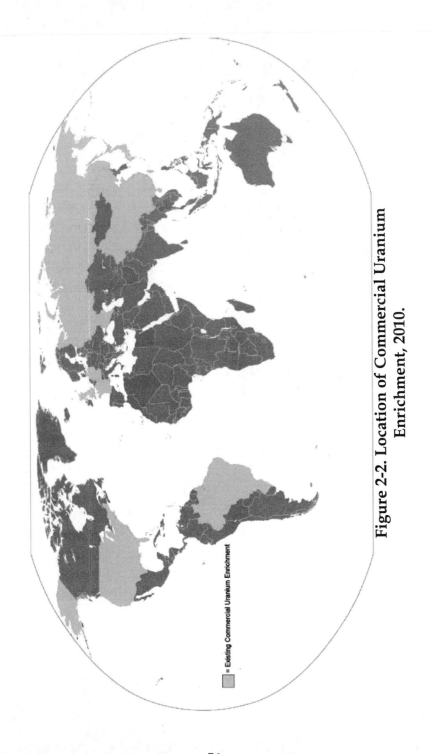

Figure 2-2. Location of Commercial Uranium Enrichment, 2010.

◻ = Existing Commercial Uranium Enrichment

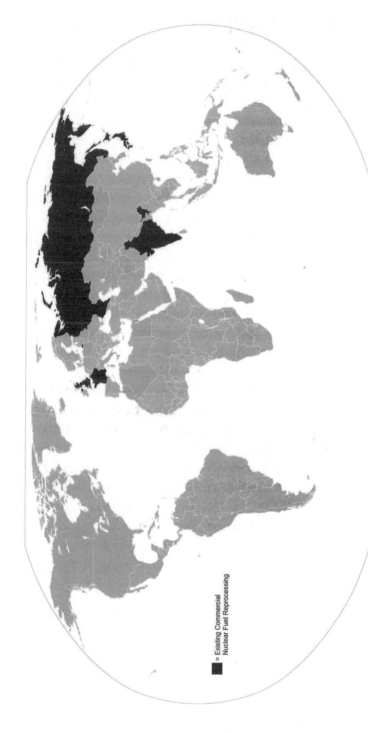

= Existing Commercial
Nuclear Fuel Reprocessing

Figure 2-3. Location of Commercial Reprocessing of Spent Fuel, 2010.

## SCENARIO I: "BUSINESS AS USUAL" GROWTH

Estimating nuclear power capacity growth out to 2030 presents some challenges. For example, will Germany and Sweden phase out nuclear power or re-think their decisions?[3] How long can the lives of older nuclear power plants be extended? According to the International Energy Agency (IEA), without significant policy changes, nuclear energy can be expected to grow to 475 GWe by 2030.[4] This amounts to an annual build rate of 4.5 reactors per year worldwide. At this rate, nuclear energy would actually decline from a 16 percent electricity market share to 11 percent, as electricity demand increases. In this business-as-usual projection, no big policy changes would be implemented and carbon emissions would rise.

Figure 2-4 depicts the first scenario of modest, or "business as usual," growth in nuclear power, using U.S. Energy Information Administration (EIA) figures. The EIA estimates 482 GWe capacity by 2030, assuming fewer retirements of older reactors in Europe.[5] In general, EIA projections factor in gross domestic product (GDP) growth, energy demand, end-use sector, and electricity supply, estimating the contribution that nuclear energy will make as a percentage of the total electricity supply. This percentage is estimated to stay even or rise slightly. See Figure 2-4.

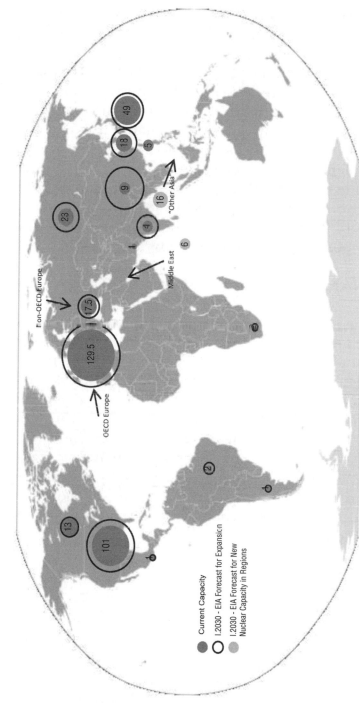

Figure 2-4. Expansion in Global Reactor Capacity, Scenario I.

In some countries, even estimates that nuclear energy's market share of electricity supply will stay level may be optimistic. For example, in the United States, a 1.5 percent rise in electricity demand each year would require 50 new nuclear power plants to be built by 2025, assuming nuclear energy maintained its 19 percent electricity generation share. (It would also require building 261 coal-fired plants, 279 natural-gas fired plants and 73 renewable projects).[6] Given that only 4 to 8 new plants might begin operation by 2015, this would require bringing 42 to 48 new plants on line in the 10 years between 2015 and 2025. While such is not impossible, it is not very likely.

## SCENARIO II: WILDLY OPTIMISTIC GROWTH

The second scenario for growth, which might be termed the "wildly optimistic" scenario, relies on countries' stated plans for developing nuclear energy. It is wildly optimistic in terms of both timing and the number of states that may develop nuclear power. Country statements were taken literally. These do not necessarily correlate to any measurable indicators (such as GDP growth or electricity demand, etc.), and in some cases the plans are unlikely to materialize. Scenario II figures, depicted in Figure 2-5, should be regarded not as projections, but as a "wish list" for many countries.

## Scenario II

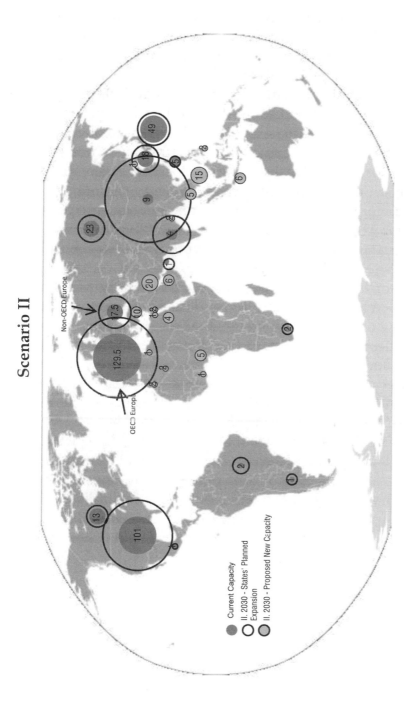

Figure 2-5. Expansion in Global Reactor Capacity According to States' Plans.

61

Some countries have modeled GDP growth, energy demand and supply, etc. Some have stated goals for specific electricity supplies. For example, the UAE has articulated a goal of diversifying its electricity production from 100 percent reliance on oil and natural gas to 30 percent liquid fossil fuels (oil and natural gas), 30 percent nuclear energy, and 30 percent renewables. The head of Brazil's nuclear association has stated that Brazil should diversify at least 30 percent of electricity generation equally into nuclear energy, natural gas, and biomass (Brazil now relies on hydroelectric power for 92 percent of its electricity). But for now, Brazil is focusing on four new nuclear power plants by 2014.

Often, countries' plans are predicated on buying one or two reactors, which would dictate how much capacity they purchase. Most of the reactors marketed today are 1,000 MWe to 1,600 MWe. However, some of these countries would be better served by much smaller reactors that would not introduce instability into their relatively small transmission grids. Some countries have not specified their plans beyond a desire to purchase nuclear power capacity. Whereas Figure 2-5 shows countries that have specified particular reactor capacities out to 2030, additional countries have articulated a need or desire for nuclear energy but have not yet been so specific. These are listed in the Appendix and shown in Figure 2-6. Some of these countries (shown with darker shading) have more detailed plans than others.

Figure 2-6. Proposed New Nuclear States, 2008.

According to the U.S. State Department, a dozen countries are "giving serious consideration to nuclear power in the next 10 years."[7] Several of this dozen, now lacking nuclear power, have plans to build nuclear reactors, including Azerbaijan, Belarus, Egypt, Indonesia, Kazakhstan, Turkey, the UAE, and Vietnam. The UAE is the furthest along in its plans, having awarded a contract to Korean firms for four reactors in December 2009. Many other countries have longer term plans — Algeria, Chile, Georgia, Ghana, Jordan, Libya, Malaysia, Morocco, Namibia, Nigeria, Bahrain, Kuwait, Oman, Saudi Arabia, Qatar, Syria, Venezuela, and Yemen number among them.[8]

If these states are serious about their plans, nuclear energy capacity could double by 2030. And if concerns over global climate change were to drive nuclear expansion, the capacity would reach 1 terawatt (or almost triple the current capacity). A more conservative estimate is that nuclear capacity could increase to 525 GW by 2030, with significant policy support. This equates roughly with the IEA's Alternative Policy Scenario from the *World Energy Outlook 2006*, which assumes that climate change policies dating from 2006 would be implemented.[9]

One of the key unknowns is how swiftly countries that are considering nuclear power for the first time will be able to implement their plans for nuclear power. The IAEA is actively providing guidance, review, and support to help them build the infrastructure for nuclear energy, and has identified 19 issues that should be addressed in building this infrastructure. The IAEA has stressed that nuclear energy is a 100-year commitment, from development to decommissioning.[10] Most developing countries would need to import reactors and, possibly, the staff to operate them. Potential suppliers will choose their business opportunities according to certainty of payment, volume of work, political stability and security, among other criteria.

There will undoubtedly be a lag between decisions to go nuclear and reactors coming on-line. The IAEA estimates about 15 years will elapse between a policy decision to develop nuclear power and the operation of a first plant.[11] By 2020, the IAEA estimates that power plant construction could begin in eight countries, and possibly in 15 more by 2030.[12] Although there is growing recognition that many of these developing countries would be better served by small and medium-sized reactors (from 300 MWe to 700 MWe) because of the capacities of their electrical grids, there will be few available options for states to purchase smaller reactors in that timeframe. For example, Westinghouse has built 600 MWe reactors in the past and has licensed the AP-600, but officials say there are no plans to market it. China has exported 300 MWe reactors, and India has built smaller reactors (from 160 MWe to 500 MWe) and has expressed the desire to get into the export market. Unfortunately, Indian reactors could pose greater proliferation risks for a variety of reasons.[13] In the meantime, most states will likely choose the reactors currently being marketed, which range predominantly from 1,000 MWe to 1,600 MWe.

Part of the challenge for many states will be adhering to international standards and conventions that have evolved over time. With no current nuclear capacity, many of these states would have had no reason to join nuclear-related conventions, or even sign comprehensive nuclear safeguards agreements. Table 2-1 shows the status of states that have declared an interest in nuclear power and certain nuclear safety, security, and nonproliferation commitments.

| Country | GWe | Target Date | Safeguards CSA AP | | Safety CNS | Security CPPNM | Waste** | Liability (Vienna Convention or CSC) |
|---|---|---|---|---|---|---|---|---|
| Turkey | 3-4? | 2014 | Y | Y | Y | Y | N | N |
| Bangladesh | 2 | 2015 | Y | Y | Y | Y | N | N |
| Jordan | .5 | 2015 | SQP | Y | N | N | N | N |
| Egypt | 1 | 2015 | Y | N | Y | N | N | VC |
| Morocco | ? | 2016 | Y | N | N | Y | Y | VC* |
| Azerbaijan | 1 | | Y | Y | N | Y | N | N |
| Belarus | 4 | 2016 | Y | N | Y | Y | Y | VC |
| Indonesia | 6 | 2016 | Y | Y | Y | Y | N | CSC* |
| Iran | 6 | 2016 | Y | N | N | N | N | N |
| UAE | 3 | 2017 | SQP | Y | N | Y | N | N |
| Vietnam | 8 | 2020 | Y | N | N | N | N | N |
| Thailand | 4 | 2020 | Y | N | N | N | N | N |
| Israel | 1 | | N | N | N | Y | N | VC* |
| Saudi Arabia | ? | | SQP | N | Y | Y | N | N |
| Oman | ? | | N | N | N | Y | N | N |
| Qatar | ? | | SQP | N | N | Y | N | N |
| Bahrain | ? | | SQP | N | N | N | N | N |
| Kuwait | ? | | SQP | Y | Y | Y | N | N |
| Kazakhstan | .6 | 2025 | Y | Y | N | Y | N | N |
| Nigeria | 4 | 2025 | Y | Y | Y | Y | Y | VC |
| Algeria | 5? | 2027 | Y | N | Y | Y | N | N |
| Ghana | 1 | 2030 | Y | Y | N | Y | N | N |
| Tunisia | .5 | 2030 | Y | N | Y | Y | N | N |
| Yemen | ? | 2030 | SQP | N | N | Y | N | N |
| Philippines | | 2050 | Y | N | N | Y | N | VC, CSC* |
| Libya | 1 | 2050 | Y | Y | N | Y | N | N |
| Venezuela | 4? | 2050 | Y | N | N | N | N | N |
| Malaysia | | 2050 | Y | N | N | N | N | N |

*= signed, not ratified.

** = Joint Convention on the Safety of Spent Fuel Management and on the Safety of Radioactive Waste Management (IN-
FCIRC/546)

CSA = Comprehensive Safeguards Agreement (IN-
FCIRC/153); AP = Additional Protocol (INFCIRC/540); CNS
= Convention on Nuclear Safety; CPPNM = Convention on the
Physical Protection of Nuclear Material; CSC = Convention on
Supplementary Compensation

## Table 2-1. States with an Interest in Nuclear Power: Status on Nuclear Safety, Security, and Nonproliferation.

Although signing conventions is an important step toward preparing for nuclear power, the real tests of responsibility may offer less tangible evidence of compliance. For example, how will vendors, regulatory agencies, and international institutions assess the maturity of nuclear safety cultures? How will states develop safety and security cultures that complement each other? Are the regulatory authorities truly independent? Many of the critical requirements will take years to develop fully.

## SCENARIO III: MAJOR GROWTH FOR CLIMATE CHANGE?

The amount of nuclear capacity needed to make a signification contribution to global climate mitigation is so large that it would inevitably be widely distributed across the globe. Such a distribution would have particular implications for nuclear proliferation. However, projected distributions of nuclear energy out to 2050 are extremely speculative. The industry itself does not engage in such projections, and countries that set nuclear energy production goals have a history of widely missing long-range targets, such as China and India. The discussion below considers a hypothetical distribution of nuclear energy for 2050, based on the 2003 MIT study, *The Future of Nuclear Power*.[14]

Scenario III, shown in Figure 2-7, uses the "High 2050" scenario described in Appendix 2 ("Global Electricity Demand and the Nuclear Power Growth Scenario") of the 2003 MIT study. Although this is not a distribution designed to achieve optimal $CO_2$ emission reductions, the level of expansion would be significant enough (1,500 GWe) to have an effect on $CO_2$ emissions. This would mean a four-fold increase from current reactor capacity.

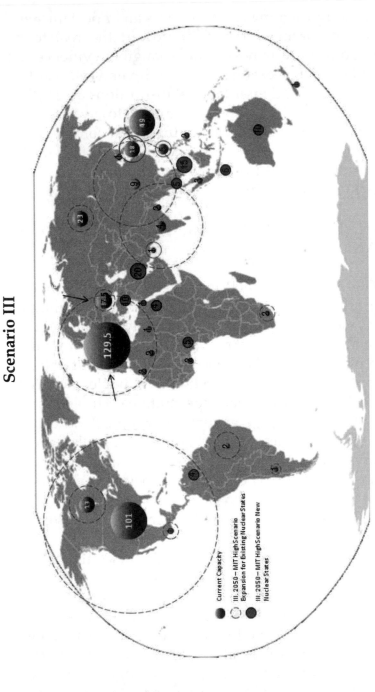

Scenario III

Figure 2-7. Illustrative Expansion to 1,500 GWe to Reduce Carbon Emissions.

The MIT study used the underlying assumption that the developed countries would continue with a modest annual increase in per capita electricity use, and the developing countries would move to the 4,000 kilowatt hour (kWh) per person per year benchmark if at all feasible (the 4,000 kWh benchmark being the dividing line between developed and advanced countries). Electricity demand was then pegged to estimated population growth. Finally, it was assumed that nuclear energy would retain or increase its current share of electricity generation. The least-well-off developing countries were assumed in the MIT study not to have the wherewithal for nuclear energy. It should be noted that MIT's 2050 projection was "an attempt to understand what the distribution of nuclear power deployment would be if robust growth were realized, perhaps driven by a broad commitment to reducing greenhouse gas emissions and a concurrent resolution of the various challenges confronting nuclear power's acceptance in various countries."[15] A few countries that the MIT High 2050 case included but that are not included here are those that currently have laws restricting nuclear energy, such as Austria.

## IMPLICATIONS FOR URANIUM ENRICHMENT

A four-fold expansion of nuclear energy would entail significant new production requirements for uranium enrichment and possibly reprocessing, as shown in Figure 8. The MIT study anticipated that 54 states would have reactor capacities that could possibly justify indigenous uranium enrichment. If a capability of 10 GWe is considered the threshold at which indigenous enrichment becomes cost-effective, more than 15 additional states could find it advantageous to engage in uranium enrichment. See Figure 2-8.

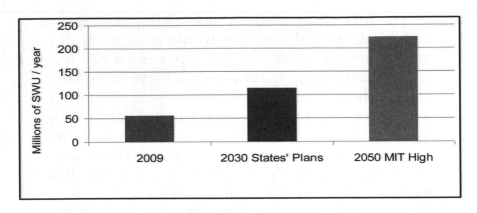

NOTE: 2030 and 2050 predict enrichment based on reactor capacity. They are based on countries' stated plans for reactor growth and the 2050 MIT "high growth" scenario, respectively. Both assume that a 1 GWe reactor requires 150,000 SWU enrichment per year.

## Figure 2-8. Enrichment Implications of Reactor Capacity Growth.

Figure 2-9 depicts what the geographic distribution of enrichment capacity might look like, based on the development of 10 GWe or more of reactor capacity. Of course, some states, such as Australia or Kazakhstan, might opt to enrich uranium regardless of domestic nuclear energy capacity, choosing to add value to their own uranium exports. In addition, states may choose to take the path of the UAE, which has formally renounced domestic enrichment and reprocessing in its domestic law, despite aspiring to reach 10 GWe of capacity. Ultimately, these decisions lie very much in the political realm, and can be reversed.

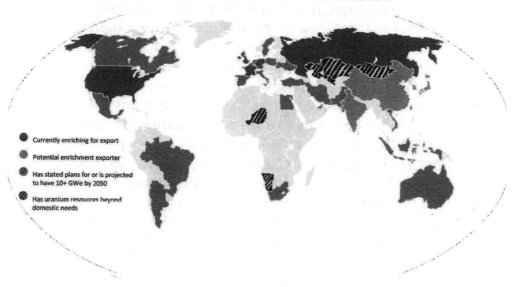

**Figure 2-9. Illustrative Uranium Enrichment Expansion Out to 2050.**

## IMPLICATIONS FOR PROLIFERATION

Proliferation experts generally fall into one of two camps—those that do not consider power reactors a cause for proliferation concern but focus on the sensitive aspects of the nuclear fuel cycle, and those that are concerned about the entire fuel cycle. Advocates of nuclear energy point out that most states that have developed nuclear weapons have used dedicated production or research reactors rather than power reactors to produce their fissile material;[16] others point to the potential for a state to use peaceful nuclear power to further a clandestine weapons program, either through technology transfer, hiding clandestine activities within a peaceful nuclear fuel cycle, or diverting lightly irradiated fuel to be further enriched. Regardless of one's views on the proliferation potential of power reactors, the recent surge of enthusiasm for nuclear energy poses several proliferation risks.

First, such enthusiasm is not limited to power reactors. On the enrichment side, President George Bush's 2004 initiative to limit capabilities to current technology holders failed, not just in strategy but also in tactics. Argentina, Canada, South Korea, and South Africa have all expressed an interest in keeping their enrichment options open. Brazil, which is commissioning a new centrifuge enrichment plant at Resende, will likely produce more low-enriched uranium than is needed for its own consumption by 2015. By and large, these countries do not produce nuclear energy on a scale large enough to make domestic enrichment capability economically sound.[17] However, they have keen national interests in maintaining their right to enrich.

Faced with allied objections to restricting future options, the Bush administration was unable to make new limits stick. The Obama administration has not fared much better. As of September 2010, the Nuclear Suppliers Group (NSG) still had been unable to agree on further detailed criteria restricting enrichment and reprocessing. Even if piecemeal efforts to limit the number of states with uranium-enrichment or spent fuel reprocessing capabilities succeed, these could ultimately further erode the Non-Proliferation Treaty (NPT) by extending the participation of nuclear weapon have-nots into the nuclear fuel cycle. In the short term, efforts to limit expansion could slow some states' implementation of the safeguards-strengthening measures in the 1997 Model Additional Protocol.

In the long term, other decisions to strengthen the NPT could be jeopardized.

At the back end of the fuel cycle, U.S. policy is very unclear. While the Bush administration supported reprocessing for the United States and key allies (including India), it did not support such capabilities for other states, and sought to provide alternatives under the Global Nuclear Energy Partnership (GNEP). This completely reversed policies adopted in the mid-1970s not to encourage the use of plutonium in the civilian fuel cycle.

Although GNEP advocates stress that the kind of spent fuel "conditioning" they favor would not result in the separation of plutonium, there are few assurances thus far that new techniques are any more proliferation-resistant than PUREX. As opponents like to point out, no future U.S. fuel conditioning technique will be more proliferation-resistant than storing spent fuel. While most countries are probably interested in having someone else solve the problem of either spent fuel storage or high-level waste storage, no commercial reprocessing service currently will store high-level waste. Neither the United States, nor Russia, nor France has committed to taking back spent fuel under GNEP. The Obama administration does not support the current generation of reprocessing techniques but will continue to fund research and development in the hopes of developing something more promising from a nonproliferation standpoint.

Two questions stand out about future sales of reactors: Who will supply them and what kind of reactors will they be? India, China, and South Korea are emerging as a second tier of suppliers interested in exporting reactors, injecting some uncertainty into assessments about the kinds of nonproliferation requirements they will place on recipients. Moreover, India actually has only heavy water reactors to sell and possibly in the future, fast reactors. Such sales would be risky from a nonproliferation standpoint. The emergence of these suppliers further points to a divide among the technological "haves" and "have-nots" — the advanced nuclear states will continue their research into advanced fuel cycles and fast reactors that may or may not be more proliferation-resistant, and the less developed states will buy what they can.

Beyond the technical realm, the widespread diffusion of civilian nuclear power raises real political questions. Do the geographic locations, the existence of terrorist groups on their soil, or other sources of political instability matter for nuclear security? Expectations will vary across regions, but in some cases, expanded nuclear infrastructure could lead neighboring countries to worry about the possibility of nuclear weapons development and possibly prepare themselves for such a development. Countries of concern include Egypt, Jordan, Indonesia, Malaysia, Morocco, Nigeria, Vietnam, and the Gulf Cooperation Council (GCC) countries.

The expansion of nuclear power would also have practical consequences for the nuclear nonproliferation regime. Additional facilities will place additional safeguards requirements on IAEA inspectors, and it is unclear how the IAEA will meet these requirements. Will more inspection days be called for, or will other approaches be used under the "integrated safeguards" program? Although reactors themselves require relatively few inspection days, there will be significant work in helping prepare new nuclear states for nuclear power programs. Already, the IAEA has conducted workshops on infrastructure requirements, including energy needs and planning considerations; nuclear security and safeguards; physical infrastructure; current and future reactor technology; experience in developing nuclear programs; human resource requirements; and public perceptions. States must also develop their own systems of accounting and control.

A nuclear expansion, in particular, that results in more states with bulk-handling facilities (enrichment and reprocessing) could place significant strain on the IAEA and the inspections system. The fact that the IAEA's goals for timely detection are clearly longer than material conversion time, that is, the time it would take for a proliferator to produce finished metal shapes, is a big concern. The largest enrichment and reprocessing plants under safeguards now are under European Atomic Energy Community (EURATOM) safeguards; the IAEA's role in verifying material balances in those plants is limited by the IAEA-EURATOM agreement. The only experience in safeguarding commercial-scale enrichment and reprocessing plants outside of EURATOM in a non-nuclear-weapon state is in Japan, where incidents with significant material losses have raised questions. British commercial reprocessing at the THORP facility also has produced recurring reports of significant materials losses.

Perhaps the largest question about a nuclear expansion is whether or not planned technological developments will outpace nonproliferation initiatives, such as fuel supply assurances and multinational fuel-cycle centers, voluntary export guidelines, and further restrictions within the Nuclear Suppliers Group. Criticism of the GNEP program had been aimed in part at the aggressive timeline for demonstrating advanced reprocessing, in contrast to developments more closely tied to nonproliferation objectives, such as supporting more proliferation-resistant reactors with sealed fuel cores that would limit handling of fuel. Already, efforts to manage expansion of the front and back ends of the fuel cycle, whether nuclear fuel assurances, fuel banks, or fuel leasing projects, have abandoned any concepts of formal restraints in favor of incentives. It is too soon to tell how compelling those incentives will be.

Finally, although there is disagreement among experts about the proliferation potential of light water reactors, it is clear that the proliferation potential of a country with no nuclear expertise is lower than that of a country with nuclear power and its associated infrastructure. The current encouraging climate for nuclear energy — new cooperation agreements between France and the UAE, Libya, and Algeria, and between the United States and Turkey and Jordan, for a few — suggests that regardless of global climate change concerns, or whether or not a significant expansion occurs, some states in the Middle East will develop nuclear energy. It is not clear whether new nuclear reactors in the Middle East would result in new enrichment or reprocessing plants in that region. In part, much depends on the outcome of negotiations with Iran on its enrichment capabilities. If states clearly renounce making nuclear fuel and allow sufficient wide-ranging inspections to verify such pledges, the proliferation implications could be significantly diminished. The hope is that more states will follow the example of the UAE, although there are few indicators so far of this.

## EXPANSION: REAL OR IMAGINED?

The largest increases in nuclear capacity in the next 20-30 years undoubtedly will occur in Asia, specifically, China, Japan, South Korea, and India. These countries are building nuclear power plants now and anticipate continued high economic growth levels. Other countries could feel the pinch of the current financial crisis more acutely, dampening demand for electricity below anticipated levels. A major expansion of nuclear power across the board, however, is not a foregone conclusion.

In addition, the traditional challenges besetting nuclear energy—cost, safety, waste, and proliferation—will likely continue to limit widespread growth. Government policies supporting nuclear energy in the future—as has been the case in the past—would be necessary to make major expansion a reality.

For many states, cost is the first and most immediate obstacle to nuclear expansion. But in those states where there is heavy involvement by the government in electricity markets, supporting nuclear energy may be as simple as providing government funding or financing. Solutions to nuclear waste tend to be deferred into the future, but policies by major suppliers to take back spent fuel could provide some incentives for growth. In states seeking nuclear power for the first time, actions to develop what some have termed the "three Ss"—safeguards, safety, and security— could improve their attractiveness to nuclear vendors. In all countries, some limits on, or costs attached to, carbon dioxide emissions could help enhance the attractiveness of nuclear power, but these should also enhance the attractiveness of renewable sources of energy as well.

## ENDNOTES - CHAPTER 2

1. Nuclear Energy Agency, *Nuclear Energy Outlook 2008*, Paris, France: Organization for Economic Cooperation and Development (OECD), p. 12.

2. In fact, the World Nuclear Association, one of the most vigorous advocates of expanding nuclear power, estimates that the total amount of nuclear power operating capacity could contract by 2030 from its current level of 435 GWe to 285 GWe. See The World Nuclear Association, *The Global Fuel Market: Supply and Demand 2007-2030*, London, UK: The World Nuclear Association, 2008.

3. In February 2009, the Swedish government called for new nuclear power plants. Following a 1980s referendum, Sweden had decided to shut all 12 reactors down, but only succeeded in shutting down two. The government's plan must be approved by Parliament. See "Sweden Changes Course on Nuclear Power," *Associated Press*, February 5, 2009.

4. International Energy Agency, *World Energy Outlook 2008*, Paris, France: Organisation for Economic Cooperation and Development (OECD), p. 92.

5. The IEA projection assumes 27 GW of reactors are retired in Europe. The EIA estimates 482 GWe for 2030, or an annual increase of 1.3 percent, but assumes planned phase-outs of nuclear power in some countries in Europe would be delayed. Note that EIA projections for nuclear energy are done "off-line" — that is, the sophisticated computer model for estimating other sources of energy is not used for the nuclear case. In addition, the estimates are aggregated into regions, with just a few country-specific breakouts. Further, retirements and the behavior of Western Europe are considered highly uncertain ("wildcards") and so estimates on those tend to be more conservative.

6. Brian Reilly, Principal Vice President, Bechtel, "Challenges of Construction Labor for New Builds," presentation to Fourth Annual Platt's Nuclear Energy Conference, February 5, 2008.

7. U.S. State Department International Security Advisory Board, "Proliferation Implications of Global Expansion of Civilian Nuclear Power," April 2008, available from *www.state.gov/documents/organization/105587.pdf*.

8. The State Department report also included Australia in this category, but the list was prepared in 2007, before Australian elections put a Labor government in power that currently has no plans for nuclear power.

9. International Energy Agency, *World Energy Outlook 2006*, Paris, France: Organization for Economic Cooperation and Development (OECD), p. 218.

10. The full list includes establishing a country's national position, legal and regulatory frameworks, financing, safeguards, energy planning, nuclear waste, nuclear safety, stakeholder involvement, management, procurement, radiation protection, human resource development, security and physical protection, the nuclear fuel cycle, environmental protection, sites and support facilities, the electrical grid, and industrial involvement. See IAEA, *Milestones in the Development of a National Infrastructure for Nuclear Power*, available from *www-pub.iaea.org/MTCD/publications/PDF/Pub1305_web.pdf*.

11. See summary of "Roles and Responsibilities of Vendor Countries and Countries Embarking on Nuclear Power Programmes to Ensure Long-Term Safety," a workshop organized by the IAEA Division of Nuclear Installation Safety in July 2008, available from *www.iaea.org/NewsCenter/News/2008/nuclnewcomers.html*.

12. Akira Omoto, Director, "IAEA support to infrastructure building in countries considering introduction of nuclear power," briefing to Division of Nuclear Power, IAEA, 2008.

13. India has built pressurized, heavy water-moderated reactors, based on Canada's CANDU reactor design. Because these reactors do not need to be shut down to be refueled, they can be easily optimized for producing weapons-grade plutonium. For the same reason, these reactors are difficult to monitor to detect military diversions. These reactors also can be fueled with natural uranium, which requires no enrichment. As a result, these

reactors and related technology are labeled as "sensitive nuclear technology" under U.S. nuclear nonproliferation law.

14. *The Future of Nuclear Power: An Interdisciplinary MIT Study*, Cambridge, MA: Massachusetts Institute of Technology, 2003, made projections of per capita electricity growth rates, assuming states would place a priority on reaching the benchmark 4,000 kWh per capita consumption level. Based on a pattern of electricity consumption, the study then estimated the proportion of nuclear power generating electricity, taking into account current nuclear power deployment, urbanization, stage of economic development, and energy resource base.

15. *The Future of Nuclear Power: An Interdisciplinary MIT Study*, Cambridge, MA: Massachusetts Institute of Technology, 2003. p. 110.

16. France, the UK, Russia, India, and the United States, however, have used unsafeguarded reactors that were designed to produce power and weapons plutonium or tritium. The reactor types they have used include heavy water reactors, gas cooled reactors, RMBKs, the U.S. Hanford Reactor, and, in the case of U.S. production of tritium, LWRs. On these points, see Zia Mian, A. H. Nayyar, R. Rajaraman, and M. V. Ramana, "Fissile Materials in South Asia and the Implications of the U.S.-India Nuclear Deal," Research Report No. 1, International Panel on Fissile Materials, September 2006; Walter Paterson, *Nuclear Power*, London, UK: Penguin Books, 1977, pp. 49-55; Lawrence Scheinman, *Atomic Energy Policy in France under the Fourth Republic*, Princeton, NJ: Princeton University Press, 1965, pp. 69, 93, 155; Alexander M. Dmitriev, "Converting Russian Plutonium Production Reactors to Civilian Use," *Science and Global Security*, Vol. 5, pp. 37-46; Arms Control Association, "Tritium Production Licenses Granted to Civilian Power Plants," *Arms Control Today*, November 2002.

17. One estimate is that indigenous centrifuge enrichment becomes cost effective at the capacity level of 1.5 million separative work units, an amount required by 10 1-gigawatt plants. Other estimates are higher, but as the price of uranium goes up, domestic production becomes more competitive with buying enrichment services on the open market. Even then, such an enrichment plant is unlikely to be competitive with larger suppliers such as Urenco.

# APPENDIX

# MAPPING GLOBAL NUCLEAR ENERGY EXPANSION

# DESCRIPTION OF SCENARIOS AND SOURCES

The maps are based on estimates of nuclear power capacity under three different scenarios. The first is a "business as usual" projection for 2030 done by the Energy Information Administration (EIA). EIA nuclear energy projections are essentially done "off-line," that is, the sophisticated computer model for estimating other sources of energy is not used for the nuclear case. This is partly because decisions about the retirement of reactors and new reactors, particularly in Western Europe, are difficult to model. In addition, the estimates are aggregated into regions, with just a few country-specific breakouts.

Scenario II is not a projection, but rather an estimate based on official statements by countries, for which a variety of sources was used. Country statements were taken at face value and do not necessarily correlate to any measurable indicators (such as gross domestic product [GDP] growth or electricity demand, etc.). In some cases, the plans are unlikely to materialize. Scenario II figures should be regarded merely as a "wish list" for many countries.

Scenario III seeks to estimate nuclear energy in 2050. It uses figures from the 2003 study by MIT, *The Future of Nuclear Power,* specifically, the "High 2050" scenario in Appendix 2, Global Electricity Demand and the Nuclear Power Growth Scenario, with some minor variations.[1] The MIT study used an underlying assumption that the developed countries would continue with a modest annual increase in per capita electricity use, and the developing countries would move to the 4,000 kWh per person per year benchmark, if at all feasible (the 4,000 kWh benchmark being the dividing line between developed and advanced countries). Electricity demand was then pegged to estimated population growth. Finally, it was assumed that nuclear energy would retain or increase its current share of electricity generation. The least-well-off developing countries were assumed in the MIT study not to have the wherewithal for nuclear energy. A final caveat regarding the MIT study is that the 2050 projection is "an attempt to understand what the distribution of nuclear power deployment would be if robust growth were realized, perhaps driven by a broad commitment to reducing greenhouse gas emissions and a concurrent resolution of the various challenges confronting nuclear power's acceptance in various countries."[2] A few countries that the MIT High 2050 case included but are not included here have laws currently prohibiting nuclear energy, such as Austria.

## CAVEATS

There is a good reason why the EIA and IEA do not make projections out to 2050—it is a highly uncertain undertaking. Some of the many uncertainties include input and construction costs, government support, and reactor operation safety. As seen from experience since Three Mile Island and Chernobyl, plans for nuclear power plant construction can be put off indefinitely in the wake of accidents.

# EXPLANATORY NOTES FOR REACTOR DATA

All figures are rounded to the nearest integer and expressed in gigawatts, electrical (GWe) (if less than 0.5 GWe, however, it has been rounded to 0.5). The organization of the data along OECD and non-OECD groupings reflects the availability of EIA projections under Scenario I. In particular, the EIA does not make projections for individual countries except where noted. Therefore, the countries are grouped by region.

In Scenario I, blank entries should not necessarily be equated with no nuclear capacity; unfortunately, the EIA does not always make individual country projections. The regional projections will include nuclear capacity for those countries that already have nuclear energy today.

In Scenarios II and III, blank entries indicate no nuclear capacity or plans, or lack of specificity in the data. There are several cases where a country has proposed power plants under Scenario II but no figure appears under Scenario III, because the MIT 2050 High Scenario did not anticipate any nuclear power development in the least developed countries, including Bangladesh, Ghana, Nigeria, and Yemen. Other states that the MIT study did not include but might build nuclear power by 2050 are the GCC states Jordan and Tunisia, and Chile.

In addition, there are several cases where a country has no current nuclear power plans, but the MIT study predicts nuclear power for them in 2050. These countries include New Zealand, Australia, Austria, Italy, Portugal, the Philippines, and Venezuela. Several countries included in the 2050 MIT projections were not included in the maps or in the data below.

Finally, there are several "placeholder" slots, where countries have expressed plans for nuclear energy, but there are no associated numbers of reactors or capacity. These include Syria (which announced it would like to generate 6 percent of its energy needs by 2020 with nuclear energy in a 2006 statement to IAEA) and Ghana (which told IAEA in 2006 it would like to introduce nuclear energy by 2020), among others.

Current:  2010 nuclear power capacity, based on Power Reactor Information Systems (PRIS), IAEA

Scenario I:  2030 — Data from Energy Information Administration, International Energy Outlook 2007, DOE/EIA-0484(2010)

Scenario II:  2030 — Proposed reactor capacities according to individual government statements. Sources are varied, but include World Nuclear Association, *Nucleonics Week,* and major trade press.

Scenario III:  2050 — MIT projection, new or expanded nuclear power capacity

| OECD | | | | |
|---|---|---|---|---|
| Country | Current | Scenario I | Scenario II | Scenario III |
| Australia | 0 | 0 | 0 | 10 |
| Canada | 13 | 17 | 22 | 62 |
| Japan | 49 | 60 | 67 | 91 |
| Korea,S | 18 | 32 | 38 | 37 |
| Mexico | 1 | 2 | 3 | 20 |
| New Zealand | 0 | 0 | 0 | 1 |
| OECD Europe (see breakout below) | 130 | 142 | 181 | 237 |
| Turkey | 0 | | 5 | 9 |
| USA | 101 | 111 | 144 | 477 |
| Regional Total | 312 | 364 | 460 | 944 |

| NON-OECD EUROPE/EUR ASIA | | | | |
|---|---|---|---|---|
| Country | Current | Scenario I | Scenario II | Scenario III |
| Non-OECD Europe | 18 | 28 | 62 | 25 |
| Russia | 23 | 46 | 43 | 52 |
| Regional Total | 41 | 74 | 105 | 77 |

## NON-OECD ASIA

| Country | Current | Scenario I | Scenario II | Scenario III |
|---|---|---|---|---|
| Bangladesh | 0 | | 2 | 0 |
| China | 9 | 65 | 200 | 200 |
| India | 4 | 24 | 63 | 175 |
| Indonesia | 0 | | 6 | 39 |
| North Korea | 0 | | 1 | 5 |
| Malaysia | 0 | | | 3 |
| Pakistan | 0.5 | 0 | 9 | 20 |
| Philippines | 0 | | 2 | 9 |
| Taiwan | 5 | | 7 | 16 |
| Thailand | 0 | | 5 | 8 |
| Vietnam | 0 | | 15 | 5 |
| Other Asia | 0 | 16 | 0 | 0 |
| Regional Total | 18.5 | 105 | 310 | 480 |

## MIDDLE EAST

| Country | Current | Scenario I | Scenario II | Scenario III |
|---|---|---|---|---|
| UAE | 0 | 5 | 6 | 0 |
| Iran | 0 | 1 | 20 | 22 |
| Israel | 0 | | 1 | 2 |
| Jordan | 0 | | 2 | 0 |
| Syria | 0 | | | 0 |
| Yemen* | 0 | | 0 | 0 |
| Regional Total | 0 | 6 | 29 | 24 |

| AFRICA | | | | |
|---|---|---|---|---|
| Country | Current | Scenario I | Scenario II | Scenario III |
| Algeria | 0 | | 2 | 5 |
| Egypt | | | 4 | 10 |
| Ghana | 0 | | 1 | 0 |
| Libya | 0 | | | 1 |
| Morocco | 0 | | 2 | 3 |
| Namibia | 0 | | 0 | 0 |
| Nigeria* | 0 | | 5 | 0 |
| South Africa | 2 | 3 | 8 | 15 |
| Tunisia* | 0 | | 1 | 0 |
| Regional Total | 2 | 3 | 23 | 34 |

| CENTRAL AND SOUTH AMERICA | | | | |
|---|---|---|---|---|
| Country | Current | Scenario I | Scenario II | Scenario III |
| Argentina | 1 | 2 | 5 | 10 |
| Brazil | 2 | 4 | 12 | 34 |
| Chile | 0 | | | 0 |
| Venezuela | 0 | | | 4 |
| Regional Total | 3 | 6 | 17 | 48 |

| WORLD TOTAL | | | | |
|---|---|---|---|---|
| World Total | 377** | 558 | 944 | 1,607 |

NOTES:

Asterisks (*) depict countries that are not included in Maps 6 or 7 but have possible GWe figures for Scenario II. These Scenario II figures were not included in the map because nuclear planning for these countries is still in the early exploratory phase.

The EIA has stated that the Africa region will produce 3 GWe of nuclear power by 2030. This table assumes this will be produced in South Africa. The country already produces nuclear power and does not face the barriers other African countries will face in developing a new nuclear power industry.

Asterisks (**) 374.6 is PRIS's world total. Our numbers do not add up to this precisely due to rounding. The EIA has stated that the Africa region will produce 3 GWe of nuclear power by 2030. This table assumes this will be produced in South Africa. The country already produces nuclear power and does not face the barriers other African countries will face in developing a new nuclear power industry.

# BREAKOUTS OF OECD EUROPE AND NON-OECD EUROPE

| OECD EUROPE | | | | |
|---|---|---|---|---|
| Country | Current | Scenario I | Scenario II | Scenario III |
| Belgium | 6 | | 0 | 11 |
| Czech Republic | 3 | | 6 | 3 |
| Finland | 3 | | 5 | 8 |
| France | 63 | | 67 | 68 |
| Germany | 20 | | 0 | 49 |
| Hungary | 2 | | 4 | 3 |
| Italy | 0 | | | 8 |
| Netherlands | 0.5 | | 1 | 4 |
| Norway | 0 | | | 5 |
| Poland | 0 | | 3 | 3 |
| Portugal | 0 | | | 1 |
| Slovakia | 2 | | 5 | 3 |
| Spain | 7 | | 7 | 18 |
| Sweden | 9 | | 9 | 16 |
| Switzerland | 3 | | 4 | 5 |
| UK | 11 | | 10 | 32 |
| Total | 129.5 | 113 | 121 | 237 |

| NON-OECD EUROPE | | | | |
|---|---|---|---|---|
| Country | Current | Scenario I | Scenario II | Scenario III |
| Albania | 0 | | 1 | 0 |
| Armenia | 0.5 | | 1 | 1 |
| Azerbaijan | 0 | | 1 | 1 |
| Belarus | 0 | | 4 | 1 |
| Bulgaria | 2 | | 4 | 3 |
| Georgia | 0 | | | 0 |
| Kazakhstan | 0 | | 0.5 | 1 |
| Kyrgyzstan | 0 | | | 1 |
| Lithuania | 1 | | 3 | 1 |
| Romania | 1 | | 3 | 2 |
| Slovenia | 1 | | 2 | 1 |
| Turkmenistan | 0 | | | 1 |
| Ukraine | 13 | | 42 | 8 |
| Uzbekistan | 0 | | | 4 |
| Total | 18.5 | 23 | 62 | 25 |

## NOTES:

Scenario I EIA projections are done primarily by region and blank spaces should not be considered to reflect zero nuclear power. Please refer to the regional totals only in Scenario I. In Scenario II, blank spaces may indicate lack of data about number or capacity of reactors, even as countries have declared interest in nuclear power.

| ENRICHMENT CAPACITIES (Millions of Separative Work Units [SWU]) | | |
|---|---|---|
| Country | 2009 | Planned (~2015) |
| Russia | 21 | 27 |
| France | 10.8 | 10.8 |
| United Kingdom | 5 | 5 |
| Germany | 2.2 | 2.8 |
| Netherlands | 3.8 | 4.4 |
| Japan | .15 | 1.5 |
| China | 1.4 | 1.9 |
| United States | 11.3 | 27.3 |
| USEC | 11.3 | 15.1 |
| URENCO US | | 5.7 |
| AREVA US | | 3.0 |
| GE Hitachi | | 3.5 |
| Brazil | 0.12 | 0.2 |
| Iran | 0.25 | 0.25 |
| India | 0.1 | |
| Pakistan | 0.2 | |

Source: Data from M. D. Laughter, "Profile of World Uranium Enrichment Programs-2009," National Nuclear Security Administration, April 2009, available from *www.fas.org/nuke/guide/enrich. pdf*.

Planned (~2015) = Countries and companies' stated plans for enrichment capacity.

Scenario III figures are estimates based on whether a state is projected to have at least 10 GWe nuclear capacity in 2050 and has expressed an interest (even if tentative) in uranium enrichment.

| ENRICHMENT REQUIREMENTS (Millions of separative work units, or SWU) | | |
| --- | --- | --- |
| Country | Scenario II (2030) | Scenario III (2050) |
| Albania | .15 | |
| Algeria | .3 | .75 |
| **Argentina** | .75 | 2 |
| Armenia | .15 | .15 |
| **Australia** | | 2 |
| Azerbaijan | .15 | .15 |
| Belarus | .6 | .15 |
| **Belgium** | .9 | 2 |
| **Brazil** | 2 | 5 |
| Bulgaria | .6 | .45 |
| **Canada** | 3 | 9 |
| **China** | 30 | 30 |
| Czech Republic | 1 | .45 |
| **Egypt** | .6 | 2 |
| Finland | 1 | 1 |
| **France** | 10 | 10 |
| **Germany** | 3 | 7 |
| Hungary | .6 | .45 |
| **India** | 9 | 26 |
| **Indonesia** | 1 | 6 |
| **Iran** | 3 | 3 |
| Israel | .15 | .3 |
| **Japan** | 10 | 14 |
| Jordan | .3 | |
| Kazakhstan | .15 | .15 |
| Kyrgyzstan | | .15 |
| Libya | | .15 |
| Lithuania | .45 | .15 |
| Malaysia | | .45 |
| **Mexico** | .3 | .3 |
| Morocco | .3 | .45 |
| Netherlands | .3 | .45 |
| New Zealand | | .15 |

| | | |
|---|---|---|
| North Korea | .15 | .75 |
| Norway | | .75 |
| **Pakistan** | 1 | 3 |
| Philippines | .3 | 1 |
| **Poland** | 2 | .45 |
| Portugal | | .15 |
| Romania | .45 | .3 |
| **Russia** | 6 | 8 |
| Slovakia | .75 | .45 |
| Slovenia | .3 | .15 |
| **South Africa** | 1 | 2 |
| **South Korea** | 6 | 6 |
| **Spain** | 1 | 3 |
| **Sweden** | 1 | 2 |
| Switzerland | .9 | .75 |
| **Taiwan** | 1 | 2 |
| Thailand | .75 | 1 |
| **Turkey** | 2 | 1 |
| Turkmenistan | | .15 |
| UAE | 1 | |
| **Ukraine** | 6 | 1 |
| **UK** | 2 | 5 |
| **USA** | 22 | 72 |
| Uzbekistan | | .6 |
| **Vietnam** | 2 | .75 |
| Venezuela | | .6 |

**NOTES:**

Numbers are rounded to nearest integer except when less than 1. Countries in bold text are projected to have 10 or more GWe capacity under at least one of the projections and might consider independent enrichment as a result.

# ENDNOTES - APPENDIX

1. *The Future of Nuclear Power: An Interdisciplinary MIT Study*, Cambridge, MA: Massachusetts Institute of Technology, 2003.

2. *Ibid.*, p. 110.

# CHAPTER 3

## NUCLEAR POWER:
## CLIMATE FIX OR FOLLY?

**Amory B. Lovins**
**Imran Sheikh**
**Alex Markevich**

Nuclear power, we are told, is a vibrant industry that is dramatically reviving because it is proven, necessary, competitive, reliable, safe, secure, widely used, increasingly popular, and carbon-free — a perfect replacement for carbon-spewing coal power. New nuclear plants thus sound vital for climate protection, energy security, and powering a vibrant global economy.

There is a catch though, the private capital market is not investing in new nuclear plants, and without financing, capitalist utilities are not buying. The few purchases, nearly all in Asia, are all made by central planners with a draw on the public purse. In the United States even the new government subsidies of 2005, which approach or exceed the total cost of new nuclear plants, failed to entice Wall Street to put a penny of its own capital at risk during what were, until autumn 2008, the most buoyant markets and the most nuclear-favorable political and energy-price conditions in history — conditions that have largely reversed since then.

This semi-technical chapter, summarizing a detailed and documented technical paper,[1] compares the cost, climate protection potential, reliability, financial risk, market success, deployment speed, and energy contribution of new nuclear power with those of its low- or no-carbon competitors. It explains why soaring taxpayer subsidies have not attracted investors. Instead, capitalists favor climate-protecting competitors with lower cost, shorter construction time, and less financial risk. The nuclear industry claims it has no serious rivals, let alone those competitors—which, however, already outproduce nuclear power worldwide and are growing enormously faster.

Most remarkably, comparing the abilities of all options to protect the earth's climate and enhance energy security reveals why nuclear power *could never deliver* these promised benefits even if it *could* find free-market buyers—while its carbon-free rivals, which won more than $90 billion of private investment in 2007 alone,[2] do offer highly effective climate and security solutions, much sooner and with higher confidence.

## UNCOMPETITIVE COSTS

*The Economist* observed in 2001 that "Nuclear power, once claimed to be too cheap to meter, is now too costly to matter—cheap to run but very expensive to build."[3] Since then, it has become even more costly to build, and in a few years, as old fuel contracts expire, it is expected to become more expensive to run.[4] Its total cost now markedly exceeds that of coal- and gas-fired power plants, let alone the cheaper decentralized competitors described below.

Worldwide construction costs have risen far faster for nuclear than for non-nuclear plants. This is not, as commonly supposed, due primarily to higher metal and cement prices: repricing the main materials in a 1970s U.S. plant (an adequate approximation) to March 2008 commodity prices yields a *total* Bill of

Materials cost only ~1 percent of today's overnight capital cost. Rather, the real capital-cost escalation is due largely to the severe atrophy of the global infrastructure for making, building, managing, and operating reactors. This forces U.S. buyers to pay in weakened dollars, since most components must now be imported. It also makes worldwide buyers pay a stiff premium for serious shortages and bottlenecks in engineering, procurement, fabrication, and construction: some key components have only one source worldwide. The depth of the decline is revealed by the industry's flagship Finnish project, led by France's top builder, which after 3 years of construction, is at least 3 years behind schedule and 50 percent over budget. An identical second unit, gratuitously bought in 2008 by the 85 percent-state-owned Électricité de France to support the 91 percent-state-owned vendor Areva (orderless 1991–2005), was bid ~25 percent higher than the Finnish plant and without its fixed-price guarantee, and suffered prompt construction shutdowns for poor quality.

The rapid escalation of U.S. nuclear capital costs can be seen by comparing the two evidence-based studies[5] with each other and with later industry data (all including financing costs, except for the two "overnight" costs, but with diverse financing models — see Table 3-1). As the Director of Strategy and Research for the World Nuclear Association candidly put it, "[I]t is completely impossible to produce definitive estimates for new nuclear costs at this time. . . ."[6]

| Date | Source | Capital Cost (2007 $/net el. W) | Levelized Busbar Cost, 2007 $/MWh |
|---|---|---|---|
| 7/03 | MIT | 2.3 | 77–91 |
| 6/07 | Keystone | 3.6–4.0 | 83–111 |
| 5/07 | S&P | ~4 | |
| 8/07 | AEP | ~4 | |
| 10/07 | Moody's | 5–6 | |
| 11/07 | Harding | 4.3–4.6 | ~180 |
| 3/08 | FPL filing | ~4.2–6.1 [3.1–4.5 overnight] | |
| 3/08 | Constellation | [3.5–4.5 overnight] | |
| 5/08 | Moody's | ~7.5 | 150 |
| 6/08 | Lazard | 5.6–7.4 | 96–123 |
| 11/08 | Duke Power | [4.8 overnight] | |

**Table 3-1. Escalating U.S. Nuclear Construction Cost Estimates (Including Interest and Real Escalation Unless [Overnight]), 2003–08 (2009–10 Continue the Trend).[7]**

By 2007, as Figure 3-1 shows, nuclear power was the costliest option among all main competitors, whether using MIT's authoritative but now low 2003 cost assessment, the Keystone Center's mid-2007 update (top of nuclear plant bar), or later and even higher industry estimates (Moody's arrow).[8] For plants ordered in 2009, formal studies have not yet caught up with the latest data, but it appears that their new electricity would probably cost (at your meter, not at the power plant) around 10–13¢/kWh for coal rather than the 9¢ shown, about 9–13¢/kWh for combined-cycle gas rather than the nearly 10¢ shown, but around 15–21¢/kWh for new nuclear rather than the 11–15¢ shown.[9] However, nuclear's decentralized competitors have suffered far less, or even negative, cost escalation, for example, the average price of electricity sold by new U.S. windfarms *fell* slightly in 2007.[10] The 4.0¢/kWh average windpower price for projects installed in 1999–2007 seems to be more representative of a stable forward market, and corresponds to ~7.4¢/kWh delivered and firmed — just one-half to one-third of new nuclear power costs on a fully comparable basis.

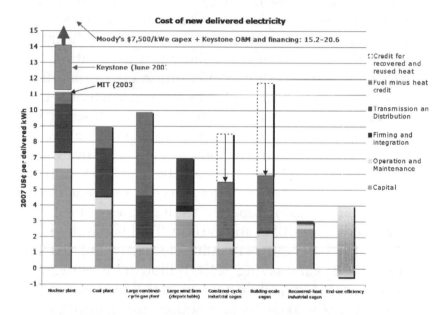

**Figure 3-1. An Apples-to-Apples Comparison of the Cost of Making and Delivering a New Firm kWh of Electrical Services in the United States Based on Empirical ~2007 Market Costs and Prices.**

**Noncentral Station Competitors.**

Cogeneration and efficiency are "distributed resources," usually located near where energy is used. Therefore, they do not incur the capital costs and energy losses of the electric grid, which links large power plants and remote wind farms to customers.[11] Wind farms, like solar cells, also require "firming" to steady their variable output, and all types of generators require some backup for when they inevitably break.[12] Figure 3-1 reflects these costs.

Making electricity from fuel creates large amounts of by-product heat that is normally wasted. Combined-cycle industrial cogeneration and building-scale cogeneration recover most of that heat and use it to displace the need for separate boilers to heat the industrial process or the building, thus creating the economic "credit" shown in Figure 3-1. Cogenerating electricity and some useful heat from currently discarded industrial heat is even cheaper because no additional fuel is needed, so no additional carbon is released—only what the factory was already emitting.[13]

End-use efficiency, by far the cheapest option, wrings more (and often better) services from each kilowatt-hour by using smarter technologies—substituting brains for dollars and carbon. That is mainly how California has held per-capita electricity use flat for the past 30 years, saving ~$100 billion of investment to supply electricity, while per-capita real income rose 79 percent (1975–2005). Its new houses, for example, now use one-fourth the energy they used to. Yet California is further accelerating all its efficiency efforts because there is so much still to save. McKinsey & Company has found that efficiency can profitably offset 85 percent of the normally projected growth in U.S. electricity consumption to 2030.[14] Just using all U.S. electricity as productively as the top 10 states now do (in terms of gross state product per kWh consumed, roughly adjusted for economic mix and climate) would save about 1,200 TWh/y—~62 percent of the output of U.S. coal-fired plants.[15]

Saving electricity costs far less than producing and delivering it, even from *existing* plants. California investor-owned utility efficiency programs cost an average of 1.2¢/kWh in 2004, and 83 Pacific Northwest utility programs cost 1.3¢/kWh.[16] The national average is about 2¢, but hundreds of utility programs (mainly for businesses, where most of the cheap savings are) cost less than 1¢.[17]

A major power engineering firm helped investment firm Lazard compare observed U.S. prices, finding that efficiency and many renewables cost less than a new central plants (see Figure 3-2). Lazard's recent comparison shows most centralized options beating all new central stations; this chart omits cogeneration, overstates wind costs, and understates nuclear costs. [18]

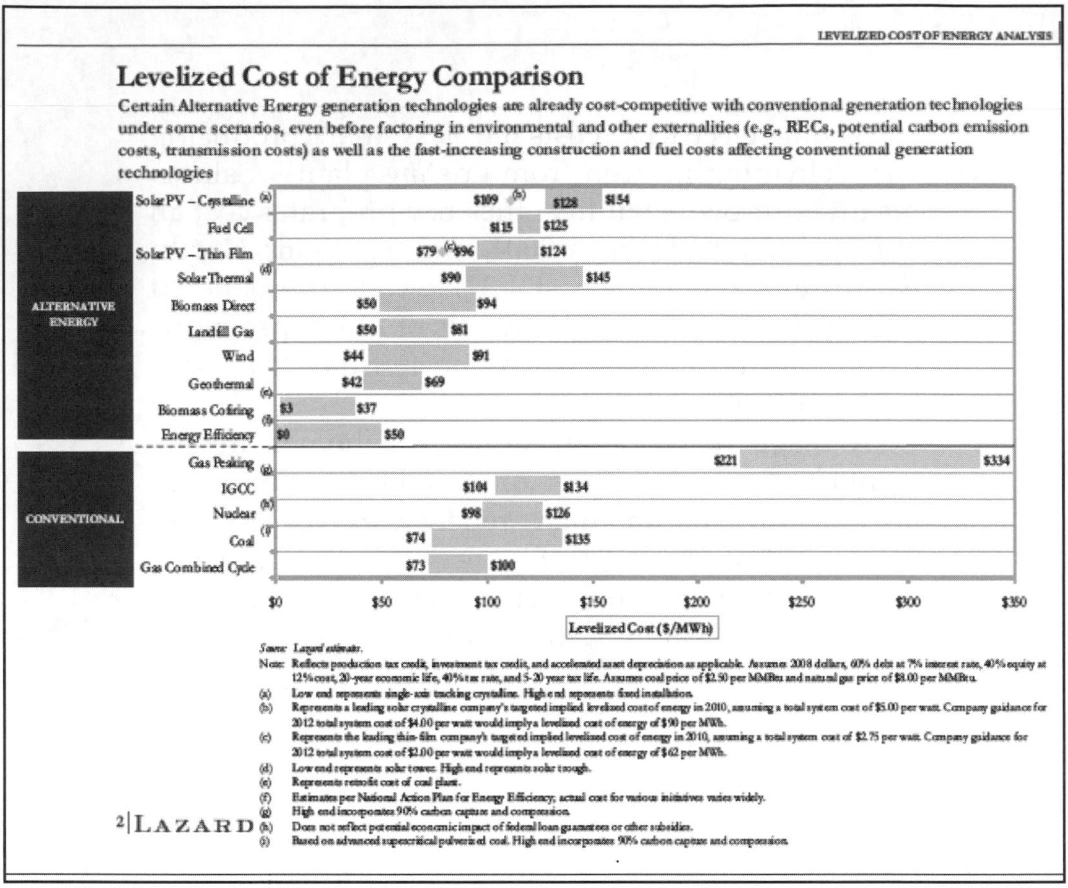

## Levelized Cost of Energy Comparison

Certain Alternative Energy generation technologies are already cost-competitive with conventional generation technologies under some scenarios, even before factoring in environmental and other externalities (e.g., RECs, potential carbon emission costs, transmission costs) as well as the fast-increasing construction and fuel costs affecting conventional generation technologies

Figure 3-2. Lazard's Levelized Cost of Energy Comparison.

## WHY THESE COMPARISONS UNDERSTATE THE LACK OF COMPETITIVENESS OF NUCLEAR POWER

These conventional results and assessments greatly understate the size and profitability of today's electric efficiency potential. In 1990, the utilities think-tank EPRI and RMI, in a joint article, assessed the potential to be as ~40–60 percent and ~75 percent, respectively, at average 2007-$ costs of about 3 and 1¢/kWh.[19] Now both those estimates look conservative, for two reasons:

1. As EPRI suggests, efficiency technologies have improved faster than they have been applied, so the potential savings keep getting bigger and cheaper.[20]

2. As RMI's work with many leading firms has demonstrated, integrative design can often achieve radical energy savings at *lower* cost than small or no savings.[21] That is, efficiency can often *reduce* total investment in new buildings and factories, and even in some retrofits that are coordinated with routine renovations.[22]

Wind, cogeneration, and end-use efficiency already provide electrical services more cheaply than central thermal power plants, whether nuclear or fossil-fueled. *This cost gap will only widen*, since central thermal power plants are largely mature and getting costlier, while their competitors continue to improve rapidly. Indeed, a good case can be made that photovoltaics (PVs) can *already* beat new thermal power plants. For example, if you start in 2010 to build a new 500-MW coal-fired power plant in New Jersey, plus an adjacent photovoltaic (PV) power plant, before the coal plant comes online in 2018, the solar plant will produce a slightly larger amount of annual electricity at lower levelized cost, but with 1.5x more onpeak output, and the PV manufacturing capacity used to build your plant can then add 750 more MW *each year*.[23] Of course, the high costs of conventional fossil-fueled plants would go even higher if their large carbon emissions had to be captured—but this coal/solar comparison assumes a carbon price of *zero*.

The foregoing cost comparison is conservative for four important *additional* reasons:

1. End-use efficiency often has side-benefits worth 1–2 orders of magnitude (factors of 10) more than the saved energy.[24]

2. End-use efficiency and distributed generators have 207 "distributed benefits" that typically increase their economic value by an order of magnitude.[25] The *only* distributed benefit counted above is reusing waste heat in cogeneration.

3. Integrating variable renewables with each other typically saves over half their capacity for a given reliability;[26] indeed, diversified variable renewables, forecasted and integrated, typically need *less* backup investment than big thermal plants for a given reliability.

4. Integrating strong efficiency with renewables typically makes both of them cheaper and more effective.[27]

The uncompetitiveness of new nuclear power is clear without these four conservatisms, and is overwhelming with them. As we will see, the marketplace concurs—and that is good news for the global climate.

## UNCOMPETITIVE CO$_2$ DISPLACEMENT

Nuclear plant operations emit no carbon directly and rather little indirectly.[28] Nuclear power is therefore touted as the key replacement for coal-fired power plants. But this seemingly straightforward substitution could be done instead by using *non*-nuclear technologies that are cheaper and faster, so they yield more climate solution per dollar and per year.

As Figure 3-3 shows, various options emit widely differing quantities of CO$_2$ per delivered kilowatt-hour:[29] Coal is by far the most carbon-intensive source

of electricity, so displacing it is the yardstick of carbon displacement's effectiveness. A kilowatt-hour of
nuclear power does displace nearly all the 0.9-plus
kilograms of $CO_2$ emitted by producing a kilowatt-
hour from coal. But so does a kilowatt-hour from
wind, a kilowatt-hour from recovered-heat industrial
cogeneration, or a kilowatt-hour saved by end-use efficiency, and all three of these carbon-free resources
cost far less than nuclear power per kilowatt-hour, so
they save far more carbon per dollar.

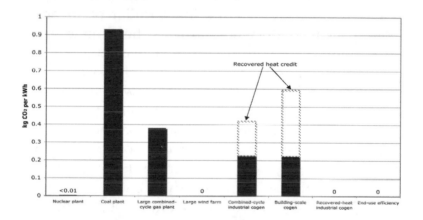

**Figure 3-3. Operating $CO_2$ Emitted Per Delivered
kWh.**

Combined-cycle industrial cogeneration and
building-scale cogeneration typically burn natural
gas, which does emit carbon (though half as much as
coal), so they displace somewhat less net carbon than
nuclear power can, around 0.7 kilograms of $CO_2$ per
kilowatt-hour.[30] Even though cogeneration displaces
less carbon than nuclear does per kilowatt-hour, it displaces more carbon than nuclear does *per dollar spent
on delivered electricity,* because it costs far less. With a

net delivered cost per kilowatt-hour approximately half of that of nuclear (using the most conservative comparison from Figure 3-1), cogeneration delivers twice as many kilowatt-hours per dollar, and therefore displaces around 1.4 kilograms of $CO_2$ for the same cost as displacing 0.9 kilograms of $CO_2$ with nuclear power.

Figure 3-4 compares the cost-effectiveness of different electricity options in reducing $CO_2$ emissions, counting both their cost-effectiveness (kilowatt-hours per dollar), and any carbon emissions. New nuclear power is so costly that shifting a dollar of spending from nuclear to efficiency protects the climate severalfold more than shifting a dollar of spending from coal to nuclear. Indeed, under plausible assumptions, spending a dollar on new nuclear power *instead of* on efficient use of electricity has a worse effect on climate than spending that dollar on new coal power! How much net carbon emissions from coal-fired power plants can be displaced by buying a dollar's worth of new electrical services using different technologies? Note that the carbon savings from realistic efficiency investments are far above the upper-right corner of the chart.

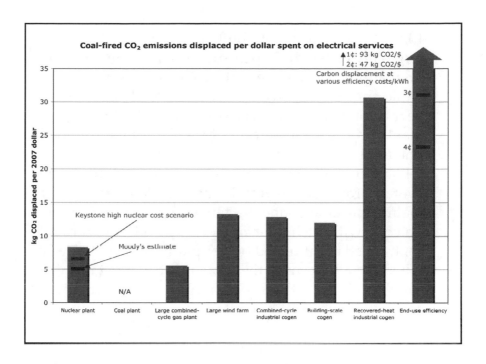

**Figure 3-4. Coal-Fired CO$_2$ Emissions Displaced per Dollar Spent on Electrical Services.**

If we are serious about addressing climate change, we must invest resources wisely to expand and accelerate climate protection. Since nuclear power is costly and slow to build, buying more of it rather than its cheaper and swifter rivals will instead reduce and retard climate protection.

## QUESTIONABLE RELIABILITY

*All* sources of electricity sometimes fail, differing only in how predictably, why, how often, how much, and for how long. Even the most reliable giant power plants are intermittent, they fail unexpectedly in billion-watt chunks, often for long periods. Of all

132 U.S. nuclear plants built (52 percent of the 253 originally ordered), 21 percent were permanently and prematurely closed due to reliability or cost problems, while another 27 percent have completely failed for a year or more at least once. The surviving U.S. nuclear plants produce ~90 percent of their full-time full-load potential, but even they are not fully dependable. Even reliably operating nuclear plants must shut down, on average, for 39 days every 17 months for refueling and maintenance, and unexpected failures do occur too. To cope with such intermittence by both nuclear and centralized fossil-fueled power plants, which typically fail about 8 percent of the time, utilities must install a roughly 15 percent "reserve margin" of extra capacity, some of which must be continuously fueled, spinning ready for instant use. Heavily nuclear-dependent regions are particularly at risk because drought, earthquake, a serious safety problem, or a terrorist incident could close many plants simultaneously.

Nuclear plants have an additional disadvantage, for safety, they must instantly shut down in a power failure, but for nuclear-physics reasons, they cannot be quickly restarted. During the August 2003 Northeast blackout, nine perfectly operating U.S. nuclear units had to shut down. After 12 days of a painfully slow restart process, their average capacity loss had exceeded 50 percent. For the first 3 days, just when they were most needed, their output was 3 percent below normal.

The big transmission lines that highly concentrated nuclear plants require are also vulnerable to lightning, ice storms, rifle bullets, cyber-attacks, and other interruptions.[31] The bigger our power plants and power lines get, the more frequent and widespread regional blackouts will become. Because 98–99 percent of power failures start in the grid, it is more reliable to bypass the grid by shifting to efficiently used, diverse, dispersed resources sited at or near the customer.

Additionally, a portfolio of many smaller units is unlikely to fail all at once because its diversity and dispersion make it more reliable even if its individual units are not.[32] The same logic applies to the two renewable electricity sources — windpower and photovoltaics — whose output varies with weather or daytime. Of course, the sun does not always shine on a given solar panel, nor does the wind always spin a given turbine. Yet, if properly firmed, both windpower, whose global potential is 35 times that of the world's electricity use,[33] and solar energy, all of which that strikes the earth's surface every ~70 minutes is equivalent to that used by humankind each year, can deliver reliable power without significant cost for backup or storage.[34] These variable renewable resources become *collectively* reliable when diversified in type and location and when integrated with three types of resources: steady renewables (geothermal, small hydro, biomass, etc.); existing fueled plants; and customer demand response. Such integration uses weather forecasting to predict the output of variable renewable resources, just as utilities now forecast demand patterns and hydropower output. In general, keeping power supplies reliable despite large wind and solar fractions may well require *less* backup or storage capacity than utilities *have already bought* to manage intermittence from big thermal stations. The renewable energy myth of unreliability has been debunked both by theory and by practical experience.[35]

## LARGE SUBSIDIES TO OFFSET HIGH FINANCIAL RISK

The latest U.S. nuclear plant proposed to be built is estimated to cost $12–24 billion (for 2.2–3.0 billion watts), much more than the industry's claims for new construction, and off the chart as shown in Figure 3-1. The utility's owner, a large holding company active in 27 states, has annual revenues of only $15 billion. Even before the current financial crisis, such high and highly uncertain capital costs made financing prohibitively expensive for free-market nuclear plants in the half of the United States that has restructured its electricity system. These high costs also make it prone to politically sensitive rate shock in the rest of the United States. For example, a new nuclear kilowatt-hour costing, say, 18 cents "levelized" over decades implies that the utility must collect ~30 cents to fund its first year of operation.

Lacking investors, nuclear promoters have turned back to taxpayers, who already bear most nuclear accident risks, have no meaningful say in licensing, and for decades have subsidized existing nuclear plants by ~1–8¢/kWh. In 2005, desperate for orders, the politically potent nuclear industry got those U.S. subsidies raised to ~5–9¢/kWh for new plants, or ~60–90 percent of their entire projected power cost, including new taxpayer-funded insurance against legal or regulatory delays. Wall Street still demurred. In 2007, the industry won relaxed government rules that made its 100 percent loan guarantees (for 80 percent debt financing) even more valuable. One utility's data indicated a cost of about $13 billion for a single new plant, which is almost equal to its entire capital cost. However, rising costs made the $4 billion of the new 2005 loan guarantees scarcely sufficient for a single reactor, so Congress raised taxpayer guarantees to $18.5 billion. Congress will soon be asked for another $30+ billion in loan guarantees, or even for a blank check (as both Houses separately approved in 2010). Meanwhile, the nonpartisan Congressional Budget Office has concluded that defaults are likely.

Wall Street is ever more skeptical that nuclear power is as robustly competitive as claimed. Starting with Warren Buffet, who recently abandoned a nuclear project because "it does not make economic sense," the smart money is heading for the exits.[36] The Nuclear Energy Institute is therefore trying to damp down the rosy expectations it created. It now says U.S. nuclear orders will come not in a tidal wave but in two little ripples — a mere 5–8 units coming online in 2015–16, then more if those are on time and within budget. Even that sounds dubious, as many senior energy-industry figures privately agree. In today's capital market, governments can have at most about as many nuclear plants as they can force taxpayers to buy. Indeed, the big financial houses that lobbied to be the vehicles of those gigantic federal loan guarantees are now largely gone; a new administration with many other priorities may be less supportive of such largesse; and the "significant" equity investment required to qualify for the loan guarantees seems even less likely to come from the same investors who declined to put their own capital at risk at the height of the capital bubble. The financial crisis has virtually eliminated private investment in big, slow, risky projects, while not materially decreasing investment in the small, fast, granular ones that were already walloping central plants in the global marketplace.

## THE MICROPOWER REVOLUTION

While nuclear power struggles in vain to attract private capital, investors have switched—and the financial crisis has accelerated their shift[37]—to cheaper, faster, less risky alternatives that *The Economist* calls "micropower"—distributed turbines and generators in factories or buildings (usually cogenerating useful heat), and all renewable sources of electricity *except* big hydro dams (those over 10 megawatts).[38] These alternatives surpassed the global capacity of nuclear power in 2002 and its electric output in 2006. Nuclear power now accounts for about 2 percent of worldwide electric capacity additions, *vs.* 28 percent for micropower (2004–07 average) and probably a good deal more in 2007–08.[39]

Despite subsidies that are generally smaller than those for nuclear power and many barriers to fair market entry and competition,[40] negawatts (electricity saved by using it more efficiently or timely) and micropower have lately turned in a stunning global market performance. Figure 3-5 shows how micropower's actual and industry-projected electricity production is running away from that of nuclear power, not even counting the roughly comparable additional growth in negawatts, nor any fossil-fueled generators under 1 megawatt.[41] Global electricity produced, or projected by industry to be produced, by decentralized low- or no-carbon resources—cogeneration ("CHP"), mostly gas-fired, and distributed renewables (those other than big hydroelectric dams). Micropower obtained over $100 billion of new private capital in 2007—roughly an eighth of the total global energy investment.

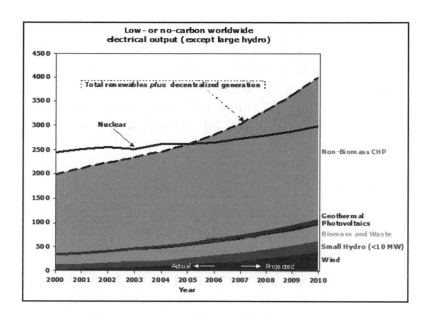

**Figure 3-5. Low- or No-Carbon Worldwide Electrical Output (Except Large Hydro).**

The nuclear industry nonetheless claims its only serious competitors are big coal and gas plants. But the marketplace has already abandoned that outmoded battleground for two others: central thermal plants vs. micropower, and megawatts vs. negawatts. For example, the United States added more windpower capacity in 2007 than it added coal-fired capacity in the past 5 years combined. By beating *all* central thermal plants, micropower and negawatts together provide about half the world's new electrical services. Micropower alone now provides a sixth of the world's electricity, and from a sixth to more than half of all electricity in 12 industrial countries, though the United States lags with ~6 percent.

In this broader competitive landscape, high carbon prices or taxes cannot save nuclear power from its fate.

If nuclear did compete only with coal, then far-above-market carbon prices might save it; but coal is not the competitor to beat. Higher carbon prices will advantage all other zero-carbon resources—renewables, re-covered-heat cogeneration, and negawatts—as much as nuclear, and will partly advantage fossil-fueled but low-carbon cogeneration as well. The nuclear industry does not understand this because it does not consider these competitors important or legitimate.

## SMALL IS FAST, LOW-RISK, AND HIGH-RISK IN TOTAL POTENTIAL

Small, quickly built units are faster to deploy for a given total effect than a few big, slowly built units. Widely accessible choices that sell like cellphones and PCs can add up to more and sooner than ponderous plants that get built like cathedrals. Small units are much easier to match to the many small pieces of electrical demand. Even a multi-megawatt wind turbine can be built so quickly that the United States will probably have a hundred billion watts of them (matching its nuclear capacity) installed before it builds its first one billion watts of new nuclear capacity, if any. As noted earlier, this speed reduces financial risk and thus makes decentralized, short-lead-time projects more financeable, especially in hard times.

Despite their small individual size, and partly because of it, micropower generators and electrical savings are already adding up to huge totals. Indeed, over decades, negawatts and micropower can shoulder the entire burden of powering the economy. The Electric Power Research Institute (EPRI), the utility think-tank, has calculated the U.S. negawatt potential (cheaper than just running an existing nuclear plant and delivering its output) to be two to three times the 19 percent share of nuclear power in the U.S. electricity market; RMI's more detailed analysis found even more. Cogeneration in factories can make as much U.S. electricity as nuclear does,[42] plus more in buildings, which use 69 percent of U.S. electricity. Windpower at acceptable U.S. sites can cost-effectively produce several times the nation's total electricity use,[43] and other renewables can make even more energy without significant land-use, variability, or other constraints. Thus, just cogeneration, windpower, and efficient use—all profitable today—can displace nuclear's current U.S. output by ~6–14 times over. This ratio becomes arbitrarily large when photovoltaics are included.

Nuclear power, with its decade-long project cycles, difficult siting, and (above all) unattractiveness to private capital, simply cannot compete. In 2006, for example, it added less global capacity than photovoltaics did, or a tenth as much as windpower added, or 30–41 times less than that added by micropower. Renewables other than big hydro dams won $56 billion of private risk capital; nuclear, as usual, got zero. China's distributed renewable capacity reached seven times its nuclear capacity and grew seven times faster. In 2007, China, Spain, and the United States each added more windpower capacity than the world added nuclear capacity. The nuclear industry trumpets its growth, yet micropower is already bigger and is growing 18 times faster.[44]

## SECURITY RISKS

President Bush rightly identified the spread of nuclear weapons as the gravest threat to America. Yet that proliferation is largely driven and greatly facilitated by the flow of materials, equipment, skills, and knowledge, all wrapped in the innocent-looking civilian disguise of nuclear power. (Reprocessing nuclear fuel, which President Bush tried to revive, greatly complicates waste management, increases cost, and boosts proliferation.) Yet by acknowledging the market failure of nuclear power and moving on to more secure, least-cost, energy options for global development would unmask and penalize proliferators by making bomb ingredients harder to get. This would make proliferation far more difficult, and easier to detect by focusing scarce intelligence resources on needles and not on haystacks.[45] The new administration has an extraordinary opportunity to turn the world away from its rush toward a "nuclear-armed crowd" by setting a good example in domestic energy policy and by helping all developing countries with the nonviolent, cheaper, faster energy alternatives that are already winning in the global market.[46]

Nuclear power has other unique challenges too, such as long-lived radioactive wastes, potential for catastrophic accidents, and vulnerability to terrorist attacks. But in a market economy, the technology could not proceed even if it lacked those issues, so we need not consider them here.

## CONCLUSION

So why do otherwise well-informed people still consider nuclear power a key element of a sound climate strategy? Not because that belief can withstand analytic scrutiny. Rather, it seems, because of a superficially attractive story, an immensely powerful and effective lobby, a new generation who forgot or never knew why nuclear power failed previously (almost nothing has changed), sympathetic leaders of nearly all main governments, deeply rooted habits and rules that favor giant power plants over distributed solutions and enlarged supply over efficient use, the market winners' absence from many official databases (which often count only big plants owned by utilities), and lazy reporting by an unduly credulous press.

Is it not time we forget about nuclear power? Informed capitalists have. Politicians and pundits should, too. After more than half a century of devoted effort and a half-trillion dollars of public subsidies, nuclear power still cannot make its way in the market. If we accept that unequivocal verdict, we can at last get on with the best buys first; proven and ample ways to save more carbon per dollar, faster, more surely, more securely, and with wider consensus. As we have seen before, the biggest key to a sound climate and a good security strategy is to take market economics seriously.

## ENDNOTES - CHAPTER 3

1. A. B. Lovins and I. Sheikh, "The Nuclear Illusion," *Ambio*, forthcoming, 2011, RMI Publ. #E08-01, preprinted at *www.rmi.org/images/PDFs/Energy/E08-01_AmbioNuclIllusion.pdf*, to be updated in 2010–11 for publication.

2. Justin Winter for Michael Liebreich (New Energy Capital, London, UK), personal communication, December 1, 2008, updating that firm's earlier figure of $71b for distributed renewable sources of electricity. The $90b is bottom-up, transaction-by-transaction and excludes M&A activity and other double-counting. Reliable estimates of investment in no-carbon (recovered-waste-heat) or relatively low-carbon (fossil-fueled) cogeneration are not available, but total global cogeneration investment in 2007 was probably on the order of $20b or more.

3. "A renaissance that may not come: this week, the Bush administration revealed an energy policy that strongly supports nuclear power," *The Economist*, Vol. 359, Issue 8222, May 19, 2001, pp. 24–26.

4. Due to prolonged mismanagement of the uranium and enrichment sectors: *Nuclear Power Joint Fact-Finding*, Keystone Center, June 2007, available from *www.keystone.org/spp/documents/FinalReport_NJFF6_12_2007(1).pdf*, estimated new fuel contracts will rise from the canonical ~0.5¢/kWh to ~1.2–1.7¢ for open or ~2.1–3.5¢ for closed fuel cycles.

5. *Ibid.*; "A renaissance that may not come."

6. Steve Kidd, "Escalating costs of new build: what does it mean?" *Nuclear Engineering International*, August 22, 2008, available from *www.neimagazine.com/storyprint.asp?sc=2050690*.

7. MIT, *The Future of Nuclear Power*, 2003, available from *web.mit.edu/nuclearpower/*. This is very conservatively used as the basis for all comparisons in this chapter, but we also show some later variants.

8. All monetary values in this article are in 2007 U.S. dollars. All values are approximate and representative of the respective U.S. technologies in 2007 except as noted. Capital and fuel costs are levelized over the lifespan of the capital investment. Analytic details are in Endnote 1, and for the underlying 2005 analysis, in A. B. Lovins, "Nuclear Power: Economics and Climate-Protection Potential," RMI Publ. #E05-14, January 6, 2006, available from *www.rmi.org/images/PDFs/Energy/E05-14_NukePwrEcon.pdf*, summarized in A. B. Lovins, "Mighty Mice," *Nuclear Engineering International*, December 2005, pp. 44-48, available from *www.rmi.org/images/PDFs/Energy/E05-15_MightyMice.pdf*.

9. Based, as in Figure 3-1, on the June 2007 Keystone findings adjusted to Moody's May 2008 capital cost, on the assumption that a somewhat stronger dollar might partly offset escalation. Anecdotal reports suggest that real capital cost escalation remains rapid in Europe and Asia, depending on exchange rates: for example, eight recent Asian plants look to end up costing ~$4/W, consistent with mid-2007 U.S. cost estimates.

10. From 4.8 in 2006 to 4.5¢/kWh, 0.9¢ higher than shown in Figure 3-1. U.S. wind turbines became 9 percent costlier during 2006–07, and may rise another ~10 percent in 2008, largely because rapid growth bottlenecked some key component supplies, but capacity factors improved, too: e.g., the average kW of Heartland wind projects installed in 2006 produced 35 percent more electricity than one installed in 1998–99, due mainly to better-designed turbines, higher hub heights, and better siting. All windpower data in this chapter are from R. Wiser and M. Bolinger, "Annual Report on U.S. Wind Power Installation, Cost, and Performance Trends: 2007," U.S. Department of Energy (DOE)/EERE, LBL-43025, May 2008, available from *www1.eere.energy.gov/windand hydro/pdfs/43025.pdf*. All windpower prices are net of some minor Renewable Energy Credit trading and of the U.S. Production Tax Credit whose levelized value is 1.0¢/kWh, far smaller than subsidies to central thermal power plants. See D. Koplow, "Energy Subsidy Links Pages," Washington DC: Earthtrack, 2005, available from *earthtrack.net/earthtrack/index.asp?page_id=177&catid=66*.

11. Distributed generators may rely on the power grid for emergency backup power, but such backup capacity, being rarely used, does not require a marginal expansion of grid capacity, as does the construction of new centralized power plants. Indeed, in ordinary operation, diversified distributed generators *free up* grid capacity for other users.

12. "Firming" is a term referring to the provision of a back-up capability to ensure continuous supply.

13. A similar credit for displaced boiler fuel can even enable this technology to produce electricity at negative net cost. The graph conservatively omits such credit (which is very site-specific) and shows a typical positive selling price. The cogeneration results shown are based on actual projects considered representative by a leading developer.

14. McKinsey & Company, "Reducing U.S. Greenhouse Gas Emissions: How Much at What Cost?," National Academies Summit on America's Energy Future, Washington DC, March 14, 2008, slide 7.

15. N. Mims, M. Bell, and S. Doig, "Assessing the Electric Productivity Gap and the U.S. Efficiency Opportunity," Snowmass, CO: Rocky Mountain Institute, January 2009, available from *www.rmi.org/rmi/Library/2009-08_AssessingElectricProductivitiyGap*.

16. C. Rogers, M. Messenger, and S. Bender, "Funding and Energy Savings from Investor-Owned Utility Energy Efficiency Programs in California for Program Years 2000 Through 2004," California Energy Commission, August 2005, available from *www.energy.ca.gov/2005publications/CEC-400-2005-042/CEC-400-2005-042-REV.pdf*; Tom Eckman, 1 May 2008 Northwest Power Planning Council memo "Conservation Savings—Status Report for 2005–07," available from *www.nwcouncil.org/news/2008/05/3.pdf*. For total societal cost, add ~30–80% depending on the sector.

17. For example, S. Nadel, *Lessons Learned*, NYSERDA 90-8, ACEEE, 1990. These 1980s results remain valid today because most U.S. utilities have invested so little in efficiency that their opportunities are more like those of the 58 firms whose 237 programs through 1988 yielded median program costs of 0.3¢/kWh for industrial savings, 0.9¢ for motor rebates, 1.2¢ for loans, and 1.4¢ for new construction rebates.

18. "Levelized Cost of Energy Analysis, v. 2.0," Lazard, June 2008, available from *www.narucmeetings.org/Presentations/2008%20EMP%20Levelized%20Cost%20of%20Energy%20-%20Master%20June%202008%20(2).pdf*.

19. A. Fickett, C. Gellings, and A. B. Lovins, "Efficient Use of Electricity," *Scientific Amererican*, Vol. 263, No. 3, 1990, pp. 64–74. The difference, analyzed by E. Hirst in ORNL/CON-312 (2001), was nearly all methodological, not substantive, A. B. and L. H. Lovins, "Least-Cost Climatic Stabilization," *Annual Review of Energy and the Environment*, Vol. 16, 1991, No. 433–531, pp. 8-11, available from *www.rmi.org/images/PDFs/Energy/E9133_LstCostClimateStabli.pdf*, for example, EPRI excluded but RMI included saved maintenance cost as a credit against efficiency's capital cost, so their respective average costs of commercial lighting retrofits (~1986 $) were +1.2 and –1.4¢/kWh; EPRI examined potential savings only to 2000 (including 9–15% expected to occur spontaneously), while RMI counted the full long-term retrofit potential;

and EPRI assumed drivepower savings 3x smaller and 5x costlier than EPRI adopted elsewhere in the same *Scientific American* article. RMI's assessment summarized a 6-volume 1986–92 analysis of ~1,000 technologies' measured cost and performance.

20. RMI estimated that during 1984–89, U.S. efficiency potential roughly doubled while its real cost fell by threefold. Since 1990, mass production (often in Asia), cheaper electronics, competition, and better technology, according to James K. Rogers PE, cut the real cost of electronic T8 ballasts by >90 percent to 2003 (while lumens per watt rose 30 percent), turned direct/indirect luminaires from a premium to the cheapest option, and cut the real cost of industrial variable-speed drives by ~83–97 percent (some vendors of midsize motors now give them away). Compact fluorescent lamps became 85–94 percent cheaper during 1983–2003; window air-conditioners got 69 percent cheaper since 1993 while becoming 13 percent more efficient; and low-emissivity window coatings became ~84 percent cheaper in just 5 years.

21. Integrative design produces these expanding (not diminishing) returns to efficiency investments: A. B. Lovins, "Energy End-Use Efficiency," Rocky Mountain Institute, 2005, available from *www.rmi.org/images/PDFs/Energy/E95-28_SuperEffBldgFrontier.pdf*, further elucidated in the senior author's five public lectures, "Advanced Energy Efficiency," delivered at Stanford's School of Engineering in March 2007 and available from *www.rmi.org/stanford*. RMI's recent redesigns of over $30 billion worth of industrial projects consistently found ~30–60 percent energy savings on retrofit, typically paying back in 2–3 years, and ~40–90 percent savings in new projects, nearly always with *lower* capital cost.

22. For example, an RMI design for retrofitting a 200,000-ft², curtainwall office building when it needed reglazing anyhow could save three-fourths of its energy at slightly *lower* cost than the normal 20-year renovation that saves nothing: A. B. Lovins, "The Super-Efficient Passive Building Frontier," *ASHRAE Journal*, June 1995, pp. 79–81, available from, *www.rmi.org/images/PDFs/Energy/E95-28_SuperEffBldgFrontier.pdf*.

23. This is simply because PVs can ride down the cost curve (they'll clearly continue to get 18 percent cheaper for each dou-

bling of cumulative global production volume, which is nearly doubling every year), they produce the most output on summer afternoons when most utility loads peak, and they can start producing energy and revenue in year one, reducing their financial risk. Many technological and institutional breakthroughs are in view that could well make the costs of PVs drop even faster than their historic cost curve. Thomas Dinwoodie, SunPower Corporation Systems (Founder and CTO), "Price Cross-Over of Photovoltaics vs. Traditional Generation," Richmond, CA: SunPower Corporation Systems, 2008.

24. For example, ~6–16 percent higher labor productivity in efficient buildings, higher throughput and quality in efficient factories, better clinical outcomes in efficient hospitals, fresher food in efficient refrigerators, better visibility with efficient lighting, etc. Just counting such side-benefits can, for example, double the efficiency gains in a U.S. steel mill at the same cost.

25. The biggest of these come from financial economics, for example, small fast modular projects have lower financial risk than big slow lumpy projects, and renewables hedge against fuel-price volatility risk. These 207 phenomena are explained and documented in an *Economist* book of the year: A. B. Lovins, E. K. Datta, T. Feiler, K. R. Rábago, J. N. Swisher, A. Lehmann, and K. Wicker, *Small Is Profitable: The Hidden Economic Benefits of Making Electrical Resources the Right Size*, Snowmass, CO: Rocky Mountain Institute, 2002, available from *www.smallisprofitable.org*.

26. For windpower in the three power pools that span the central United States from Canada to Texas: J. Traube, L. Hansen, B. Palmintier, and J. Levine, "Spatial and Temporal Interactions of Wind and Solar in the Next Generation Utility," *Windpower 2008*, June 3, 2008, available from *www.rmi.org/rmi/Library/2008-20_WindSolarNGU*.

27. For example, an integrated retrofit of efficiency, demand response, and 1.18 MW of PVs at the Santa Rita Jail in Alameda County, CA, easily met a 10 percent/year IRR hurdle rate—the $9-million project achieved a present-valued 25-year benefit of $15 million and hence would have made money even without its $4-million state subsidies—because on the hot afternoons when the PVs produced the most power, the efficient jail used little,

leaving a bigger surplus to resell to the grid at the best price. Or my own household can run on ~120 average W (a tenth the U.S. norm), obtainable from 3 m² of PVs—a system cheaper than connecting to wires 30 meters away. If built today, my household would need only ~40 average W, from 1 m² of PVs—a system cheaper than connecting to wires already on the side of the house. Both these comparisons assume free electricity; the point is that superefficient end-use can make the breakeven distance to the grid, beyond which it is cheaper to go solar than to connect, drop to about zero.

28. We ignore here the modest and broadly comparable amounts of energy needed to build any kind of electric generator, as well as possible long-run energy use for nuclear decommissioning and waste management or for extracting uranium from low-grade sources and restoring mined land afterwards. B. K. Sovacool, "Valuing the greenhouse gas emissions from nuclear power: A critical survey," *Energy Policy*, Vol. 36, August 2008, pp. 2490–2953, surveyed these issues. He screened 103 published studies of nuclear power energy inputs and indirect carbon emissions; excluded the 84 studies that were older than 10 years, not in English, or not transparent; and found that the other 19 derived $gCO_2e$/busbar kWh figures ranging from 1.4 to 288 with a mean of 66, which is roughly one-seventh the carbon intensity of combined-cycle gas but twice that of photovoltaics or seven times that of modern onshore windpower. This comparison, or its less favorable dynamic equivalent described by A. B. Lovins and J. Price, *Non-Nuclear Futures*, Part II, Cambridge MA: Ballinger, 1975, however, is scarcely relevant, since the unarguable *economic opportunity cost* shown in this section is far more important and clear-cut.

29. Conservatively, assuming industry claims that nuclear power indirectly emits about one-seventh as much carbon as the mean of the 19 studies analyzed by Sovocool's literature review (Endnote 28), and similarly omitting the probably even smaller carbon footprint of renewables, recovered-heat cogeneration, and efficiency.

30. Since its recovered heat displaces boiler fuel, cogeneration displaces more carbon emissions per kilowatt-hour than a large gas-fired power plant does.

31. A. B. and L. H. Lovins, report to DoD republished as *Brittle Power: Energy Strategy for National Security*, Andover, MA: Brick House, 1981, posted with summaries #S83-08 and #S84-23, available from *www.rmi.org/sitepages/pid114.php*; Defense Science Board, *More Fight, Less Fuel*, February 13, 2008, available from *www.acq.osd.mil/dsb/reports/2008-02-ESTF.pdf.*

32. These arguments are elaborated and documented in Lovins *et al.*, *Small Is Profitable*.

33. C. L. Archer and M. Z. Jacobson, "Evaluation of global windpower," available from *www.stanford.edu/group/efmh/winds/global_winds.html*, calculated at 80 m hub height. A later National Renewable Energy Laboratory study, published February 19, 2010, found three times the originally assessed profitable U.S. wind potential, totaling 10 TW or 37 PWh/y on available categories of land. Available from *www.windpoweringamerica.gov/filter_detail.asp?itemid=2542.*

34. Wiser and Bolinger, p. 27, document 11 recent U.S. utility studies showing that even variable-renewable penetrations up to 31 percent generally cost <0.5¢/kWh to "firm" to central-plant reliability standards. The two studies that found costs up to 0.8¢ didn't assume the sub-hourly market-clearing that most grid operators now use.

35. The nuclear industry's claim that because a modern economy needs highly reliable electricity, and therefore it also needs "24/7" power *stations* of billion-watt scale is absurd. *No* power source is 100 percent reliable; that is why utilities must use redundancy and elaborate operating techniques to ensure reliable supply despite unpredictable failures, which are especially damaging when the failed units are large. The same proven techniques apply similarly, but more easily, to large numbers of diverse renewables whose variable elements can be readily forecast. Without exception, more than 200 international and 11 U.S. studies have found this (see Lovins and Sheikh, pp. 22–27). Wind-rich regions of Germany, Spain, and Denmark have already proven it by meeting 20–39 percent of all annual electrical needs (and at times over 100 percent of regional needs) with variable renewables, without encountering instability nor significant integration costs.

36. Scott DiSavino, "MidAmerican drops Idaho nuclear project due to cost," *Reuters*, January 29, 2008, available from *www.reuters.com/article/idUSN2957446620080129*.

37. New Energy Finance found only a 4 percent drop in 3Q08 renewables financing, and recent data suggest a robust, even growing, solar sector despite grave financial distress and accelerating decline in the central-station business.

38. A term originated by *The Economist*'s then-energy correspondent Vijay Vaitheeswaran and publicized in his book, *Power to the People: How the Coming Energy Revolution Will Transform an Industry, Change Our Lives, and Maybe Even Save the Planet*, New York: Farrar, Straus and Giroux, 2005.

39. A thorough database of industry and official data sources is posted and available from *www.rmi.org/sitepages/pid256.php#E05-04*. Similar renewable energy data are available from *www.ren21.net*.

40. A policy agenda for removing many of these obstacles is in the last section of Lovins *et al.*, *Small Is Profitable*.

41. Data for decentralized gas turbines and diesel generators exclude generators of less than 1 megawatt capacity.

42. O. Bailey and E. Worrell, "Clean Energy Technologies: A Preliminary Inventory of the Potential for Electricity Generation," LBNL-57451, Berkeley, CA: Lawrence Berkeley National Laboratory, April 2005, available from *repositories.cdlib.org/lbnl/LBNL-57451/*.

43. *20% Wind Energy by 2030*, Washington, DC: USDOE, Chap. 2, p. 2, available from *www.20percentwind.org/20p.aspx?page=Report*.

44. Lovins and Sheikh.

45. A. B. and L. H. Lovins and L. Ross, "Nuclear power and nuclear bombs," *Foreign Affairs*, Vol. 58, No. 5, Summer 1980, pp. 1137–1177, available from *www.foreignaffairs.org/19800601faessay8147/amory-b-lovins-l-hunter-lovins-leonard-*

*ross/nuclear-power-and-nuclear-bombs.html* or *www.rmi.org/images/ other/Energy/E05-08_NukePwrEcon.pdf*; and also found in *Foreign Affairs*, Vol. 59, 1980, p. 172. Had that paper's market-driven strategy been adopted 30 years ago, the world would not be worrying about Iran and North Korea today.

46. This would satisfy the intent of the "nuclear bargain" in Article IV of the Non-Proliferation Treaty. See also C. A. Ford (Hudson Institute), "Nuclear Technology Rights and Benefits: Risk, Cost, and Beneficial Use under the NPT's Article IV," Conference on Comparing Electricity Costs, held at NPEC/Carnegie Corporation of New York, December 1, 2008.

# CHAPTER 4

# THE CREDIT CRUNCH AND NUCLEAR POWER

## Stephen Thomas

Since the decline following nuclear power's golden era of the mid-70s, there have been frequent predictions of an imminent nuclear revival, but all came to nothing. The latest revival, widely known as the "nuclear renaissance" and dating from 2002-03, is being pursued with greater determination than its predecessors. However, after 5 years, the absence of any new orders in key markets like the United States and the United Kingdom (UK), and unresolved issues (for example, on finance) have led to increasing doubts — even before the extent of the impact of the "credit crunch" on the world economy was apparent — as to whether the renaissance will again be still-born. While the credit crunch will not be good for most large scale projects, will it be the last straw for the prospects of a nuclear renaissance?

## FINANCE

The most obvious place to start is at the heart of the credit crunch itself, the banking system, in particular, the ability of electric utilities to borrow the money needed to build nuclear plants. It is clear that one of the legacies of the credit crunch will be that banks will be more risk-averse and will also be more careful with their risk assessment procedures.

A nuclear power station is the most capital-intensive way to generate electricity and, based on its past record, the most economically risky. So it is clear that unless ways can be found to insulate the banks from this risk, the impact on the prospects for the nuclear renaissance will be very severe. There are two main ways that banks can be insulated, at least in part, from this risk: by electricity consumers, or by government credit guarantees.

## DEREGULATION AND INVESTMENT RISK

In the past, while electricity was still a regulated monopoly, obtaining cheap finance to build nuclear power plants was made easier by the fact that consumers effectively guaranteed the loans. If costs escalated, performance was worse than expected, alternatives proved cheaper, or electricity demand had been overestimated, the plant owners simply increased electricity prices to recover the additional costs they had incurred. When this assurance broke down, either because competition had been introduced to electricity or, as in the United States in the late 1970s, because regulators were no longer prepared to make consumers pay for the errors of electric utilities, finance became a thorny issue. When regulators began to disallow part of the cost of imprudent investments, in short, utilities were made to pay for the plants from their profits, the ordering of new facilities ground to a halt, and many existing orders were cancelled.

Some have suggested that new nuclear units would most likely be built in states where the electricity industry is not deregulated and is still regulated under cost-of-service procedures. But this assumes that regulators will be willing to pass on whatever costs the utility incurs; a risky assumption to make.

The poor record of nuclear plants being built on time and on budget, and the mixed record of reliability has always made nuclear power a risky option, but now the risks are falling directly on the utility building the plant. If, as a result, the utility failed, financiers would not be repaid. This has been proved to be more than a theoretical risk more than once. In 2002, the privatized British nuclear generating company, British Energy, collapsed because its costs were higher than the wholesale electricity price of the electric power that it sold. In this case, the British government chose to rescue the company using taxpayer money and the banks did not lose, but this will not always be the case. The Olkiluoto project in Finland, the only Generation III+ design on which substantial construction work has been completed, is acknowledged to be 50 percent over budget and 3 years late after only 3 years of construction.[1] The owners, Teollisuuden Voima Oy (TVO), expect to be covered for the cost escalation by a turnkey construction contract. Whether this contract will stick is now far from clear.[2] But most of the costs of late completion — buying the replacement power from a potentially tight Nordic wholesale electricity market — will fall on the owners.

Negligible quantities of new power generation have been built since the Nordic market was created in the late 1990s, and already dry winters, which reduce the availability of hydro-power, have led to large increases (up to 6-fold) in the wholesale electricity price. So for the period 2009-12, when Olkiluoto should have been producing 12 terawatts (TWh) per year, the owners will have to buy that power from the wholesale market, assuming that amount of power is available. The economic studies on which Olkiluoto was based assumed the generation cost would be €24/megawatt

(MWh). If the Nord Pool price was three times that, far from unusual in recent years, the extra cost of purchasing this power from the market would be in the order of €2 billion over 3 years.

However, TVO is owned by its customers, energy intensive industries such as paper and chemicals, for which electricity purchase is likely to be one, if not the largest of their input costs. So passing these extra costs on to consumers has serious repercussions. While the owners of TVO would not want to cause its failure, their first priority must be to ensure that the cost of the power they buy is not so high as to make their products uncompetitive. It is not hard to imagine a utility with less financial and contractual back-up than TVO collapsing under the strain of the cost and time overruns suffered at Olkiluoto. If cost escalation at the site continues, perhaps even TVO will collapse, with a resulting long-lived impact on the financeability of nuclear projects.

## GOVERNMENT GUARANTEES

Even before the credit crunch, the risk premium involved in nuclear projects, as discussed above, was a severe barrier to new orders. At the top of the utilities' wish list for government support were credit guarantees, which shift this risk to taxpayers. One of the factors that made the Olkiluoto order financeable was export credit guarantees from the French and Swedish government. This made loans at only a 2.6 percent interest rate possible. At the time, the guarantees were shocking and looked extensive but in comparison with what U.S. utilities are asking for, they now seem small.

In the United States, Congress has made $18.5 billion in federal loan guarantees for new nuclear plants available for 2008-09.[3] This is part of the Bush Nuclear Power 2010 initiative, which was based on the premise that some federal subsidies and guarantees to a handful of new plants would overcome barriers to new ordering and lead to a flow of new, unsubsidized orders. The Department of Energy estimates that loan guarantees could reduce total generation cost by about 40 percent: "A new merchant nuclear power plant with 100 percent loan guarantee and 80/20 debt to equity ratio could realize up to a 39 percent savings in the levelized cost of electricity when compared to conventional financing with a 50-50 debt to equity ratio."[4]

There are restrictions on the type and number of plants that would be eligible for loan guarantees. The Congressional Budget Office stated:

> The Department of Energy has indicated that it will deny a utility's application for a loan guarantee if the project is not deemed to be both innovative (essentially, in the case of nuclear technology, a plant design that has not been built in the United States) and commercially viable, and that no more than three plants based on each advanced reactor design can be considered innovative.[5]

If three units of each of the five plant designs under consideration were built, 15 units would be built. But while utilities have been keen to stand in line for these handouts, with 30-40 plants now at various stages of planning, it seems increasingly likely that only plants with loan guarantees will be ordered. If the new U.S. administration really wants to get a significant proportion of the 30-40 reactors proposed built, the $18.5 billion will not go very far.

If we assume that a new plant will cost no more than $7-9 billion and that industry gets its wish that 80 percent of this cost is covered by federal loan guarantees, guarantees worth about $100 billion would be needed to build just the 15 innovative units. To build 35 units, guarantees of $230 billion would be needed. By October 2008, 17 power companies had already applied for $122 billion in federal loan guarantees.[6] If, as argued by Standard & Poor's,[7] skills and component bottlenecks mean that only a few units can be supplied per year to the U.S. market, the need for this very large number of guarantees may not arise.

There has also been speculation that the French and Japanese governments would offer loan guarantees for plants supplied by their national companies.[8] Areva NP is controlled by French interests, indeed, it is majority-owned by the French state, and the French government has already proved itself willing to offer loan guarantees, for example to Finland and South Africa.

The Japanese government is much less experienced with supporting Japanese vendors. Despite the extensive nuclear program in Japan as well as large exports of nuclear components, this is the first time Japanese vendors have tried to win foreign orders as a main contractor. Nevertheless, Japanese vendors are involved in four out of five of the designs being considered in the United States — the Franco-German engineering, procurement, and construction project, European Pressurized Water Reactor (EPR) is the fifth.

Mitsubishi has its own design, the U.S. Advanced Pressurized Water Reactor (US-APWR). Hitachi is collaborating with General Electric (GE) to offer the Economic Simplified Boiling Water Reactor (ESBWR) and, perhaps, the Advanced Boiling Water Reactor

(ABWR). Westinghouse, which is offering the AP-1000, although largely based in the United States, is now owned by Toshiba, which is also offering the ABWR. Standard & Poors believes the Japanese government will provide finance for orders from Japanese vendors through the Japan Bank for International Cooperation.

Providing guarantees for one order, like Olkiluoto, which was seen as opening up the market for French exports might be acceptable to French and Japanese taxpayers. However, if such guarantees are a condition for all orders to be placed, taxpayers will see this as a blank check and, especially if the Olkiluoto order does lead to a default, a highly risky one.

For U.S. orders, if public opinion remains that failures of the U.S. banking were at the root of the credit crunch, the idea of foreign banks supporting U.S. financial institutions to again make risky investments will be even more unpopular.

This is an issue that the new Obama administration will need to look at urgently. The U.S. Government seems to have three choices:

1. Abandon the program;

2. Build 3-4 totemic plants within an $18.5 billion budget; or

3. Cave in to the nuclear industry's demands for blank check support.

The first option is more feasible for a new administration at the start of its term and would be the logical choice if it was judged that orders without loan guarantees would not be feasible. It would face huge opposition from those who stood to gain from nuclear orders. The second option would be politically less contentious by avoiding the opposition that abandoning the program would lead to, but would put $18.5 billion of public money at risk.

For other countries, especially the UK, the government has not faced up to the prospect that loan guarantees will be necessary if orders are to be placed. It is one thing for taxpayers to be forced to find this sort of sum to save the global banking system, it is a very different thing to volunteer this level of taxpayer money simply to get nuclear power plants built when there are non-nuclear alternatives that would not need this level of support. The public opposition to the U.S. Government $700 billion bail-out of the banking sector demonstrated that the public is not prepared to risk its money on what appear to be ill-thought-out policies.

## KEYNESIAN STIMULATION

With governments desperately looking for measures that will prevent their economies from slipping too deeply into recession, there is bound to be some pressure for Keynesian measures to stimulate the economy through government or government-inspired investment in infrastructure. Building nuclear power plants might seem to be a good way to do this. To some extent, any major infrastructure project will stimulate the economy because it will employ labor and use materials but that does not avoid the need for governments to choose projects that have long-term value to the economy so choices still have to be made. The other relevant issue is how quickly can the chosen project have an impact, and this is the major weakness for nuclear projects. Even in the countries where the process of restarting nuclear ordering is most advanced, notably the United States and the UK, no orders can be realistically placed for 4-5 years.

If an immediate stimulus is needed in the energy sector, energy efficiency measures, which have a short lead-time, which employ a large number of workers with varying skills and which have a huge long-term welfare benefit would seem likely to be far more effective. It is therefore particularly surprising that the British government is cutting funding for its flagship energy efficiency program, Warmfront.[9]

## NUCLEAR CONSTRUCTION COSTS

### Cost Estimates.

One of the most bewildering aspects of the nuclear debate over the past few years has been the escalation in forecast nuclear costs, even before any new plants have been built. The figure of $1,000/kW (so that a 1,000 MW plant would cost $1 billion) was toted by the nuclear industry in the late 1990s as an achievable cost for the new Generation III+ nuclear plants then being designed. This figure was seen by many outside the industry as a target rather than a realistic forecast. So when the first order for a Generation III+ plant was placed for Olkiluoto in 2004, the size of the contracted cost, €3 billion or $3000/kW — three times the figure that the nuclear industry had forecast — was not a surprise to experienced industry watchers. It was seen as a loss-leader, although given that the vendors would have to pay for any cost overruns, there was an expectation that it was at least of the right order of magnitude.

It is now clear that construction at the Olkiluoto is going very badly, and that the project is 50 percent over budget and 3 years late, and additional cost increases are expected. Even companies as big as Areva

NP's owners (Areva and Siemens) cannot easily take losses on this scale without expecting serious repercussions from their shareholders.

How much of the cost overrun is the result of the problems at the site and how much is because the price was an underestimate will probably never be known. Areva, in its attempt to pass these costs on to TVO, will have a strong incentive to argue it is due to specific site problems.

However, prices continued to escalate rapidly even after the Olkiluoto price was announced. By 2008, the estimated construction cost from a range of sources for a Generation III+ unit seemed to be settling at around \$4,000-6,000/kW, double the Olkiluoto price and often double the estimates made by the same utilities a year or two previously. These cost estimates are not extrapolations by anti-nuclear activists, they are from credible organizations with no apparent motive for overestimating costs such as experienced nuclear utilities and financial institutions like Standards & Poor's. The figures need to be treated with some care partly because the projects are still at an early stage of development and partly because it is not always clear what is included in the estimates. In particular, some estimates may include finance costs, while others, e.g., Duke Power, Progress, and Florida Power & Light, are overnight costs. See Figure 4-1.

| Organization | Plant | Estimate ($/kW) | Date |
|---|---|---|---|
| Duke Poweri | Lee (AP-1000) | 4700 | November 2008 |
| Progress Energy/Harrisii | Harris (AP-1000) | 4000 | October 2008 |
| Standard & Poorsiii | n/a | 3000-5000 | October 2008 |
| E.ONiv | n/a | 6000 | May 2008 |
| Florida Power & Lightv | Turkey Point | 5400-7900 | February 2008 |

Sources:

i. "Duke Doubles Cost Estimate for Nuclear Plant," The *Business Journal of the Greater Triad Area*, November 4, 2008.

ii. "Reactors Likely to Cost $9 Billion; Progress Energy Doubles Estimate," *The News & Observer*, Raleigh, NC, October 17, 2008.

iii. "Construction Costs To Soar For New U.S. Nuclear Power Plants," *Standard & Poor's*, October 15, 2008.

iv. "Reactors Will Cost Twice Estimate, Says Chief," *The London Times*, May 5, 2008.

v. "FPL Says Cost of New Reactors at Turkey Point Could Top $24 Billion," *Nucleonics Week*, February 21, 2008.

## Figure 4-1. Recent Estimates of Nuclear Construction Costs.

A variety of explanations can be suggested for this escalation.[10] These include:

- Rapidly rising commodity prices driven by China's demands for them which makes all power plants more expensive, but affects nuclear plants particularly severely because of their physical size;
- Lack of production facilities, which is means that utilities hoping to build nuclear plants are taking options on components like pressure vessels;
- Shortages of the necessary nuclear skills as the nuclear work force ages and is not replaced by younger specialists; and,

- Weakness of the U.S. dollar.

All of these deserve consideration in light of how the credit crunch will impact them.

**Commodity Prices.**

If the recession triggered by the credit crunch does bite hard, commodity prices (including fossil fuels) could drop steeply in the short-term and this might at least help check the growth in estimates for nuclear construction costs—it will also tend to reduce the price of other types of power plants, albeit to a lesser extent. In the longer term, whether lower prices can be maintained will depend on resource issues. If the price of commodities rose because of resource issues, e.g., the marginal reserves that were being exploited had much higher costs than the main resource base, prices will tend to remain high. Advocates of the peak oil theory would probably argue this was the case for oil.[11] If the high prices are simply the result of a short-term supply-demand imbalance, as new capacity is built, prices will drop back sharply. This may be the case for steel and concrete, where there does not appear to be any basic resource problem. Note that some of the escalation in commodity prices may also be due to the decline from the end of 2005 to mid-2008 of about a third in the value of the U.S. dollar. Much of this decline had been recovered by November 2008.

## COMPONENT BOTTLENECKS AND SKILLS SHORTAGES

Standard & Poor's[12] places great emphasis on the issue of shortage of component manufacturing facilities. It identifies in particular pressure vessels, circulating water pumps and turbine forgings as particularly problematic. While a large demand for these products would undoubtedly lead to an increase in capacity, the certification requirements for nuclear components will make this a slower process than it would be for less demanding technologies and companies will be reluctant to commit the investment needed to build such production facilities until they see solid evidence of long-term demand. Standard & Poor's also notes skills shortages as a major constraint and, again, such skills shortages cannot quickly or easily be overcome.

## CURRENCY INSTABILITY

Currencies values have been particularly volatile in the past 2 years with the dollar hitting historic lows against European currencies. Between November 2005 and July 2008, the value of the dollar against the Euro had fallen from €1=$1.17 to €1=$1.57. Yet by November 2008, the dollar had recovered most of this ground to €1=$1.27. It seems likely that at least some of the cost escalation was related to the decline of the U.S. dollar making some inputs more expensive in dollar terms. For the future, this currency instability represents a particular risk to all sides. For example, a Japanese company selling plants or components for which the contract is denominated in dollars would lose substantial amounts of money if the value of the dollar was to fall back sharply again.

A fifth factor, greater awareness among utilities that if the estimates they make are not accurate that there will be serious financial consequences for them, which is difficult to quantify. Experience with Olkiluoto and awareness that regulators and the public are likely to be much less indulgent to cost overruns than they were in the past will be a strong incentive for utilities to build in ample contingencies.

Given that the current costs estimates are based on minimal actual construction experience and that such estimates have, in the past, seriously underestimated actual costs, the figure of $6,000/kW may yet turn out to be grossly inaccurate.

## TURNKEY CONTRACTS

The financial assurance that a turnkey contract seemed to give was an important element in Areva NP winning the Olkiluoto contract and also for the French and Swedish governments to offer loan guarantees. However, it was surprising that Areva NP was so desperate for the order that it was prepared to take the massive financial risk a turnkey contract involves. There have been few (if any) genuine whole plant (as opposed to individual component) turnkey contracts since the notorious 12 turnkey orders that launched commercial ordering in the United States in 1964-66.[13] These lost the vendors massive amounts of money although they did achieve one of their aims, which was to convince utilities that nuclear power was little more challenging than, say, a coal-fired plant and could be ordered with confidence as a proven technology. Turnkey orders for nuclear plants are much more risky compared to other power plants because so much of the work in nuclear construction is on-site engineering and construction, a process that is notoriously difficult to control. It is also not easy for the vendor to control the quality of work because of the large number of contractors involved.

Standard & Poor's were clear in a recent report that turnkey contracts would not be offered. "We expect no EPC [engineering, procurement, and construction] contracts to be fully wrapped through a fixed-price, date certain mechanism."[14]

## COMPETITIVENESS AND DEMAND

Nuclear power is just one of many possible ways of meeting electricity demand and, if it is not competitive or demand does not justify it, in the long term plants will not be built. Going back 30 years, large numbers of U.S. orders were cancelled when it became clear that either demand did not warrant them or that the cost of meeting demand with nuclear plants would have been prohibitive.

### Competitiveness.

Even though estimated costs have escalated rapidly in the past 3 years, this seems to have had little impact on the enthusiasm of governments for nuclear power. One explanation for this was the rapid rise of fossil fuel prices and insecurity in their markets. As in 1975, after the first oil crisis, the notion that fossil fuels could ever be cheap again seemed unimaginable. But now, as then, while fossil fuel markets are far from perfect, they do respond and by autumn 2008, this response was already apparent. Sharp declines in electricity demand were also becoming apparent.

High oil prices led in the short term to recession, and the credit crunch is likely to deepen this recession. This will reduce energy demand in the short term because of the reduction in economic activity. In the longer term, there will be a more significant demand and supply side response. This is clearly illustrated by the marketing of new cars, which for the first time in 30 years are being sold based on their fuel consumption. On the supply side, higher oil and gas recovery rates will be justified, exploration efforts redoubled and previously uncommercial reserves, especially for gas, will become more economically viable.

The competitiveness of renewables will be improved, but it might be energy efficiency that is the real winner. Fuel poverty, as defined in the UK as a household spending more than 10 percent of disposable income on energy, has become a major issue and the forecast indicates that by the end of 2008 a quarter or more of British households will be fuel poor. Building nuclear plants might help keep the lights on in the long term, but even its most committed advocates cannot claim it will reduce the price of power. Spending money on energy efficiency to reduce demand will not only keep the lights on and replace fossil fuels, it will also permanently lift households out of fuel poverty with huge health and welfare benefits as well as reducing the strain on the social security system. Few policies pay off so handsomely and in so many ways.

**Capacity Need.**

In the past when the economic case for nuclear power was not so strong, nuclear programs were justified by the nuclear industry on the basis of needed capacity. Without a nuclear power program, they argued, the lights will go off, a prediction usually based on a projection of high electricity demand growth. High energy prices and the credit crunch are likely to cause a recession and a strong demand side response on energy efficiency, so electricity demands will be much lower than earlier forecasts.

## Other Markets.

While most eyes are on the U.S. and UK nuclear programs, other countries' programs are also being affected. South Africa has, for the past decade, been trying to commercialize Pebble Bed Modular Reactor technology, but progress has been slow and the publicly-owned South African utility, Eskom, is now prioritizing orders for conventional nuclear plants, either the Areva NP EPR or the Westinghouse AP-1000. It has a budget of R343 billion ($34 billion) to build 16GW of capacity from new coal and nuclear plants by 2017. In the longer term, it plans to build 20GW of nuclear plant capacity by 2025. But at $6000/kW, its budget would provide less than 6GW of new nuclear capacity. Eskom's credit rating is falling, in August 2008, Moody's reduced their rating to Baa2. It is also deeply unpopular because of numerous blackouts over the past 2 years so its priority must be to deal with power shortages and strengthen the grid in order to ensure that these blackouts are a thing of the past. New nuclear plants which, realistically, will not be on line before 2020 will do nothing to achieve this. So South Africa's ability to proceed with any nuclear program now looks questionable.[15]

Berlusconi has been vocal in his support for nuclear power and is trying to overturn the 1987 referendum verdict that required the phase-out of nuclear plants in Italy.[16] However, the practical difficulties of relaunching the program, such as rebuilding skills and capabilities, were always underestimated and the credit crunch may make finance, even for a utility of the size of ENEL, difficult, especially given the financial strain on ENEL from its purchase last year of over €40 billion of the Spanish utility, Endesa.

## Decommissioning Funds.

While the credit crunch could have an immediate impact on the prospects for new nuclear orders because of its impact on finance, construction, demand, and competitiveness, it could also have a long-term impact on funding for decommissioning.

Under the polluter pays principle, the responsibilities for decommissioning should be clear. Those that consume the electricity should be responsible for paying for the clean-up of the site. The consensus now emerging is that this is best ensured by setting up segregated funds that are only invested in low risk investments. In practice, funds have not always been segregated and decommissioning cost estimates have been largely underestimated so funds have been lost or are inadequate. While for long-term investments, the return will fluctuate over time, the credit crunch may well lead to large shortfalls in these funds which will not be repaired simply by the next economic upswing. Only a few examples have surfaced so far, but if these prove to be the tip of an iceberg, more extensive ways of ensuring adequate funds are available when needed. The decommissioning fund of the Vermont Yankee plant was reported to have lost 10 percent of its value in a matter of weeks.[17] This plant is licensed until 2012 but the license may be extended for another 20 years, in which case, there will be time to make up the shortfall. Decommissioning of the Zion plant (already closed) had to be delayed because its fund also lost 10 percent of its value.[18] On average, U.S. decommissioning trusts are 60 percent equity and 40 percent debt (bonds). Given that indices like the S&P 500 lost more than a third of their value in 2008, it is not difficult to see how losses could be as high as 20 percent. If plants are reaching the end of their life with inadequate funds for decommissioning there may well be a need for further assurance mechanisms. For example, it could be required that utilities take out financial instruments (insurance policies) so that if there is a shortfall, it will be covered by the insurers.

## CONCLUSIONS

Even before the scale of the impact of the credit crunch began to be appreciated, the cracks in the Nuclear Renaissance were becoming clear. The designs were unproven; costs were escalating sharply; obtaining finance was problematic; and there were skills shortages and component supply bottle-necks. The credit crunch has done nothing to lessen these problems.

There are likely to be many unexpected developments before business-as-usual for the world economy resumes, but two changes are clear:

1. The scrutiny by banks of the projects that they lend money to will be far more rigorous in the future so that the mistakes that led to the credit crunch can never be repeated; and,

2. Public appreciation of risk will be sharpened, and where risk is being passed to taxpayers (or electricity consumers), government will need a strong case for such support to be obtained.

The implications for nuclear power concerning these changes are severe, and it is clear that governments and utilities will no longer be able so easily to pass the risk of nuclear programs on to taxpayers and electricity consumers. Nuclear power has demonstrated extraordinary resilience over the past 2 decades, still remaining on the policy agenda despite its failings. So it would be unrealistic to assume that in a decade that powerful interests will no longer be lobbying for more nuclear orders. But the current conditions may be the best and perhaps the last chance for the nuclear industry. The external factors, such as fossil fuel prices, the need to act on climate change and the geopolitical situation are as favorable as they are likely to get. So if the nuclear industry cannot take advantage of these, will it get another chance? The nuclear workforce is aging and is not being replaced, and if a whole generation of new designs, which in a decade will be looking a little dated, has remained largely on paper, will there really be the appetite among private companies to spend the money necessary to bring another generation of designs to the market? Olkiluoto will continue to be the marker for the industry. At best, if there are no more delays and cost overruns, it will be a warning to potential investors, but if things keep going wrong and TVO fails financially, the ability to finance any future nuclear project will be put in doubt.

## ENDNOTES - CHAPTER 4

1. **Finland's Olkiluoto Plant.** The Olkiluoto construction project in Finland has become an example of all that can go wrong in economic terms with a new nuclear plant build. It demonstrates the problems of construction delays, cost overruns, and hidden subsidies. A construction licence for Olkiluoto was issued in February 2005, and construction started that summer. As it was the first reactor ordered in a liberalized electricity market, it was seen as proof that nuclear power orders were feasible in these markets and as a demonstration of the improvements offered by the new designs. To reduce the risk to the buyer, Areva NP offered

the plant under "turnkey" terms, which means that the price paid by the utility (TVO) is fixed before construction starts, regardless of what actually happens to costs. The contract allows for fines on the contractors if the plant is late. The schedule allowed 48 months from pouring of first concrete to first criticality.

**Finance**. The European Renewable Energies Federation (EREF) and Greenpeace France made complaints to the European Commission in December 2004 that the financial arrangements contravened European State aid regulations. The Bayerische Landesbank (owned by the German state of Bavaria) led the syndicate that provided a loan of €1.95 billion, about 60 percent of the total cost, at an interest rate of 2.6 percent. France's Coface provided a €610 million export credit guarantee covering Areva NP's supplies, and the Swedish Export Agency SEK provided €110 million. In October 2006, the European Commission finally announced it would investigate the role of Coface. Subsequently, in what was seen as an eccentric judgement, it found that the guarantees did not represent unfair state aid. Regardless of this, it is clear that the arrangements for Olkiluoto are based on substantial state aid that will not be available to many plants. The interest rate on the loan is far below the levels that would be expected to apply for such an economically risky investment.

**Construction Problems**. In August 2005, the first concrete was poured. In September 2005 problems with the strength and porosity of the concrete delayed work. In February 2006, work was reported to be at least 6 months behind schedule, partly due to the concrete problems and partly to problems with qualifying pressure vessel welds and delays in detailed engineering design. In July 2006, TVO admitted the project was delayed by about a year and the Finnish regulator, STUK, published a report which uncovered quality control problems. In September 2006, the impact of the problems on Areva started to emerge. In its results for the first 6 months of 2006, Areva attributed a €300 million drop in first-half 2006 operating income of its nuclear operations to a provision to cover past and anticipated costs at Olkiluoto. The scale of penalties for late completion was also made public. The contractual penalty for Areva is 0.2 percent of the total contract value per week of delay (past May 1, 2009) for the first 26 weeks, and 0.1 percent each week beyond that. The contract limits the penalty to 10 percent, about €300 million. In December 2006, after only 16 months of construction, Areva announced the reactor was already 18 months behind schedule, which seems to assure that the full penalty will be due. In late 2007, the cost overrun was reported to have increased to €1.5 billion and in October 2008, the estimated delay was increased to 3 years.

Relations between Areva NP and TVO are near a breaking point, with Areva NP now appearing to want to renege on the turnkey contract, claiming that TVO had not fulfilled its part of the deal. The turnkey contract is now in dispute and seems likely to be settled acrimoniously. In December 2008, Areva announced it had initiated a second arbitration against TVO to recover €1 billion in compensation for the delays, which it attributes to failings on the part of TVO, in particular, slowness in processing technical documentation. TVO countered in January 2009 by demanding €2.4 billion in compensation from Areva NP for delays in the project. These cases are likely to take several years to settle and will hang over both TVO and Areva NP until they are resolved.

**Implications**. The scale and immediacy of the problems at Olkiluoto has taken even sceptics by surprise. It remains to be seen how far these problems can be recovered, what the delays will be, and how far these problems will be reflected in higher costs (whether borne by Areva or TVO). However, a number of lessons do emerge:

- The contract value of €2000/kW, which was never a cost estimate due to the turnkey nature of the contract, now appears likely to be a significant underestimate.
- Turnkey contracts may well be required by competitive tenders in liberalized electricity markets. Or, regulators may impose caps on recoverable nuclear construction costs, which would have the same effect. The willingness of vendors to bear the risk of cost over-runs in the light of the Olkiluoto experience is open to serious question.
- The skills needed to successfully build a nuclear plant are considerable. Lack of recent experience of nuclear construction projects may mean this requirement is even more difficult to meet.
- There are serious challenges to both safety and economic regulatory bodies. The Finnish safety regulator had not assessed a new reactor order for more than 30 years and had no experience of dealing with a first-of-a-kind design.

2. "Target date for operating Olkiluoto-3 again delayed, this time until 2012," *Nucleonics Week*, October 23, 2008.

3. "Nuclear Energy Institute president says Congress needs to boost loan guarantees," *Platt's Global Power Report*, October 16, 2008.

4. Prepared remarks of Deputy Energy Secretary Dennis Spurgeon at the second annual "Nuclear Fuel Cycle Monitor Global Nuclear Renaissance Summit," Alexandria, Virginia, July 23, 2008.

5. "Nuclear Power's Role in Generating Electricity," Washington, DC: Congressional Budget Office, 2008, p. 33, available from *www.cbo.gov/ftpdocs/91xx/doc9133/05-02-Nuclear.pdf*.

6. Available from *www.lgprogram.energy.gov/press/100208.pdf*.

7. "Construction Costs To Soar For New U.S. Nuclear Power Plants," Standard & Poor's, 2008.

8. "US Working with Allies to Change Global Rules for Nuclear Financing," *Nucleonics Week*, October 23, 2008.

9. "Extra Fuel Poverty Funding 'Will Last a Single Winter'," *Observer*, November 9, 2008.

10. For more discussion on these factors, see "Construction Costs To Soar For New U.S. Nuclear Power Plants."

11. Peak Oil advocates claim we are near the maximum oil production level that can be achieved, and that as reserves become further depleted and supply cannot fulfill existing demand, prices will rise rapidly.

12. "Construction Costs To Soar For New U.S. Nuclear Power Plants."

13. I. C. Bupp and J. C. Derian, *Light Water: How the Nuclear Dream Dissolved*, New York, Basic Books, 1978.

14. "Construction Costs To Soar For New U.S. Nuclear Power Plants."

15. "South Africa; Funding Costs 'Will Rise for Eskom'," *Business Day*, September 3, 2008.

16. "Credit crunch may slow down Italy nuclear relaunch," *Reuters*, October 17, 2008.

17. "Entergy Balks at Document Requests," *Brattleboro Reformer*, November 1, 2008.

18. "Economy Delays Dismantling of Zion Nuclear Plant," *Chicago Tribune*, October 17, 2008.

# CHAPTER 5

# TAXPAYER FINANCING FOR NUCLEAR POWER: PRECEDENTS AND CONSEQUENCES

## Peter A. Bradford

## INTRODUCTION

In recent years, Presidents George W. Bush and Barack Obama, as well as Congress, have offered extraordinary incentives for the building of new nuclear power plants in the United States. For a while, these efforts seemed to stimulate a nuclear "renaissance." By late 2008, applications for more than 30 reactors jostled for position in the nuclear subsidy queue. Even now, as cancellations, delays, and cost overruns dominate the nuclear trade press, many in Washington behave as if the "renaissance" were a great success, deserving further subsidy to produce further marvels. Their clamor evokes Hans Christian Anderson's brilliant ending to the fable, The Emperor's New Clothes:

> "But he has nothing on!" everybody shouted at last. And the emperor shivered, for it seemed to him that they were right; but he thought within himself, "I must go through with the procession." And so he carried himself still more proudly, and the chamberlains walked along holding the train which wasn't there at all.

But could the industry's quest for further taxpayer and customer subsidy in the face of demonstrated economic illogic possibly succeed? Sure it could.

The Energy Policy Act of 2005 offered a production tax credit of 1.8¢ per kWh to some 6,000 MW of new nuclear capacity for 8 years. In addition, the first six plants were offered insurance against various types of delays. The U.S. statute limiting accident liability to an inflation-adjusted $10.5 billion and spreading it across all nuclear power plants was extended to new units for the next 20 years.[1] The U.S. Government remains committed to taking the waste fuel rods eventually, another valuable benefit for which other industries need not apply. As yet it has no place to put them.

In December 2007, Congress responded by extending some $18 billion in loan guarantees for new nuclear plants. The process by which this was done was sufficiently irregular and cumbersome that the extent of the benefit remains uncertain,[2] but Congress's bipartisan determination to override the 30-year market verdict against new nuclear power in the United States could not be much clearer.

In addition, the Bush administration undertook to pour taxpayer financing into the reprocessing of spent fuel, an activity even more uneconomic than new nuclear power plants, and one which does not diminish the waste disposal problem appreciably. U.S. reprocessing was suspended by President Ford in 1976 on account of its potential connection to the proliferation of nuclear weapons. President Carter terminated the suspended programs because he shared President Ford's proliferation concerns and because he saw no economic justification for reprocessing. The latter conclusion was validated when President Reagan 4 years later withdrew government objections to reprocessing done by the private sector, and the private sector showed no interest.

This chapter discusses this history in the context of current government efforts to assure the construction of new nuclear power facilities. This chapter also describes some consequences of subsidies in terms of patterns of growth and economic activity that are demonstrably unsustainable today. It shows that hiding the costs of megaprojects in order to improve their competitiveness against more sustainable and less dangerous alternatives can have seriously adverse long-term consequences. As to nuclear power, these consequences are not yet as clearly visible as they are in the water and agriculture sector, which makes the water/agriculture cases a valuable light to shine on the nuclear.

The chapter begins with a review of the effects of using federal loan guarantees to further particular forms of energy infrastructure development. It then reviews some past energy developments to assess their potential to misallocate resources and expose the taxpayer to liability in the event of default. It includes an overview of the package of federal programs that combines water resource development with energy facilities. Because these programs involve urgently needed resources in two separate realms, the opportunities for subsidy and misallocation were compounded. The chapter concludes with a comparison of past uses of federal credit support with the proposed efforts in support of new nuclear units. It suggests that all of the ingredients of past resource misallocations are aligned in such a way as to create high potential for similar results if a similar course is followed with regard to new nuclear units.

## SOME CONSEQUENCES OF TARGETED CREDIT SUBSIDIES WHEN TAXPAYERS SHOULDER INVESTORS' RISKS

In response to the energy crisis of the 1970s, the Ford administration proposed an "Energy Independence Authority" to extend loans and loan guarantees to projects making a significant contribution to the energy independence or the energy security of the United States. The necessary legislation was not enacted. However, the legislation did serve as a precursor to the more limited Synthetic Fuels Corporation Act of 1980. This legislation established the Synthetic Fuels Corporation (SFC), with initial authority to provide up to $18 billion in loans, grants, and price guarantees to support coal gasification and oil shale development. The SFC was abolished 5 years later in the face of collapsing oil prices. The only facility it ever built was the Great Plains coal gasification plant, constructed with the aid of $2 billion in federal loan guarantees. Great Plains went bankrupt in 1988, was sold for $88 million, and emerged to sell overpriced synthetic gas on the basis of federally required purchases for the next 2 decades.[3]

Although it was never enacted, the more ambitious Energy Independence Authority legislation did give rise to an insightful 1978 critique of government credit subsidies in the context of energy facility development, prepared by Murray Weidenbaum (then soon to be the first Chairman of President Reagan's Council of Economic Advisors) and Reno Harnish.[4]

This thirty-year-old Weidenbaum/Harnish critique invalidates the 2007 nuclear loan guarantee legislation and subsequent proposals to create a government-run public-private bank to provide financial support (loans, guarantees and equity capital) for U.S. clean energy projects, including nuclear power.

In assessing the potential of subsidies to misallocate energy resources, the Weidenbaum report makes the following points:[5]

- Federal credit programs merely shift funds from one borrower to another. They do not increase the amount of funds available to the economy. Rather, to the extent they succeed, they take capital away from the unassisted sectors of the economy, leading them to request aid (pp. 17-18).
- New and small businesses, school districts, smaller local governments and individuals, private mortgage borrowers not under the federal umbrella — generally the weaker borrowers — are the ones squeezed out. The unsubsidized private borrowers wind up paying higher interest rates (pp. 52-53).
- Federal credit programs put the government in the position of holding assets of questionable quality or limited use, making it difficult to recover the original value of the loans in the case of default, and complicating the process of liquidating the agency (p. 17).
- A basic function that credit markets are supposed to perform is that of distinguishing credit risks and assigning appropriate risk premiums. This function is the essence of the ultimate resource allocation of credit markets. As an increasing proportion of issues coming to the credit markets bears the guarantee of the federal government, the ability of the market to differentiate credit risks inevitably diminishes. Theoretically the federal agencies issuing or

guaranteeing debt perform this role, charging as costs of the programs differing rates of insurance premiums. In practice, all of the pressures are against such differential pricing of risks (p. 13).

- [Quoting MIT Professor Henry Jacoby, a supporter of limited loan guarantees] "The problem with loan guarantees is that they tend to hide the true cost of the technology that is being demonstrated. . . . If I thought this bill was a prelude to a massive program of loan guarantees for new energy facilities, for multiple plants with known technology and not just for a limited set of demonstrations, then I would oppose it. I think it would be a terrible mistake to embark on a large scale program of hidden subsidies for energy supply from new capital intensive technologies. . . . The disadvantage of the widespread use of loan guarantees is that they will obscure the true cost to the economy. . . . More important, they hide the true cost from consumers and encourage wasteful consumption practices" (pp. 41-42).

- [Quoting the General Accounting Office] "The bill is not neutral on conservation options. Actually, it would hamper conservation efforts rather than simply fail to promote them. . . . Its guarantees would make projects it assists financially more attractive to private capital than conservation projects not backed by federal guarantees. Thus both its loans and its guarantees will siphon private capital away from those conservation projects which might have been able to obtain private financing" (p. 12).

- The size of the undertaking in itself does not necessitate governmental assistance; large commercial energy projects, such as the $7 billion ($35 billion in 2007 dollars) Alaska pipeline project, are proceeding with private finance (p. 49).

## NUCLEAR POWER AND FEDERAL CREDIT SUPPORT

Uniquely among major industries and energy sources, nuclear power was created by federal expenditures. Controlled nuclear fission was developed as part of the Manhattan Project during World War II. Fission was first used as a nonexplosive energy source in the propulsion of nuclear submarines for the U.S. Navy.

While these expenditures were not targeted to the benefit of the nuclear power industry, they certainly had the effect of bringing nuclear power closer to commercial reality than private capital would have been likely to do during the same period. In addition, they created a pool of skilled labor, a supportive national laboratory capability, and an industrial infrastructure that were readily convertible to the needs of the civilian nuclear power program.

This initial support did not take the form of loan guarantees of the sort reviewed in the Weidenbaum study, but there was more to come.[6] Consider, for example, the case of the West Valley reprocessing plant, which operated sporadically in upstate New York from 1966 until 1972 when it closed for "retrofitting" and "expansion." It never reopened. During its 6 working years, it achieved the equivalent of about 2 full years of operation.

At the 1963 groundbreaking, Governor Nelson Rockefeller announced the entire cost of the project to be $28 million (about $165 million in today's dollars), including $20 million from W. R. Grace Company to build the facility and $8 million from New York for support facilities. Governor Rockefeller's speech captures in unusually pure form the extraordinary marriage of free enterprise imagery to government largesse that is a staple of such occasions:

> We are launching a unique operation here today, which I regard with pride as a symbol of imagination and foresight on the part of your state government — an operation that will make a major contribution toward transforming the economy of western New York and indeed the entire state. . . .

> I would like to express my appreciation of the leadership and imagination of W. R. Grace and Company, in the best tradition of the American free enterprise system, for its decision to pursue this pioneering undertaking in New York State. The company will find here an understanding and congenial home....The project is illustrative of the vigor, farsightedness, and boldness, which is characteristic of free enterprise in New York state. . . .[7]

> The presence in the state of the nuclear fuel reprocessing industry will, as time passes, have an increasingly favorable impact on the economics of energy production and utilization in the State with a resultant stimulation of over-all industrial development.

> In short, this state-sponsored project, operating through private enterprise with federal cooperation, places New York in the forefront of the atomic industrial age now dawning — "to the benefit of the health, safety and prosperity of this generation and many generations to come."[8]

Today, however, every hope has ended in disappointment. Expectation has given way to irony; pretense is exposed as nonsense. What was really guaranteed in this unique operation, and what did it cost?

The state of New York was the "landlord," meaning that it owned the site and built a number of the support facilities, including those for waste storage. In 1976, the "tenant" — Nuclear Fuel Services Corporation (NFS), a subsidiary of W. R. Grace and American Machine and Foundry Company until it was sold to Getty Oil Company in 1969 — notified the state that it would not renew the lease when it expired in 1980. NFS thereby turned the entire contaminated facility plus considerable unreprocessed spent fuel over to New York. The taxpayers of New York had — through a lease arrangement that left them with the cleanup responsibility — guaranteed that the private "tenant" would be indemnified against cleanup costs, an open-ended obligation whose full extent remains unclear 36 years after the facility closed.

Luckily for New York, a federal takeover of the cleanup responsibilities was arranged in the form of the 1980 West Valley Demonstration Project Act, which provided for a Department of Energy (DoE) cleanup that is still not complete. The New York share of the cleanup costs was set at 10 percent of the total. That amount had reached $250 million in 2006, so the cleanup of that one facility to date has cost federal and state tax payers $2.5 billion in unadjusted dollars.

# THE ORIGINS OF FEDERAL CREDIT SUPPORT
# IN MAJOR ENERGY AND WATER PROJECTS

The U.S. Government's first major involvement in the electric power sector evolved out of efforts to support farmers in the western states through federally provided irrigation dams. Some of these dams also generated electricity. Indeed, the revenues from the hydroelectric dams provided one of the funding sources for federal credit support for the irrigation farmers. The complex accounting for costs and benefits of dams that provided hydroelectricity, irrigation, flood control, and urban drinking water provided opportunity for subsidy and favoritism of many sorts and made effective oversight difficult.

Because these projects have existed for nearly a century, their consequences—benign and otherwise—are now relatively clear. Their history shows both the potential and the pitfalls of using federal credit support on a long-term basis to underwrite established industries and economic patterns. The parallels to the potential misallocations resulting from using such support on behalf of an established nuclear industry are imperfect but often compelling.

## DEMAND FORECASTING, CLIMATE SCIENCE, AND MYTHOLOGY

Much of the western United States between the Mississippi River and the Rocky Mountains was marked as desert on the maps of the 19th century. Not until the 1870s did early experiments in irrigation enable significant settlement based on farming. The 1870s were a decade of exceptional rainfall in the arid regions, resulting in heavier settlement than the normal climate could sustain. While some urged caution in federal policies subsidizing settlement, others felt that it was America's "Manifest Destiny" rapidly to settle the nation from the Atlantic to the Pacific.

The believers in Manifest Destiny and their allies in local real estate and finance found support in a theory of human-induced climate change based on the proposition that "rain follows the plow." According to this theory, the rainfall that coincided with the initial settlements was, in fact, produced by those settlements. Professor Cyrus Thomas exemplified this view:

> Since the territory has begun to be settled, towns and cities built up, farms cultivated, mines opened, and roads made and traveled, there has been a gradual increase in moisture....I therefore give it as my firm conviction that this increase in moisture is of a permanent nature, and not periodical, and that it has commenced within 8 years past, and that it is in some way connected to the settlement of the country, and that as population increases the moisture will increase.[9]

Politicians, newspaper editors, believers in American expansion, and promotional land development policy combined to lure refugees from American and European cities and rocky eastern soil westward with visions of endless easily farmed land. The truth was quite different.

John Wesley Powell, who had headed the first expedition successfully to raft and map the Colorado River, wrote a warning document entitled *A Report on the Arid Lands of the United States,* in which he forecast that, even with irrigation, only a small portion of the land on which settlement was pouring could be sustainably farmed. He recommended major reforms in land grant practices and the development of carefully sited reservoirs to assure that the best land received adequate water to maximize its productivity. He cautioned that the subsurface waters available in the West were not likely to be a sustainable basis for farming in the long term.

When he explained his recommendations to the Congress, Powell was vilified by the representatives of the Western states that he sought to protect. As Wallace Stegner described the scene in his biography of Powell:

> They clamored to know how their states had got labeled "arid". . . . What about the artesian basin in the Dakotas? What about irrigation from that source? So he gave it to them: artesian wells were and always would be a minor source of water. . . . If all the wells in the Dakotas could be gathered into one county, they would not irrigate that county.

Senator Moody thereupon remarked that he did not favor putting money into Major Powell's hands when Powell would clearly not spend it as Moody and his constituents wanted it spent. "We ask you," he said in effect, "your opinion of artesian wells. You think they're unimportant. All right, the hell with you. We'll ask someone else who will give us the answer we want."[10]

Powell, then the head of the U.S. Geological Survey, was defunded and forced into retirement, replaced by successors whose opinions were more congenial. But time was to prove him far more right than wrong. Only federal assistance for water and energy projects on a scale that turned a blind eye to both economic logic and the laws of nature could maintain the settlement flows across the Great Plains and into the Rocky Mountains for a while.

## DEVELOPING A FEDERAL ROLE

Droughts in the 1890s made nonsense of the proposition that settlement increased rainfall. Many farmers and developers faced ruin. Combinations of private citizens and state governments failed at the task of organizing and financing broader irrigation projects. Despite strong belief in the importance of preserving private enterprise and individual initiative against government encroachment, western state representatives acquiesced in the passage of the Reclamation Act of 1902.

The Reclamation Act established the Reclamation Service, whose projects were to be financed by a federal Reclamation Fund. Monies for this fund were to come from the sale of federal land. The fund would be replenished from the sale of water to farmers. However, the farmers were to be excused from paying any interest on this money, the first of many substantial subsidies.

The Reclamation Service attracted idealistic graduates of the country's finest engineering schools, who headed west in a fog of idealism ready to take on the most implacable foe of mankind, the desert. . . . The engineers who staffed the Reclamation Service tended to view themselves as a godlike class performing hydrologic miracles for grateful simpletons who were content to sit in the desert and raise fruit. About soil science, agricultural economics, or drainage they sometimes knew less than the farmers whom they regarded with indulgent contempt. As a result, some of the early projects were to become painful embarrassments and expensive ones.[11]

More aggressive subsidies were needed to prevent the embarrassing failure of the initial subsidy program. First came a $20 million loan from the Treasury to the Reclamation Service in 1910. In addition to the Treasury loan, Congress extended the repayment period for the farmers from 10 years to 20. Still, by 1922 only 10 percent of the money paid from the Reclamation Fund had been repaid, and 60 percent of the irrigators were in default on their obligations.

Congress responded by doubling the repayment period again, to 40 years. However, crop prices fell following the end of World War I. Farmers continued to default. The Reclamation Service (renamed the Bureau of Reclamation) rarely cut off the water. Instead, monies from oil production and potassium mining on federal lands were channeled into the Reclamation Fund rather than into the federal treasury, a further subsidy from the U.S. taxpayer to the Reclamation Fund.

All of this might have ended in a relatively modest financial loss had it not been for the election of Franklin Roosevelt to the U.S. presidency and the onset of the Great Depression, a combination of need and visionary hope that was to elevate reclamation project expenditures to an entirely new level, based on the concept of river basin development.

## COMPLEX VARIANTS OF FEDERAL CREDIT ASSISTANCE

The concept of river basin development got its start on the Colorado River in the 1920s. The Colorado—far from the largest U.S. river—begins in the Rocky Mountains of central Colorado and flows southwest through Utah and Arizona, becoming the border first between Arizona and Nevada and then between Arizona and California, before crossing into Mexico. It drains mountains whose snow pack pours prodigious spring runoffs into desert lands with few other water sources. It was also the only river whose water could be diverted in sufficient quantity to meet the growing demands of urban Los Angeles and agricultural southern California, whose earlier grab of the entire but ultimately insufficient Owens River in southeastern California is loosely commemorated in the movie, *Chinatown*.

Because the Colorado flows were so seasonal, massive storage was required to meet year-round demands. So in 1935 the Bureau of Reclamation completed the Hoover Dam in Nevada, at the time the world's largest hydroelectric project and reservoir. The Hoover Dam was completed just as the Midwestern drought that turned the center of the United States into the Dust Bowl entered its final stages. Hundreds of thousands of farmers fled westward from Oklahoma, Kansas, Nebraska, and the Dakotas, potentially overwhelming the ability of the West Coast states to absorb them. One essential part of the response fashioned by President Roosevelt was the building of more dams to create more farmland.

The Bureau of Reclamation and the Army Corps of Engineers combined several river basin projects in central California into the Central Valley Project, vastly increasing the agricultural potential of a large part of the state. Even this was inadequate to cope with the dislocations caused by the Dust Bowl and the Depression. The Roosevelt administration responded with an even greater river basin development on the Columbia River in the Pacific Northwest.

The centerpiece of the Columbia projects was the Grand Coulee Dam in Washington state, three times larger than the Hoover Dam. But the Grand Coulee was just one of many dams built on the Columbia River and its tributaries over the next 3 decades, dams that provided the cheapest power in the United States.

With the completion of the Hoover and Grand Coulee dams, as well as the Shasta Dam in California's Central Valley Project, considerable political pressure developed for additional Colorado River projects to serve the "upper basin" states of Colorado, Utah, and Wyoming. However, these states—at higher altitude with colder climates—lacked the agricultural potential of Southern California. Smaller crops of lower value were all that could be grown, even with water from expensive projects. Repayment potential was nonexistent. Yet the Bureau of Reclamation and the elected officials wanted more projects. The answer to their financial dilemma was the "cash register dam."

The cash register dams had their roots in a Bureau of Reclamation creation called "river basin accounting." In the world of river basin accounting, the profits generated by the sale of electricity from a dam could be used to offset losses from other projects, such as irrigation, rather than going into the federal budget. The concept differed subtly but crucially from the accounting by which electric sales paid off nearly all of the bonds issued to build the Hoover Dam, even though that reservoir was essential for both electricity and for irrigation. Under river basin accounting, electric revenues could be used not just to offset such common costs, but also to offset the costs of irrigation trenches and other expenses that had nothing to do with electricity, as long as those expenses were Bureau of Reclamation outlays on the same river basin.

Thanks to this accounting, water was provided at little or no cost to grow products and raise livestock in the late 1940s and the 1950s, when the nation had a surplus of both, a surplus that would have lowered prices ruinously but for the fact that the government was paying farmers elsewhere not to grow or to raise the commodities that it was subsidizing in the river basins that it was developing.

The leading congressional opponent of these practices was Senator Paul Douglas of Illinois. He pointed out that the cash register dams were producing electricity considerably more expensively than fossil fuels might have done, and far more expensively than the dams of the Tennessee Valley Authority and the Bonneville Power Administration in the Pacific Northwest, dams that had been the original justification for the federal government's going into the electric power business. He noted the irony of the cash register dams being championed by politicians who had opposed the TVA and Bonneville Dams as "creeping socialism." But his greatest scorn was leveled at the economics of the irrigation projects enabled by the cash register dams.

The original projects tended "to be at low altitudes and in fertile soil, and to involve low costs. . . . Now we are being asked to irrigate land in the uplands, at altitudes between 5,000 and 7,000 feet, where the growing season is short. . . .

In my state of Illinois, the price of the most fertile natural land in the world is now (1955) between $600 and $700 per acre. In the largest [irrigation] project of all, the Central Utah Project, the cost [of supplying water] would be nearly $4,000 per acre—six times the cost of the most fertile land in the world. . . . We are being asked to make an average expenditure [on 16 projects under consideration] of $2,000 an acre on land which, when the projects are finished, will sell for only $150 per acre.[12]

Despite the cogency of Senator Douglas's analysis, we hear this cynical contemporaneous pep talk that Commissioner of Reclamation Michael Straus gave to his Montana employees: "I don't give a damn whether a project is feasible or not. I'm getting the money out of Congress, and you'd damn well better spend it. And you'd better be here early tomorrow morning ready to spend it, or you may find someone else at your desk."[13]

As the economically preferable projects were gradually taken care of, the Bureau resorted to ever more outlandish accounting to justify the less desirable projects. Low discount rates understated the costs. Comparisons to alternatives never included options based on resource conservation. Benefits were overstated, as was demand for power and for irrigation. But at least the Bureau was required to subject its projects to some semblance of cost/benefit analysis. It had a rival far less subject to such awkward limitations.

The details of the dam building rivalry between the Bureau of Reclamation and the U.S. Army Corps of Engineers are fascinating but beyond the scope of this chapter. Suffice it to say that for more than 3 decades—from California's Central Valley to the length of the Missouri River and its tributaries to the remote vastness of central Alaska—the two bureaucracies competed with one another to build increasingly uneconomic and often destructive projects. Citizen booster groups, engineering firms, and contractors became adept at playing one off against the other, as did the different congressional committees to which each agency was accountable.

By the time their rivalry had played itself out in the early 1970s, they had done much to discredit their projects among fiscal conservatives and even among some farm groups. When the Bureau proposed two dams that would flood beautiful canyons on the Colorado River as well as part of the Grand Canyon (and defended the latter by saying that tourists would have improved access by motorboat), environmentalists defeated them with an ad campaign whose centerpiece asked, "Should we also flood the Sistine Chapel so that tourists can get nearer the ceiling?"

A few more examples will serve to illustrate both the realities and the dreams that have emerged from this century-old transformation of idealism and social engineering in the best sense into the largest of congressional pork barrels.[14]

**The Central Arizona Project.**

The Central Arizona Project (CAP) is, in essence, a 330-mile channel—the Granite Reef Aqueduct—to bring water from the Colorado River uphill to the cities of Phoenix and Tucson, as well as to store and distribute it. Because the aqueduct had to lift the water 1,000 feet, considerable electricity was needed. Hence the project included the two cash register dams on the Colorado that were ultimately defeated by environmentalists. To replace the power, the Bureau bought an interest in a large coal plant.

Because the Colorado River was overallocated as a result of optimistic forecasts of average flows, because California had succeeded in obtaining a guarantee that its share would be provided regardless of hardship to other states, and because Mexico had eventually succeeded in obtaining a guaranteed allotment of reasonably pure water instead of the salty soup that was its lot for most of the 1960s,[15] CAP could not be assured of its full water allocation in dry years. As a result, the cost of the water was unpredictable but likely to be more than Arizona farmers could afford, even with the customary subsidies.

The astonishing answer to the Arizona shortage was—in the 1960s—a planned diversion into the Colorado basin from the Pacific Northwest. The Bureau publicly admitted to designs on a river or two in northern California, but its real aim was the much larger Columbia River, further north in Washington state. However, the Northwest would not hear of such a plan, and the pumping costs might well have been insurmountable without the cancelled cash register dams in any case.

Sam Steiger, an Arizona congressman who had been a major CAP supporter, had second thoughts in retirement. Describing a process in which cities would be forced to take large quantities of CAP water in return for long-term supply assurances on which their growth depended while farmers' water continued to be subsidized to whatever extent was needed to make it affordable, he summarized:

> They'll skin the cat twenty ways if they have to, but they're going to make the water affordable. Congress will go along, because it will be goddamned embarrassing for Congress to have authorized a multi-billion dollar water project when there's no demand for the water because no one can afford it. The CAP belongs to a holy order of inevitability. . . .
>
> There are hundreds of thousands of acres of good farmland right along the Colorado River . . . but the farmers got established in the central part of Arizona

because of the Salt River Project (a smaller and earlier Bureau undertaking). The cities grew up in the middle of the farmland. The real estate interests, the money people — they're all in Phoenix and Scottsdale and Tucson. They didn't want to move, so we're going to move the river to them. At any cost.[16]

## The Teton Dam.

Built in the early 1970s in Idaho, the Teton Dam was in most ways just another uneconomic and environmentally unsound Bureau of Reclamation project. When realistic discount rates were used, the costs were twice the benefits, but the costs were dispersed to the taxpayers while the benefits flowed to a powerful local constituency that already had a groundwater supply 10 times the amount used in dry lands elsewhere. As one project critic described these farmers, "Mormons get burned up when they read about someone buying a bottle of mouthwash with food stamps, but they love big water projects. They only object to nickel-and-dime welfare. They love it in great big gobs."[17]

Assistant Interior Secretary Nathaniel Reed, a dam opponent, went to Idaho to dedicate the Snake River Birds of Prey Natural Area. Also attending was Idaho Senator Len Jordan, the leading dam proponent. Reed said later: "As soon as the photogs went off, Jordan got crude and angry. He yanked me aside and said 'Listen, Nathaniel Reed, we're going to build this . . . dam and you're going to come out and dedicate it. I've used every chip I've got on Teton Dam. What do you think I'm doing here dedicating this goddamned vulture site?"[18]

As the Nixon White House aide John Ehrlichman recalled later, "The economics of a bad federal project did not matter all that much in the larger equation. At the time, Nixon was about to open the gates to China. Then there was the international monetary agreement, the SALT talks, détente with the Soviets. He could not get anywhere on those without congressional support, and Congress knew that and the Idahoans in Congress wanted that dam."[19]

Once in a while, when political imperatives repeatedly trump prudence, the laws of physics will provide sterner oversight than will the processes of Congress. The Teton Dam site was geologically unsound. Another Bureau dam on a problematic site had nearly failed just 5 years earlier, and some engineers within the Bureau doubted the wisdom of building the Teton Dam at all. They were overridden. The dam was completed in 1976, and the reservoir filled rapidly as springtime melted the snow from the mountains.

At this point, the Bureau took a series of actions that foreshadowed the Chernobyl, Ukraine, plant operators 10 years later. Not wanting to lose the water from the snow melt, the project engineer (30 years old and supervising his first big project) received permission to allow filling at twice the normal rate for a new dam despite the discovery during construction of unusually large fissures in the right-hand canyon wall. Grouting the fissures had been shoved aside in order to avoid further expense and delay. In addition, the main outlet through which water could be spilled was not yet complete. The emergency outlet was complete but sealed off by a huge metal barrier because it was being painted.

In 36 hours in early June 1976, the dam went from an initial leak to complete failure. Because the failure occurred visibly during the day, it allowed some minimal time for warnings. Nevertheless, the flood obliterated two towns and badly damaged a third.

Thousands of acres that were to have received water from the dam were stripped of topsoil and ruined; 11 people and 13,000 cows died. Had the failure occurred at night, the human death toll would have been at least in the hundreds.

**The Texas Water Project.**

In some ways the most grandiose of the Bureau of Reclamation projects, this 1960s scheme would have moved an amount of water equivalent to the lower Colorado River 1,200 miles from the Mississippi River below New Orleans across Louisiana and the lowlands of east Texas before pumping it up 3,000 feet to the high plains of west Texas.[20] The aqueduct would have had to go under four major rivers, while more than 100 smaller streams would have had to be tunneled under the aqueduct. The Texas Water Project was thought to be needed because the High Plains farmers had been pumping water from the gigantic Ogallala aquifer on which the region depended at a rate well above sustainability ever since pumping technology improved in the 1930s to make such excess possible.

To pump the needed Mississippi River water up 3,000 feet, 12 new power plants providing extremely cheap energy would be needed. The Bureau thought it knew just how to get it: "We took the most pie-eyed projections we could find from the Atomic Energy Commission. We figured the plants would cost $250 million apiece. The plan required about 12 of them. . . . You couldn't build one nuclear plant in 1985 for what we thought we were going to pay for 12 in 1971."[21] The Texas Water Project ultimately sank under the weight of cost and hostility from the state of Louisiana, though not before a politician from an adjoining state remarked, "If those Texans can suck as hard as they can blow, they'll probably build it."[22] Depletion of the Ogallala aquifer has slowed as pumping costs have risen and usage has become more efficient, but the aquifer is generally thought to be unable to meet the potential demands on it much after 2025.

### President Carter's Quest for Reform.

Shortly after taking office in 1977, President Jimmy Carter announced that he wanted to terminate funding for 18 water projects, including the Central Arizona Project, because none was remotely cost effective. A furious Congress responded with an appropriations bill that restored all but one of the 18 and included several new projects that Carter had not asked for. A story told by Congressman Bob Edgar, Paul Douglas's successor in lonely opposition to wasteful water and energy projects, illustrates the dominance of the energy and water appropriations committees and their business constituents during this period:

> We are a tyranny presiding over a democracy. Congressman Floyd Fithian of Indiana has a water project planned for his district that he doesn't want. . . . But he hasn't been able to remove the project from the appropriations bill. Congressman John Meyers sits on the Appropriations Committee and its Energy and Water Development Subcommittee. He has some big construction people in his district, which is next door to Floyd's, who would get some big contracts if the project is built. So every time Fithian tries to remove the project, Myers puts it back in.

The struggle over the terminated projects continued for 2 years. In 1979, President Carter was forced to sign a bill that continued funding for all of the terminated projects in order to secure the votes necessary to implement his agreement to return the Panama Canal to Panama, avoiding a major foreign policy embarrassment.

President Carter's successor, Ronald Reagan, had served as governor of California, a state that had benefited as much as any other from federal water and energy projects. The Bureau and its allies hoped that President Reagan would champion their projects. Instead, the fiscal conservatives in his administration worked in parallel with environmentalists to force more rigorous repayment and state contribution terms which killed many projects.

From that time forward, far fewer energy and water projects — and none of the grand river diversions — have been undertaken. The same cannot be said for the misallocations of resources that the more poorly considered projects have set in place. For example, the growing of water intensive crops in desert climates through heavily subsidized irrigation has created constituencies that stymie wiser water, energy, and agricultural policy. Some of these constituencies, rather than face a reality of diminishing supply and higher price, still talk longingly of the North American Water and Power Alliance (NAWAPA) — the greatest water and energy project of all.

NAWAPA would build high dams, pumps, and tunnels in Alaska and western Canada to route immense wild rivers south. Most would move through the Rocky Mountain Trench in British Columbia to unite with some reversed flow from the Columbia River and pour south into the Colorado River basin and California, alleviating any fears of drought in those regions for decades.

A considerable amount would also move east into Lake Superior and the Great Plains, there to make its way south to Texas to relieve pressure on the Ogallala aquifer. Variants involve damming off the southern end of James Bay in Quebec, Canada, turning it into a giant fresh water reservoir from which water could be pumped south and west to join the NAWAPA water in the Great Lakes and western Canada and to rescue the drought-prone southeastern United States.

## NUCLEAR REVIVAL, LOAN GUARANTEES, AND THE LESSONS OF HISTORY

With Congress having already allocated $18 billion dollars in loan guarantee authority for new nuclear units and with the industry and its congressional champions already complaining that this will not be nearly enough if nuclear power is to play a major role in combating climate change, it seems important to put nuclear credit support in the context of the knowledge that we have acquired over a century of federal credit support for major projects.

We know at least the power of self-interested myth to ride roughshod over fundamental economics. The historical evidence reveals beyond doubt the ineffectuality of mere proof of waste and risk when it comes to dissuading an eager Congress from lavishing credit support on a favored technology.

The evidence to date suggests that few if any new nuclear units will be built if they must obtain private capital either in power markets or under the regulatory treatment normally afforded new investment. But it seems equally clear that a major scaling up of nuclear power, while potentially helpful in combating climate change if it were truly a low cost approach, is not essential to doing so. Indeed, something on the order of building three times the existing nuclear capacity in the world is needed to provide 10-15 percent of the necessary carbon reduction.[23] At current rates of new construction, this goal cannot be attained. Indeed, world nuclear capacity will decline as plants reach the end of their operating lives.

However, many ways to reduce green house gas emissions exist. Principles will be more important than prophecy when it comes to choosing wisely among them. Among the reasons for preferring technology neutral options such as a carbon tax or a cap-and-trade mechanism is the likelihood that entrusting the federal government to achieve optimal results by manipulating access to capital is no more likely to produce sensible results now than it has in the past.

After all, the ways in which today's nuclear industry might echo the water and energy history set forth above are compelling:

- Nuclear power, too, was born in idealism and nurtured in government agencies that believed in it fervently. The Atomic Energy Commission was every bit as promotional as the Bureau of Reclamation and the Corps of Engineers, especially in the national mood that followed President Dwight Eisenhower's "Atoms for Peace" speech in 1953. Today's DoE is no less enthusiastic, not only for conventional nuclear power but for reprocessing and advanced reactors whose economics are outlandish and whose technical feasibility is unproven. Today's DoE has repeatedly shown itself to be incapable of sound economic analysis of the potential risks of nuclear development. Yet, it will have responsibility for screening the applicants for nuclear loan guarantees and for setting fees that reflect the risks of default inherent in such guarantees. Its record with similar programs in the past suggests strongly that it will err in ways that impose excessive risk on the public.

- Nuclear power also has been championed by powerful congressmen and senators whose states were home to major nuclear development. Initially, nuclear oversight was housed in a unique joint committee of both houses of the U.S. Congress. The Joint Committee on Atomic Energy was so unabashed in its promotion of nuclear power and so indifferent to public concern that it was abolished in 1975. In its place, Congress assigned primary responsibility to the same committees that oversaw the Corps of Engineers and the major hydroelectric development entities of the federal government. Initially more vigorous in their oversight of nuclear matters, these committees have over time become increasingly supportive of the economic interests involved with the technology. Senator Pete Domenici in the recent past has had influence over federal nuclear policy and the application of federal support comparable to that of champions of the dam projects of the last century. In short, the allocation of federal credit to nuclear power is every bit as subject to political influence as were the dam projects a generation ago.

- Nuclear power development too was accelerated by competition among two rival developers. Some in the Congress championed public ownership and wanted the Atomic Energy Commission itself to build the plants. Others wanted private ownership and arranged for subsidies of several sorts to make the plants attractive to investor-owned utilities. When President Eisenhower's Atomic Energy Commission (AEC) chair held out the vision of nuclear power "too cheap to meter," he was presiding over a program of unprecedented government assistance designed to lower costs and risks to companies interested in building the first power plants. The term came to haunt nuclear power as it contributed to a tripling of U.S. electric rates between 1970 and 1980.

- Under congressional and vendor company pressure to push ahead, nuclear power grew too fast for its own good, just as the Bureau and the Corps of Engineers pushed the water program into unwise and uneconomic development after the best sites had been developed by mid-century. Operating mishaps of several sorts, including the Brown's Ferry fire and culminating at Three Mile Island, caused vast nuclear cost increases and brought on an environmental backlash. The potential for a similar over acceleration of nuclear development exists again if nuclear power is assumed to be essential to dealing with climate change and is promoted accordingly. At present neither the regulatory process nor the nuclear industry infrastructure is adequate to handle rapid expansion, so the pressure to cut corners will once again be substantial.

- Rather than face up to the fact that a dozen new nuclear units will call for more than $100 billion in credit support from taxpayers and/

or customers (more that $300 per U.S. citizen), nuclear proponents blame the U.S. nuclear licensing process for nuclear power's trouble. But this is nonsense. The Bush administration nuclear regulators devoted themselves to compressing the licensing process to an extent such that the public has few meaningful rights left. The current schedule for pending applications calls for a review lasting some 3 years. The industry, delighted with the new process, has no idea how it could be cut further, especially since even the former, slower U.S. licensing process licensed more nuclear capacity than the next four countries combined. The problem was that more than half of the licensed plants proved unnecessary and were cancelled, some after billions had been spent on them.

- Nuclear power also involved "big government" approaches to choosing and building a particular technology. It also has enjoyed the support of many politicians who normally describe themselves as strong proponents of free enterprise, small business, and minimal government. Indeed, many of nuclear power's strongest supporters are small government champions from states that have hosted a large share of the big water resource projects—New Mexico, Washington, California, Idaho, Texas, and Tennessee come to mind. In the early years of nuclear development, the Democrats in this group tended to favor government ownership and used this possibility to push the investor-owned utilities to move faster. By the early 1970s, the distinction between investor-owned and government-owned nuclear development had become unimportant. However, it returned in a new form as the Bush administration put the Tennessee Valley Authority and even the Air Force[24] in the forefront of those willing to host new nuclear units while putting the DoE in a position to sponsor and perhaps build a new reprocessing plant, an advanced recycling reactor, and an advanced fuel cycle research facility. For the time being, only U.S. Government entities can raise funds for new nuclear projects. Meanwhile, the technology-neutral approaches to the problems of climate change and energy security—the approaches most compatible with reliance on private enterprise—are shunned by the conservatives most rhetorically eager to rely on markets rather than government.

- Both sets of projects also depended heavily on cost-benefit and environmental impact studies that were distorted in important ways. Demand or need for projects was overstated. Potential for shortage, even catastrophe, without them was exaggerated. Fictitious discount rates were used. Costs were understated and benefits overvalued. Risks and uncertainties were ignored. Alternatives chosen for comparison purposes were the most expensive and objectionable. Yet when the projects were delayed or cancelled, the forecasted shortages never occurred because more efficient usage and/or different combinations of alternatives filled the forecasted void.

- Both sets of projects followed a strategy of maximizing the number of states with an economic interest in their programs to maximize political support in Washington, DC. The current nuclear industry approach to its asserted renaissance reflects this approach, with many more nuclear power plants and Global Nuclear Energy Partnership projects announced than can conceivably be built anytime soon under current levels of federal support. As a result, congressmen will come under pressure to expand the funding and other support to cover a larger population of new plants.

- President Carter failed to constrain the most uneconomic aspects of both sets of policies when he took office. As with the water projects, the nuclear industry went around him to Congress and — in the case of nuclear power — also went overseas to urge defiance of the Carter effort to curtail the breeder reactor and reprocessing (never mind that both had originally been suspended by President Gerald Ford). And in both cases, the proponents took great comfort from the election of President Reagan, only to be disappointed when — despite his supportive record and rhetoric — his actual refusal to put the federal treasury at their disposal doomed their prospects.
- Finally, most fundamentally, when the economic justification for new projects is lacking, coerced capital remains the option of last resort for both sets of projects. Not only does this approach make capital available, but it permits charging a lower price for the output of the facilities even though they have not become cheaper. Instead, cost and risk have shifted from the investor to the taxpayer. Providing water to irrigation farmers who sometimes paid less than 5 percent of the cost of supplying

them may have been the most extreme example, but calculations of the impact of the loan guarantee program about to be offered by the DoE suggest that it may cut some 30 percent off the price of nuclear electricity by shifting risk from investors to taxpayers.[25]

## CONCLUSION

This chapter does not argue that all federal credit support is undesirable. But it does argue that proponents of such support confront a heavy burden to show that the program they have put forward has built-in checks against the pitfalls described in this chapter. In particular, they need to provide reason to believe that analysis of need and of alternatives will be more rigorous than has been the case in the past. They need also to show insulation from political pressure, a capacity to charge participants an amount commensurate with the benefits that they are receiving, and a determination to hold participants liable in case of default. For the nuclear industry, this will require coming to terms with its brush with economic catastrophe in the 1990s, when only the willingness of state regulators to allow extraordinary surcharges for the excess costs of the last generation of nuclear units avoided massive write-offs for many utilities.

Proponents should have to establish also that the problem that they seek to address cannot be solved in the absence of federal credit support. The shortcomings of such programs are clear enough and persistent enough such that they should be an option only of last resort, one that is turned to upon a showing that the capital needed to solve a major problem cannot be raised in any other way.

If Congress does not insist on more rigorous analysis in the face of the nuclear industry's current loan guarantee claims, it risks sitting through a sordid melodrama it has seen before. In the name of urgent societal necessity, we have literally moved mountains to deliver resources that the private sector alone would not put forward. But we have also seen what happens as these programs develop privileged constituencies who become expert at corralling strong political support. In particular, we have seen that the use of federal credit support can indeed hide costs, but federal credit support does not make those costs go away. It assures only that they will not have to be paid in the prices of the favored projects—the irrigation tunnels, nuclear power plants, or federally built reprocessing infrastructure.

## ENDNOTES - CHAPTER 5

1. The Price-Anderson Act was originally enacted in 1957, with a liability limit of $500 million to encourage private industry to build the first few nuclear power plants.

2. In December 2007, Congress enacted a provision in its report on the Energy and Water Appropriations Act for Fiscal 2008 which purported to approve $18.5 billion in loan guarantees for new nuclear plants in Fiscal Year (FY) 2008 and 2009 based on nonlegally binding report language. The provision fails to resolve major congressional conflicts over law and policy, such as the applicability of the Federal Credit Reform Act and the role of congressional oversight. It also creates more hurdles for the DoE to overcome before it can execute its program. Even if DoE decides it has sufficient legal authority and is willing to overcome a likely legal challenge, it will be difficult to issue the loan guarantees in the near term.

3. Economists Linda Cohen and Roger Noll were to conclude that "the entire synfuels program had a quality of madness to it.

Project after project failed. Cost estimates were connected to the price of substitutes rather than to the program itself. Goals were unattainable from the start. . . ." *The Technology Pork Barrel*, Washington, DC: The Brookings Institution, 1991.

4. Murray Weidenbaum and Reno Harnish, with James McGowen, "Government Credit Subsidies for Energy Development," Washington, DC: American Enterprise Institute for Public Policy Research, 1978. For a more amusing and recent critique, see "Nusubsidies Nuclear Consortium: Where the Taxpayer is Our Favorite Investor," Cambridge, MA: Earthtrack Institute, 2005, available from *www.earthtrack.net/earthtrack/library/NNC_Overview.pdf*.

5. Murray L. Weidenbaum and James McGowen, *Government Credit Subsidies for Energy Development*," Washington, DC: American Enterprise Institute for Public Policy Research, 1976.

6. This paper is concerned only with support taking the form of a commitment to assume liabilities, whether in the form of loan guarantees or other obligations. It does not, for example, deal with the grants made toward reactor development by the Atomic Energy Commission, the limitation of liability in the Price-Anderson Act, the assumption of responsibility for waste disposal in the Nuclear Waste Policy Act, or the production tax credits in the Energy Policy Act of 2005.

7. Remarks of Governor Nelson A. Rockefeller at the groundbreaking ceremony for the reprocessing plant of Nuclear Fuel Services at West Valley, New York, June 13, 1963.

8. *Ibid.*

9. Quoted in Marc Reisner, *Cadillac Desert: The American West and Its Disappearing Water*, London, UK: Penguin Books, Rev. Ed., 1993, p. 36. This book, from which most of the history of water resource development is drawn, is a remarkable review of the American water development saga. As water resource issues take on increasing importance around the world, it is worth the attention of anyone with an interest in that field.

10. Wallace Earle Stegner, *Beyond the Hundredth Meridian: John Wesley Powell and the Second Opening of the West*. With an

introduction by Bernard De Voto, Boston, MA: Houghton, Mifflin, 1954.

11. *Ibid.*, p. 114.

12. *Ibid.*, pp. 142-143.

13. David Ogilvy, "Should we flood the Sistine Chapel so that tourists can get nearer to the ceiling?" *Ogilvy on Advertising,* ad campaign written by Howard Gossage for the Sierra Club.

14. "Pork barrel" is an Americanism describing the process by which congressmen obtain financial support for projects that lack economic justification. It derived from the old Southern custom of putting out a barrel of pork that slaves often fought over.

15. Here is Reisner's discussion of the government's solution of the Mexican problem, which resulted largely from agricultural practices in Arizona's Wellton-Mohawk district:

> The solution of choice at Wellton-Mohawk has been construction of a reverse osmosis desalination plant — 10 times larger than any in the world — which, while consuming enough electricity to satisfy a city of forty thousand people, will treat the waste water running out the drain canal. . . . What Congress has chosen to do, in effect, is purify water at a cost exceeding $300 an acre foot so that upriver irrigators can continue to grow surplus crops with federally subsidized water that costs them $3.50 an acre foot.

> "If the farmers at Wellton-Mohawk adopted efficient irrigation methods" says Jan van Schilfgaarde (of the Department of Agriculture's Salinity Control Laboratory), "you could solve the problem without even retiring the lands. . . . I'm not even talking about installing drip irrigation. I'm talking about laser-leveling fields and reusing water on salt tolerant crops and not doing stupid things like irrigating at harvest time. . . . A lot of these guys are actually absentee owners farming by telephone from their dentists' offices in Scottsdale. . . . They're not in this business to farm crops, or even to make a profit. They're

farming the government. They're growing tax shelters. But even if you do have a highly competent farmer who wouldn't mind reducing his wastewater flows, he has no incentive to conserve. Federal water is so cheap it might as well be free. What's the point of hiring a couple of additional irrigation managers to save free water? He's being forced to consume water.

In fact, the desalination plant was completed in the late 80s but operated for just a few months before shutting down because some wet years, coupled with rising energy costs, made it too expensive. It may soon be reopened in light of increased demands for Colorado River water. Marc Reisner, who died in 2000, would not have been surprised.

16. Reisner, pp. 304-305.

17. *Ibid.*, p. 386. Jerry Jayne, the nuclear engineer who was also president of the Idaho Environmental Council, a dam opponent, said of the farmers, "I can talk to the loggers. I can talk to the ranchers. I can talk to the mining companies. I can say nothing to the irrigation farmers. They're not reasonable. They don't listen. They're true believers. They're like communists — only in reverse."

18. *Ibid.*, p. 394.

19. Reisner.

20. A benefit undervalued at the time was the use of the earth that would have been removed in digging the aqueduct to enhance the levee system protecting New Orleans from hurricanes.

21. Jim Casey, Deputy Chief of Planning, Bureau of Reclamation, quoted in Reisner, p. 447.

22. Reisner.

23. For a thoughtful analysis of carbon reduction options, see Robert Socolow and Stephen Pacala, "A Plan to Keep Carbon in Check," *Scientific American*, September 2006, p. 50. See also S. Pa-

cala and R. Socolow, "Stabilization Wedges: Solving the Climate Problem for the Next 50 Years with Current Technologies," *Science*, August 13, 2004, pp. 968-972, available from *fire.pppl.gov/energy_socolow_081304.pdf*.

24. "U.S. Air Force May Lease Land for Nuclear Plant," *Nucleonics Week*, March 6, 2008, p. 1.

25. Doug Koplow, "Nuclear Power in the U.S.: Still Not Viable Without Subsidy," presentation at Airlie House, November 2005, p. 6, available from *www.earthtrack.net/earthtrack/library/NuclearSubsidies2005_NPRI.ppt*.

# PART II

# EXPANDING NUCLEAR POWER
# IN EXISTING AND TO FUTURE NUCLEAR
# STATES

# CHAPTER 6

## NUCLEAR POWER MADE IN FRANCE: A MODEL?

**Mycle Schneider**

## INTRODUCTION

Not long ago, french fries were renamed freedom fries in the United States as part of an intense anti-French campaign following the French government's refusal to join the Iraq war. But the rage against France now seems forgotten. Not only are french fries politically correct again, but France has become a model for nuclear power in the United States and beyond. "It's time to look to the French," *New York Times* columnist Roger Cohen wrote. "They've got their heads in the right place, with nuclear power enjoying a 70 percent approval rating."[1] Similarly, former Republican presidential candidate John McCain has wondered, "If France can produce 80 percent of its electricity with nuclear power, why can't we?"[2]

The current Sarkozy-Fillon government, acting presidency of the European Union's (EU) Council of Ministers, has chosen to massively promote nuclear power even to newcomer countries like Algeria, Jordan, Libya, Morocco, Tunisia, and United Arab Emirates. As President Sarkozy put it in a speech in Marrakech: "We have it in France, why shouldn't they have it in Morocco?"[3] The French president travels the world as a salesman for the glittering nuclear industry; from the Middle East to China; from Brazil to India. On September 29, 2008, even before the U.S. Congress had given the green light for the U.S.-India nuclear deal, France signed a similar cooperation agreement with India.

The international credit crunch will not make it easier for nuclear planners to implement their projects. Finance is rare and will be more expensive. On November 19-20, 2008, the participants to the 1978 Arrangement on Guidelines for Officially Supported Export Credits met in Paris under the auspices of the Organization for Economic Cooperation and Development (OECD). The main goal was to extend the granted payback period for nuclear credits from 15 years to up to 30 years. The key expected sources of funding of export credit agencies are Japan and France. These two countries, in addition to the United States, recently agreed to fund a study within the World Bank to reassess the cost-competitiveness of nuclear power. The move is a further step to increase pressure on multilateral development banks that generally have an outspoken or implicit ban on nuclear financing. For example, the World Bank has not financed nuclear projects in decades, while the Asian Development Bank has never provided funding for nuclear. "I don't understand and I don't accept the ostracism of nuclear [power] in international financing," French President Sarkozy commented in March 2010.[4]

The state utility, Electricité de France (EDF), has amplified its own international strategy with the recent takeover of British Energy, investment in the U.S. utility Constellation, and the creation of the Guangdong Taishan Nuclear Power Joint Venture Company with the purpose of building and operating two European Pressurized Water Reactors, in which EDF holds a 30 percent interest for 50 years.

The general message is clear: in France nuclear power works. In 2007, nuclear power provided 77 percent of the electricity in France and 47 percent of all nuclear electricity in the EU. "The requests by countries that wish to profit from that clean and cheap source of energy are legitimate," claims French Foreign Minister Bernard Kouchner.[5] But does it really work that well, and is it all that clean and cheap in France?

France is among the top economic powers in the world, and it has considerable political influence on the international level. The country had the seventh largest Gross Domestic Product (GDP) in 2006, the eighth largest primary energy consumption in 2007, and by far the most visits by foreign tourists worldwide. With over 63 million inhabitants, France has the second largest population in the EU, behind Germany.

French energy policy has considerable international influence, in particular through a constant strong representation at the Directorate General of Transport and Energy (DG TREN) of the European Commission and through other organizations like the International Energy Agency (IEA) of the OECD. The IEA has remarkably increased its pro-nuclear stance since 2003 when the term of Claude Mandil as Executive Director began. Mandil is a member of the *Corps des Mines*, a French State elite of engineers that has designed, pushed through, and implemented the nuclear program in France, with its members holding key positions in ministries, industry, and State agencies.

Industry and utility representatives, diplomats, and civil servants have been highly successful in depicting the nuclear program as a great achievement, leading to a great level of energy and oil independence and carbon free power.

With nuclear power gaining increasing acceptance in the EU and elsewhere, it is worthwhile to have a closer look at the "French model." To understand the overall impact of the French nuclear energy strategy, it is necessary to look beyond the number of kilowatt hours produced. Many of the impacts are system effects that are not obvious at first sight.

## HISTORICAL ASPECTS

In 1946, the French government nationalized "the production, transport, distribution, and the import and export" of electricity and natural gas and created Electricité de France (EDF) and Gaz de France (GDF) as state energy monopolies.[6] The legislation stipulated that 1 percent of the companies' turnover go to the Central Fund for Social Activities (CCAS),[7] a fund to be managed by a board composed of representatives from the different trade unions according to the previous union election outcome. Since the Confédération Générale de Travail (CGT), close to the French communist party, won the absolute majority of votes every single time from the start, CGT was in a position to manage a huge budget, about €450 million in 2006, in principle on a large number of social and associated issues (vacation facilities, restaurants, child care centers, etc.). The CCAS employs over 5,600 people. It has been suspected for a long time of constituting a convenient and abundant source of subsidies for the French communist party. In a 2006 confidential report, the French Court of Auditors accused the Fund of "total lack of transparency on resources and employment . . . and insufficient internal control."[8]

However, more importantly, the arrangement constituted a long-term guarantee for social peace. The extraordinary advantages for EDF employees funded by the CCAS were and are complemented by preferential power tariffs. It comes as no surprise that EDF has been hit significantly less by strikes than many other French companies (including public ones), and that only on rare occasions have union activities led to power cuts.

In addition to the average preferential electricity tariff for each EDF employee, during the project planning and construction phases, EDF practiced lower tariffs in the vicinity of nuclear power plant sites. The practice has been declared illegal on the grounds that it violates the principle of equal treatment. However, the court case, initiated by consumer and environmental protection organizations, took over 5 years, time enough for the construction sites to get into an advanced stage. Incitement to acceptance had done the job prior to the method being declared illegal because it was obviously violating the equality principle.

The relationship between trade unions and nuclear sector has been instrumental to the implementation of the various phases of the nuclear program. While EDF was pacified by the historical social fund deal, the history of the French Atomic Energy Commission (CEA) was slightly different. After what was later termed the "reactor line war" (la guèrre des filières), the CEA remained responsible for the implementation of the nuclear fuel chain. Nevertheless, the CEA had lost the war at the beginning of the 1970s. Its own gas-graphite reactor line was abandoned in favor of the Westinghouse Pressurized Water Reactor (PWR). By 1972, nine gas-graphite reactors had been started up, of which eight were producing power. At least four of them were used to generate plutonium for the French nuclear weapons program. The last one was shut down in 1994.

In the same year that the last unit of the CEA natural uranium reactor line started up, the European

Gaseous Diffusion Uranium Enrichment (EURODIF) consortium was created with the intention to provide low enriched uranium to a group of participating countries for their light water reactors.[9] In 1974 the first large-scale 16-unit nuclear power program was launched exclusively on the basis of a Westinghouse PWR license held by Framatome until 1982. At that point, 50 of the now 58 operating units were already in operation or under construction.

By the middle of the 1970s, the French trade union Democratic Confederation of Labour (CFDT), close to the Socialist Party, had gained considerable influence within the CEA. The CFDT was very critical about the plutonium program and the health and safety conditions at the La Hague reprocessing plant. In 1976, the Compagnie générale des matières nucléaires (CO-GEMA) was established as a 100 percent subsidiary, under private law, of the public CEA with the clear strategic orientation to build up over time a powerful nuclear fuel group that would master everything from uranium mining to plutonium fuels, civil and military. The "private" company COGEMA had been given the industrial uranium fuel and reprocessing facilities. The change broke up the powerful position of the La Hague section of the CFDT, which had led a successful strike there the same year. Union leaders left the site to stay within the public CEA in order to protect their status. In 1981, the newly elected President François Mitterrand entrusted top CFDT leaders and engineers from the CEA and EDF with the development of the French Energy Efficiency Agency (AFME). The deal was basically that they would get full government support on the condition that they would no longer intervene on nuclear issues. The only existing nuclear expertise federated on the national level was decapitated. In the following years, the CFDT lost most of its credibility in the sector. Later, EDF, CEA, and CO-GEMA were cleared of potentially costly trade union concessions.

## DECISIONMAKING

Until 1991, France did not have any specific nuclear legislation, and the 1991 law was limited to the question of research and development on high-level radioactive waste. It was only in 2006 with the "Law relative to transparency and security in nuclear matters"[10] that specific legislation was introduced. There has never been a vote in parliament on the launch of the nuclear power program. A "great energy debate" promised by François Mitterrand prior to his election in 1981 never materialized. It took until 1989 for anyone to finally discuss national energy policy in the National Assembly: the discussion lasted 3 hours and was followed by 24 members.[11] The energy debates organized in various French cities in 2004 and after, like the events organized by the National Public Debate Commission, did not influence the decisionmaking in any way. Major decisions like the construction of a first Generation III European Pressurized water Reactor (EPR) at Flamanville were made before a parliamentary debate even took place. Greenpeace accused the government of demoting the members of parliament to individuals "elected for nothing" (Elus Pour Rien).

This is no coincidence. The elected representatives always had and have a very minor influence on the development, orientation, design, and implementation of energy and nuclear policy in France. The issue is entirely under control of the elite technocrats of the state Corps des Mines.[12] Officially, the governing body of the Corps des Mines is the General Mining Council and is presided over by the Minister of Industry. However, ministers change, but the "corpsards" remain. Therefore the most powerful position in reality is the vice-president of the Council who, as are all its other members, is a member of the Corps des Mines. It is remarkable to what extent the Corps has managed to lockup all of the key positions linked to the nuclear issue. The nuclear advisors to the President of the Republic, the Prime Minister, the Ministers for Economy, Industry, Environment and Research, the chief executive officers of the CEA, AREVA, Framatome, and the safety authorities, all have historically been members of the Corps des Mines.[13] If there is an "Energy Mission" to advise on policy, it is headed by the Corps des Mines.

This state organized elite clan has made it possible to push through long-term policy orientations, like the nuclear program, entirely outside the election process. The mechanism provides a huge advantage for long-term planning and the implementation of large infrastructure projects. It also constitutes a significant disadvantage for democratic decisionmaking, which is entirely cut off. It is a serious handicap for any significant policy adaptation or reorientation.

Georges Vendryes, who represented France at the IAEA's Board of Governors for 23 years and who is considered the "father of Superphénix" (the fast breeder reactor at Creys-Malville) summed up the French exception this way:

> Since 40 years the big decisions concerning the development of the French nuclear program are taken by a very restricted group of personalities that occupy key

positions in the government or in the top administration of EDF, CEA, and the few companies involved in the program. The approach remains unchanged in spite of the change of ministers thanks to the permanence of these personalities that occupy the same position generally for some 10 years.[14]

## ACCESS TO INFORMATION

Access to information on the nuclear sector in France has been restricted since the launch of the industry in 1946. Trust in the information provided by state and industry were entirely demolished in the aftermath of the Chernobyl accident. On May 6, 1986, 1 week after the disaster, the Ministry of Agriculture issued a press statement declaring that "the French territory, due to the distance, has been totally exempted from radioactive fallout after the accident at the Chernobyl plant."[15] While vegetables were destroyed systematically on the other side of the borders with Germany and Italy, the French government did not take any precautionary measure except for the destruction of a single load of spinach. The head of the radiation protection agency SCPRI declared in a telex "to be distributed to medical doctors and the public," that one would have to imagine "levels 10,000 or 100,000 times higher" in order to justify precautionary measures.[16]

In reality, airborne radioactivity from the Chernobyl accident had triggered the alarm systems on many nuclear sites in France. The reaction was to modify the set-point accordingly rather than to inform the public. An extensive environmental measuring campaign (over 3,000 samples) carried out by the independent laboratory CRIIRAD, that started up in the aftermath of Chernobyl and published in 2002, revealed current cesium contamination levels of up to 50,000 Bq/m². In

1999, a small group of people that had contracted thyroid cancer started an association that by 2002 developed into a group complaint by over 400 people with thyroid sicknesses. They accused the government of misleading statements and lack of precautionary measures. The case is still under active investigation.

On June 13, 2006, the French Nuclear Transparency and Safety Act was passed. This is the first more comprehensive piece of legislation on the nuclear industry. Until then, the sector had been legally managed by a 1963 truncated act and specific regulations. The new law stipulates in section 2:

> Every person shall be entitled, on the conditions laid down by the present act and its implementing decrees, to be informed about the risks related to nuclear activities and their impact on the health and safety of persons and on the environment, and on the release of effluents from installations.

It is seen as a major change that

> while the government remains responsible for informing the public about the risks related to nuclear activities and their impact all operators and persons in charge of transport now have obligations to disclose information too, which considerably broadens the range of enterprises concerned.[17]

However, the author's experience since the passing of the Transparency and Safety Act illustrates that much needs to be done, in particular when it comes to cost figures. For example, none of the three main nuclear operators—EDF, CEA, and AREVA—answered a questionnaire on decommissioning and waste management funds, developed in the framework of an official study on behalf of the European Commission.

On the other hand, a more constructive attitude has been adopted by the Nuclear Safety Authority (ASN) and its technical back-up, the Institute for Radiation Protection and Nuclear Safety (IRSN). For example, even prior to the Nuclear Transparency Act, ASN began publishing all of the letters that are sent to nuclear operators following inspections. On request, it has also released very detailed emissions data and transmitted a database on nuclear events.

Following an incident that was discovered on July 7, 2008, that involved the spill of uranium into groundwater at a nuclear maintenance facility at Tricastin in the south of France, the Minister of Ecology asked the High Committee for Transparency and Information on Nuclear Security (HCTISN) to elaborate an opinion on the radio-ecological monitoring of all nuclear sites and the management of former nuclear waste storage facilities. The minister requested that particular attention be given to the quality of information, the level of transparency and the modalities allowing better involvement of stakeholders in the process. On November 7, 2008, the HCTISN published its report, with 18 recommendations concerning access to information, including "the development of an expert capacity that is diversified and independent of the organizations currently implicated in the evaluation of nuclear dossiers."[18] However, nothing has happened since.

## CIVIL-MILITARY LINKS

Unlike the United States, which has attempted to a large extent to separate civil and military uses of nuclear power, France has never divorced the administration of nuclear energy and nuclear weapons, this has remained the underlying rational until today. As the latest official French report on the protection and control of nuclear materials states: "In fact, France is a civil and military nuclear power but does not have two separate [fuel] cycles."[19]

The state-owned CEA, created in 1945 with the explicit, though secret, task of developing the French nuclear bomb program, has since implemented the military-civilian nuclear link. Until today, the same number of people, 4,500 each, work under the military applications and civil energy departments. The CEA has a wide area of responsibility in nuclear matters, which includes everything from fundamental research in physics to research and development for radioactive waste management. Its Direction des Applications Militaires (DAM) was responsible for warhead testing at Moruroa. Its former subsidiary, COGEMA (Compagnie générale des matières nucléaires) (now AREVA NC), is responsible for the production and maintenance of nuclear materials, including plutonium. The CEA built the plutonium production plants at Marcoule and La Hague.

The French civil nuclear program has largely profited from the military program and vice-versa. The 1973 CEA annual report explains the French approach: "The CEA must, within the framework of a rigid budget and strictly limited possibilities of expansion, adapt the production of military nuclear material to rapidly changing needs by taking advantage of technical progress and civilian programs (which themselves have greatly benefited from military programs) in order to limit the costs."[20]

The following quote stems from a document dated 1964, which introduces defence-planning legislation for the years 1965 to 1970. The chapter on nuclear materials production explains unambiguously:

> This chapter deals with the investment and operational expenditures connected with the production of nuclear materials for military use. This expenditure relates to:
> — the completion, the start-up, and the operation of the plutonium extraction plant at La Hague, since that plutonium, extracted from fuel irradiated in EDF reactors, will be used for military purposes.
> — the costs incurred from the production of military grade plutonium in EDF reactors.[21]

It is only consequent that the original La Hague reprocessing plant was financed in equal shares by the civil and military budgets of the CEA. Civil military cross subsidizing has been a principle throughout the entire French nuclear program.

## INTERNATIONAL NUCLEAR SAFEGUARDS IN FRANCE

The fact that international nuclear safeguards are in force in France frequently leads to the belief that military and civil nuclear activities are separated, since France deals with so many foreign nuclear products from countries that do not want to participate in the French nuclear weapons program and as such have signed bilateral or multilateral agreements with France. This is not so. International nuclear safeguards in France are a delicate compromise between defense needs and peaceful end-use obligations. A part of the nuclear installations in France co-processes civil and military nuclear materials. At the same time public opinion in client countries that calls for a clear separation of French nuclear weapons from civilian programs needs to be satisfied.

Spent fuel from Australia, Belgium, France, Germany, Italy, Japan, the Netherlands, Spain, and Switzerland has been reprocessed at La Hague. Australia, Switzerland, and Japan are not members of the EU. La Hague facilities are subject to two international safeguards regimes, the EURATOM, and IAEA Safeguards. Nevertheless, France is a nuclear weapons state and, as such, is allowed to withdraw French-labeled material from international safeguards control as often as it likes, thus freeing it for military use. In fact, the rules are even less stringent than that.

**EURATOM Safeguards in France**.

The EURATOM treaty is, first of all, a treaty meant to facilitate and to structure the development of the nuclear industry in Europe. One of the tasks of EURATOM is "ensuring, through appropriate control and verifications, that civil nuclear materials are not diverted to uses other than those for which they were intended."[22] The EURATOM agreements forbid a non-nuclear weapon state to develop a nuclear capacity, but accept that nuclear weapon states possess the necessary production and maintenance installations for their nuclear weapons.

The EURATOM Safeguards implementation scheme requires continuous inspection for the duration of the operation of certain facilities. The materials for which EURATOM Safeguards apply are materials subject to a Community commitment for peaceful use, materials subject to a bilateral agreement between France and another country, and nuclear materials which are free for any use, i.e., not subject to any constraint. EURATOM Safeguards do not apply to materials for which no attribution has yet been decided, nor to nuclear materials directed towards any military use by France.

**IAEA Safeguards in France**.

Some non-European countries like Japan, Canada, Australia, and Sweden (before it joined the EU) have requested from France that their materials there be placed under IAEA Safeguards.[23] Nuclear safeguards are meant to allow detection of any diversion of nuclear materials or installations to any undeclared activity, particularly military uses. This applies in principle to the spent fuel sent to La Hague for reprocessing, as well as to uranium and plutonium separated during reprocessing. Can the IAEA guarantee the peaceful end-use of these materials?

According to COGEMA[24] and the IAEA itself,[25] the only nuclear installation in France that the IAEA has selected for inspection is the spent fuel storage pools at La Hague. The 13,700 ton capacity pools contain the spent fuel from the different client countries awaiting reprocessing. Safeguards of storage pools cannot measure the plutonium content of the fuel. The amount of plutonium contained in spent fuel assemblies is estimated by calculations based on the characteristics of the fuel. According to the IAEA,[26] poor accuracy of these calculations limits their value as a safeguards reference. The IAEA does not indicate any figure for error margins. No alternative solution to establish the plutonium content of spent fuel more accurately seems technically possible at the moment. Plutonium and uranium contained in spent fuel can only be measured accurately once the fuel assemblies have been sheared, dissolved, and transferred into an account-ability vessel.

## THE TRIPARTITE EURATOM/IAEA/FRANCE AGREEMENT

A tripartite agreement on safeguards was signed on June 20, 1978, between France, the IAEA, and EUR-ATOM. The IAEA was supposed to be able to control nuclear materials from different foreign countries for which France had agreed to conform to IAEA Safe-guards, while at the same time not intervening in the French military program. By 1993, 103 facilities in which this foreign material could be processed could also contain nuclear materials for French military uses. The tripartite agreement was meant to diminish IAEA Safeguards costs while facilitating communication between IAEA and EURATOM.

The first article states:

(a) France shall accept the application of safeguards, in accordance with the terms of this Agreement, on source or special fissionable material to be designated by France. . . .

(b) France shall provide the Community and the Agen-cy with a list (herein-after referred to as 'the facilities List') of the facilities or parts thereof which contain the nuclear material referred to in paragraph (a)...France shall keep the Facilities List up to date and may at any time make deletions to it. . . . [27]

According to Article 14, "if France intends to make any withdrawals of nuclear material from the scope of this Agreement . . ., it shall give the Community and the Agency advance notice of such withdrawal."[28] In other words, France is entirely free to use any of its installations for military purposes, and all it has to do is declare it to the IAEA and EURATOM.

A French Industry Ministry report on physical protection and safeguards of nuclear material describes the scope and the structure of the EURATOM safeguards in France.[29] The following information concerning the excluded material is particularly illustrative:

> On the contrary, are excluded from Safeguards materials which are free of use and which are declared by France as affected for its defence needs, as well as eventually those for which the affectation has not yet been decided. In all cases, France always has full control over materials which are free of use and may at any time, by a simple accountancy movement, transfer them from an area under EURATOM Safeguards to an area not under EURATOM Safeguards and reciprocally.[30]

According to the same report, following the 1978 tripartite agreement, France communicated a list of 116 facilities which contain nuclear material where community Safeguards are liable to be enforced. Also according to the same report, 265 facilities in France contain nuclear material. Therefore, at that point, France did not allow access for EURATOM inspectors to more than half of its facilities which contain nuclear materials. The list of these 149 unsafeguarded facilities is called the negative list of facilities. Furthermore, 30 of the 116 facilities which have been declared to EURATOM "reflect [France's] only nuclear fuel cycle," and "are under a mixed status, since they may contain alternatively or simultaneously material under or not under Safeguards. In this last case, their access is temporarily closed to EURATOM."[31]

In its latest report, the French Government states that it has transmitted a list of 171 facilities where EURATOM safeguards are "susceptible to be carried out."[32] However, it does not specify the number of facilities that are under civil-military mixed status.

The scope of IAEA Safeguards in France is even more limited than the scope of EURATOM safeguards. According to the French Industry Ministry, the agreement which was concluded was a compromise: France named eight installations, internally called the "gas meter" facilities, to be subject to IAEA Safeguards. The identity of these facilities is secret. According to the terms of the agreement, the nuclear materials contained in these facilities shall be at least equivalent to foreign nuclear materials officially subject to Safeguards, in quantity and in quality.[33] According to the International Safeguards Application Division of the French government that follows international Safeguards inspections in France, the IAEA would be allowed to inspect about eight installations in which nuclear materials equivalent to the foreign materials are stored. "Eight is a number which varies," but "like everyone, the IAEA has budget problems" and therefore only inspects the La Hague spent fuel storage facility. However, "nothing forbids the IAEA from making inspections in the other installations."[34]

This is confirmed by an IAEA statement: "The IAEA applies limited safeguards at THORP and at the UP2 and UP3 facilities in the UK and France, respectively. The IAEA has not been given the financial resources necessary for full coverage of civil nuclear installations in [Nuclear Nonproliferation Treaty] NPT nuclear weapon states allowed under voluntary safeguards agreements currently in force."[35] According to its International Safeguards Application Division, France sends accountancy reports on foreign safeguarded materials at La Hague to the IAEA. Therefore, the 1978 agreement allows France to manage according to its civil and military nuclear needs the nuclear materials in the majority of its facilities, even if those materials are in contact with or consist physically of foreign materials under Safeguards.

IAEA and EURATOM Safeguards are controls, which are meant to verify that no nuclear material declared for peaceful use has been diverted to a military use or to a use different from the one declared. France, as any other nuclear country, also implements national physical protection measures, to prevent any nuclear material from being diverted from the facilities. However, this scheme is not designed to prevent the direct or indirect use of foreign nuclear materials in the French defense programs.

## PLUTONIUM SWAPS

The plutonium coming out of reprocessing is never identical to in-going plutonium since a certain amount stays in the piping system and in the waste, and plutonium in the piping from earlier campaigns might come out in later campaigns. Theoretically, physical tracking of the plutonium of a certain origin would be possible to a fairly high degree of accuracy. But this would mean that the whole piping of the reprocessing plant would have to be cleaned out and rinsed each time before the following batch of fuel of a different client is introduced into the system. This would raise the economic costs of plutonium separation to impracticable levels. Even if this was not done, it would be possible after each reprocessing campaign to allocate to the customer a given amount of plutonium of the corresponding age and of a certain quality based on the isotopic composition of the plutonium delivered in the spent fuel. Even if atoms do not carry a "flag," this procedure would allow for at least a share of the physical identity of the plutonium to be allocated to the original owner, and it would allow for the tracking of this material from then on.

In practice, however, the plutonium in store does not have a label that identifies its origin or the client to whom it is allocated. The plutonium is identified only according to its quality. The plutonium is allocated on paper to each customer according to an unknown set of parameters, without regard to the actual physical origin of the material. One batch of plutonium can be replaced by another. Therefore, not only is there always a certain mixing of plutonium of different origins during the reprocessing process, but also, under current practice, there is the conscious exchange (swapping) of plutonium of a given origin/allocation for plutonium from another origin/allocation.

Plutonium swaps can have different purposes. Plutonium separated through reprocessing spent fuel can have very different isotopic compositions, which vary notably with the burn-up rate of the spent fuel, the delay before reprocessing, and the storage time after reprocessing. A higher burn-up rate increases the radiotoxicity of the plutonium and diminishes the fissile plutonium content of the resulting plutonium. A long delay after reprocessing increases the americium-241 content (plutonium-241 decay product), which increases the radiotoxicity of the fuel and diminishes its fissile properties. It is therefore useful for plutonium fuel production to use plutonium with a low burn-up rate, and which has not been stored too long since reprocessing. Since it is in everybody's interest and there are large amounts of plutonium in stock, it is quite logical that AREVA NC actually sends back to client countries rather "fresh" plutonium, whatever its origin.

The 1978 agreement between France, IAEA, and EURATOM is a de facto flag swapping authorization. It allows France to exchange foreign safeguarded nuclear materials and materials considered to be equivalent in a chosen facility. Therefore, once foreign nuclear materials are sent to France, AREVA NC processes them as it desires, and sends back materials considered to be equivalent.

In civil facilities, this exchange of equivalent material can enable the use of embargoed material, as happened during the embargo of South African uranium. In this case, in the late 1980s Finland got what was actually South African uranium that was physically located in France and was transformed, "flag-swapped," by the German uranium broker NUKEM into uranium of nonembargoed Niger origin. One of reasons that France has developed Niger as a major uranium provider is precisely because it was not only cheap, but free of end-use obligations.

The consequences for the French facilities are different, since there is no separation between civil and military installations. The 1978 agreement allows AREVA NC and the CEA to consider that once foreign nuclear materials under safeguards enter French facilities, they lose their country specificity, i.e., their "flag." As former EURATOM Safeguards director Wilhelm Gmelin put it: "We do not have any obligation to follow-up the origin of material."[36]

In a May 13, 1983, working group organized by the Nuclear Control Institute in Washington, DC, on "Nuclear Explosives Control Policy," Bertrand Barré, nuclear attaché at the French Embassy in the United States, made the following statement: "As a nuclear weapons State respectful of its international commitments, France would never use for military purposes any fissile material which, directly or through 'filiation', would be subject to a civilian use pledge."[37] The meaning of "filiation" in French is radioactive decay, so this covers any radioactive nuclear substance, which derives by natural decay directly from a given material. Indirectly, Barré's declaration states that France does not use civil nuclear material for military purposes if the civil material has not been subject to any irradiation. Thus civil nuclear material irradiated in a reactor is not influenced by this statement. Plutonium produced in French fast breeder reactors from peaceful-end-use-labeled fuel, and civil plutonium irradiating blankets in the fast breeder reactors, could accordingly be used in military programs. To make things very clear, Barré stated further: "Beyond that, it is not France's policy to disclose which, if any, of its nuclear facilities are used for military purposes."[38]

A EURATOM spokesperson, asked by the science journal, Nature, to comment on the findings of a WISE-Paris report on the Japanese-French Plutonium Connection,[39] confirmed that "the possibility that foreign nuclear waste might end up in military programmes cannot be discounted . . . given the practice of 'flag-swapping' equivalent nuclear materials."[40]

## THE PLUTONIUM INDUSTRY[41]

France initiated a spent nuclear fuel reprocessing program to provide plutonium for its nuclear weapons programs in Marcoule in 1958. Later, the vision of the introduction of plutonium-fueled fast breeder reactors drove the large-scale separation of plutonium for civilian purposes starting with the opening of the La Hague plant in 1966.

Military plutonium separation by France ceased in 1993, but civilian reprocessing continues. Virtually all other European countries, apart from the United Kingdom, have abandoned reprocessing. France's last foreign reprocessing customers for commercial fuel are the Netherlands and Italy, with negligible quantities under contract, and provide no more than a few months of activity to the La Hague reprocessing complex.

France abandoned its fast breeder reactor program in 1998 when the only industrial-scale plutonium fueled breeder in the world, the 1,200 MW Superphénix in Creys-Malville, was officially shut down permanently. Superphénix was a financial disaster. Started up in 1986, it produced electricity only in 6 of the 12 years it was officially in operation. Its lifetime load factor was less than 7 percent. Plagued by technical problems and a long list of incidents, the cost of the adventure was estimated by the French Court of Auditors at FRF60 billion (close to €9.15 billion) in 1996. However, the estimate included only FRF5 billion (€0.760 billion) for decommissioning. That figure alone increased to over €2 billion by 2003. At a lifetime power generation of some 8.3 TWh, Superphénix produced the kWh at about €1.35 (to be compared with the French feed-in tariff of €0.55 per building integrated solar kWh).

**Marcoule**.

France's first reprocessing plant was the Usine de Plutonium 1 (UP1, Plutonium Factory 1) at Marcoule. Thirteen thousand tons of reactor fuel from gas-graphite plutonium production and power reactors were reprocessed there between 1958 and late 1997. Today the site hosts a huge decommissioning and clean-up effort. In 2003, the cost of clean-up, including waste management, was estimated to eventually reach about €6 billion. The clean-up is currently expected to last till 2040. In 2005, these costs and liabilities were transferred from AREVA NC to the CEA.

## La Hague.

Between 1966 and 1987, about 5,000 tons of gas graphite reactor (GGR) fuel and, between 1976 and the end of 2007, about 24,000 tons of light water reactor fuel (LWR) fuel were reprocessed in the UP2 and UP3 plants at La Hague. Small batches of breeder reactor and LWR mixed uranium-plutonium oxide (MOX) fuel also have been reprocessed. Over the last few years, together, the two reprocessing lines have processed about 1,100 tons annually.

Until around 2004, close to half of the LWR spent-fuel put through at La Hague was foreign-owned spent fuel. Almost all of the foreign spent fuel under contract has been reprocessed. At the end of 2007, the total quantity of foreign fuel awaiting reprocessing was so small, in total about 6 tons, that AREVA NC indicated the quantity per client country in kilograms. It should be noted that, at the same time, the total quantity of spent fuel awaiting reprocessing at La Hague was 8,849 tons, and thus 99.8 percent of French origin. At the end of 2009, there was practically no foreign fuel left —3.4 tons in total, including the countries that had contracted research reactor fuel (Australia and Belgium)—and the total quantity of spent fuel in storage at La Hague was 9,421 tons, with a tendency on the rise.

EDF has a large backlog of about 12,000 tons of spent fuel, three-quarters of which are stored at La Hague, the equivalent of over 10 years of throughput at the current rate of reprocessing. Since 1987, France has also built up a large backlog of almost 55 tons of its own unirradiated plutonium in various forms, of which more than half is stored as separated plutonium at La Hague. Plutonium is being used in MOX fuel in 20 900-MWe LWRs that are operating with up to 30 percent MOX fuel in their cores. While there was no plutonium stock when the MOX program started in 1987 (see Figures 6-1 and 6-2), stockpiling has increased every year since. In addition to the French stocks, AREVA's foreign clients currently store more than 30 tons of separated plutonium in France.

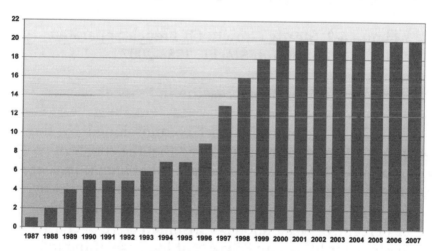

Sources: Mycle Schneider, and Yves Marignac, *Spent Nuclear Processing in France*, Research Report No. 4, Princeton, NJ: International Panel on Fissile Materials, April 2008, available from *www. fissilematerials.org*.

**Figure 6-1. Number of French Reactors Loaded With Mox Fuel, 1987-2007.**

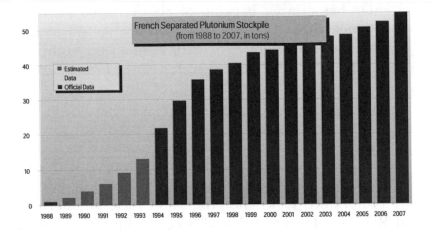

Sources: For Estimated Data, see WISE-Paris, unpublished. For Official Data, see French Government INFCIRC-549, declarations to the IAEA, 1988-2007.

### Figure 6-2. Steady Growth Of French Separated Plutonium Stocks, 1988-2007.

### Economic Costs of Reprocessing in France.

In 2000, an official report commissioned by the French Prime Minister concluded that the choice of reprocessing instead of direct disposal of spent nuclear fuel for the entire French nuclear program would result in an increase in average generation cost of about 5.5 percent, or $0.5 billion per installed GWe over a 40-year reactor life, or an 85 percent increase of the total spent fuel and waste management ("back-end") costs.

Recent projected costs by the industry and the Ministry of Industry show that, in addition to a number of other favorable assumptions, the investment and operating costs of a future reprocessing plant would need to be half of the costs for the current La Hague facilities in order for reprocessing to cost no more than direct disposal. Since 1995, EDF has assigned in its accounts a zero value to its stocks of separated plutonium, as well as to its stocks of reprocessed uranium.

With the liberalization of the electricity sector in the EU, the pressure to lower costs has increased significantly. EDF's massive subsidy of AREVA's plutonium industry is becoming unbearable, and the EDF management has not yet signed on to a follow-up agreement that should replace the reprocessing/MOX fabrication contract that ended in 2007. In an unusual press statement, AREVA's CGT trade union section alleges that "in the difficult year 2007 EDF has not respected its contractual engagements. . . . The CGT is concerned that EDF's posture, including the request for drastic cost reductions in reprocessing-recycling, will not be without consequences on safety, security, and working conditions."[42] A few days later, the negotiations collapsed. The two parties signed a provisional agreement for 1 year to avoid the worst case, which would have been the closure of the MELOX plant because there was simply no contractual basis any longer. Apparently the initial positions were very far apart: while EDF wished a 30 percent reduction over previous prices, AREVA wanted a 30 percent increase. The French government, majority shareholder of both companies, requested the managers to target the signature of an agreement by the end of 2008. On December 19, 2008, both companies announced the signature "in extremis" of an unpublished "framework agreement" that apparently reaches from 2008 to 2040! The only information released is that the amount of EDF spent fuel to be reprocessed as of 2010 is to be increased from around 850 to 1050 tons per year and, accordingly, the amount of MOX fuel fabricated should be expanded from 100 to 120 tons per year.

**Waste Volumes.**

A major argument for reprocessing is that it would dramatically reduce the volume of radioactive waste. However, a number of serious biases have been found in official comparisons made by EDF, AREVA and ANDRA, the organization responsible for radioactive waste disposal in France. These include:

- Exclusion of decommissioning and clean-up wastes stemming from the post-operational period of reprocessing plants.
- Exclusion of radioactive discharges to the environment from reprocessing. Their retention and conditioning would greatly increase solid waste volumes.
- A focus on high-level waste (HLW) and long-lived intermediate-level waste (LL-ILW), leaving aside the large volumes of low-level waste (LLW) and very low-level wastes (VLLW) generated by reprocessing.
- Comparison of the volumes of spent fuel assemblies packaged for direct disposal with those of unpackaged wastes from reprocessing, which overlooks, for instance, the fact that packaging reprocessing waste is expected to increase its volume by a factor of three to seven.
- Failure to include the significantly larger final disposal volumes required for spent MOX fuel, because of its high heat generation, unless it is stored on the surface for some 150 years instead of the 50 years for low-enriched uranium spent fuel.

**Radiological Impact.**

The global, collective dose over 100,000 years — due primarily to annual releases to the atmosphere from La Hague of the low-level but long-lived emitters, krypton-85 (half-life of 11 years), carbon-14 (5,700 years) and iodine-129 (16 million years) — have been recently recalculated at 3,600 man Sv,[43] which is in excess of the estimated impact of the 1957 Kyshtym nuclear accident that led to widespread contamination. Continuing discharges at this level for the remaining years of La Hague's operation theoretically could cause over 3,000 additional cancer deaths.

## RESEARCH AND DEVELOPMENT

There is no comprehensive overview and analysis of public support for nuclear research in France. It is definitely one area where the overlap between civil and military applications played a significant role. An independent analysis carried out on research and development (R&D) expenditures on nuclear issues in France from 1960 to 1997[44] illustrates that there is a fundamental lack of public statistical data on nuclear research expenditures, but that most likely at least half of the research has been carried out under CEA public funding.

According to OECD-IEA figures exclusively based on data transmitted by the French government, between 1985 and 2001 nuclear fission has constituted between 75 percent and 86 percent (93 percent including nuclear fusion) of public energy research expenditures in France. It is only over the last few years that more resources have been allocated to other energy technologies, but mainly to fossil fuels (18-22 percent). While efficiency and all renewable energies combined have increased from less then 1 percent each in 1997 to 8 percent and 5 percent, respectively in 2005, the French research efforts in these areas remain remarkably low.

## OIL, ENERGY DEPENDENCE, AND NUCLEAR POWER

Thus far, France has been able to achieve a relative energy independence and acquire competitive electricity favorable for the development of industry and employment.

The disproportionate public research effort for the nuclear sector becomes even more obvious if one considers the fact that nuclear power only provides about 16 percent of final energy in France, while fossil fuels continue to cover three-quarters (73 percent) of the demand. In 2007, after 3 decades of major nuclear power development, oil alone provides almost half (48 percent) of the final energy consumed in France.

Gaining energy independence through the massive development of nuclear power! That was the message in 1974 when the French government launched the first large-scale nuclear power program. The so-called oil crisis of 1973 had impacted on collective consciousness. The oil price skyrocketed, supply shortages appeared, and neighboring Germany even invented the car-free-Sunday. (See Figure 6-3.)

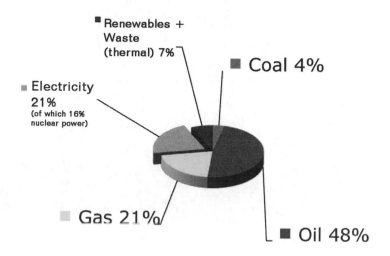

Renewables +
Waste
(thermal) 7%

Coal 4%

Electricity
21%
(of which 16%
nuclear power)

Gas 21%

Oil 48%

Source: "Bilan énergétique de la France pour 2007" Commissariat Général au Développement durable, June 2008, available from www.developpement-durable.gouv.fr/spip.php?page=article&id_article=2369.

**Figure 6-3. Final Energy Supply in France in 2007
(Per Fuel, in Percent).**

However, the French government's announcement that it intended to render France independent from oil through the development of nuclear energy remains astonishing considering the fact that electricity generation accounted for less than 12 percent of the oil consumption in the country in 1973 (see Figure 6-4).

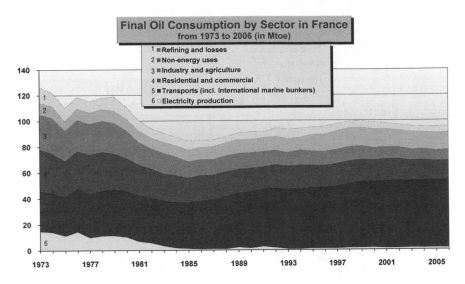

Sources: *"Les Bilans Energétique de la France"*, 1973-2006, *Ministère de l'Ecologie*, available from *www.developpement-durable. gouv.fr/.*

**Figure 6-4. Oil Consumption in France.**

It is remarkable to what extent the oil consumption in the country followed the oil price rather than the electricity supply policy. The key sector for oil consumption in the early 1970s was already the transport sector. The substitution of oil for nuclear power in the electricity sector was very successful and brought the electricity sector share in oil consumption down to 1.5 percent by 1985. At the same time, overall oil consumption hit a long time low. Between 1973 and 1985, the industry and residential/commercial sectors had saved double the amount of oil that the electricity sector had saved essentially through substitution. Four years of work by the French Energy Efficiency Agency (AFME) was harvesting spectacular results. But the 1985 counter oil shock, combined with a dramatic shift in government policy in 1986, led to the radical downsizing of the AFME and the energy conservation and efficiency policy. The result was the immediate resurgence of overall oil consumption. By the end of the 1990s, oil consumption corresponded once again to the level of the early 1980s—in spite of the closure of some of the oil consuming heavy industries, and the nuclear program.

In 2007, per capita oil consumption in France of 1.5 tons was higher than the EU average and higher than consumption in non-nuclear Italy and in nuclear phase-out Germany where the average is about 1.4 tons per person.[45] It is a clear historical lesson that if independence from oil imports had been really the driving force behind French energy policy, the transport sector would have long ago been the target for reform.

Since oil consumption is sensitive to the price of oil, the $CO_2$ emissions are relative to oil consumption. The respective graphs almost have the same shape (see Figures 6-4 and 6-5). While direct per capita emissions remain significantly lower than in most neighboring countries, there is hardly any identifiable structural emission reduction. France's total emissions of the six main greenhouse gases were 2 percent below 1990 levels in 2005. This had little to do with the power sector. In fact, the emissions of public electricity and heat generation were 5 percent above the 1990 levels.[46] But large reductions were achieved for example in $N_2O$ emissions from the adipic acid production.[47]

In 2009, energy related $CO_2$ emissions decreased by an estimated 5.7 percent (Germany reduced by 8.2 percent) compared to 2008 and were 6.1 percent below the 1990 level. In other words, in 2008, $CO_2$ emissions were roughly identical to 1990. The Ecology Ministry explains the (yet provisional) 2009 outcome: "The development of renewable energy but especially the economic crisis have strongly contributed to this result."[48] Indeed, the plunge in industrial activity decreased the sector emissions by 10.7 percent. The energy sector emissions dropped by 12.8 percent, mainly because of the crash in oil refining and marginally because of fuel shifting in fossil fuel plants (-3.8 percent).

Per capita direct greenhouse gas emissions have decreased continuously since 1999, and in 2006, France registered the eighth lowest emissions in the EU27 — not the lowest though, as often assumed. However, as new figures published by the French government illustrate, taking into account the net carbon content of imported goods (minus the carbon content of exported items), per capita greenhouse gas emissions (2005) increase from 8.7t to 12t of $CO_2$ equivalent and thus likely reaches the level of coal-based Germany.[49] France has a large trade deficit while Germany has been the world's leading export nation until China took over in 2009.

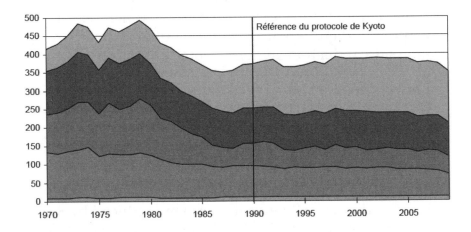

Sources: "*Bilan énergétique de la France pour 2009*" *Commissariat Général au Développement durable, June 2010, available from www. developpement-durable.gouv.fr/Bilan-energetique-de-la-France,17214. html.*

**Figure 6-5. $CO_2$ Emissions in France 1970-2006 (in Million Tons).**

## ELECTRIC HEAT AND POWER TRADE

By the middle of the 1980s — when most of the currently operating nuclear plants in France were either already operating or in advanced building stage — it had become clear that the dimension of the nuclear generating capacity had been vastly oversized. France was not the only country whose energy technocrats had got it wrong. In most industrialized countries, the dogma was to plan on the basis of a doubling of the consumption every 10 years. Instead, there was a clear decoupling of economic development and energy consumption during the 1970s. However, the energy establishment did not adapt its planning, and phenomenal overcapacities were built up in the power sector as well as in refineries and nuclear fuel industries all over Europe and beyond. This was the death knell for any significant intelligent energy initiative based on efficiency and conservation.

In France, the nuclear overcapacities were and remain tremendous. They were estimated already by the middle of the 1980s at 12 to 16 nuclear reactors. While in the United States, 138 units in total were cancelled in various stages of planning and construction, in France the state owned EDF did not abandon any project. Between 1977 and 1999, EDF started up 58 PWRs[50] with a total capacity of 63 GWe (net). At of the end of 2007, France had a total installed power generating capacity of some 116 GW (see Figure 6-6), adding 24 GW of fossil fueled capacity (coal 8 GW, oil 6 GW, gas, and others 10 GW) and 25 GW of hydro to the nuclear plants. Other renewable energy sources have remained marginal, with less than 5 percent of installed capacity (mainly wind) but now have a strong growth rate exceeding 33 percent in 2009.

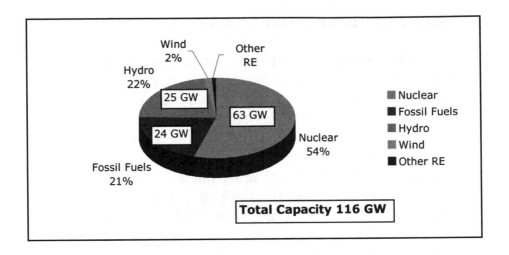

Mycle Schneider Consulting
Source: RTE 2008.

Source: *"L'empreinte carbone de la demande finale intérieure de la France", Ministère de l'Ecologie, de l'Energie, du Développement durable et de la mer, August 2010, available from www.stats.environnement. developpement-durable.gouv.fr.*

**Figure 6-6. Figure 6-6. Electricity Generating Capacity in France in 2007.**

In 2007, nuclear plants generated 76.9 percent of the electricity in France, fossil fuel plants (coal, gas, oil) produced 10.1 percent, hydro plants 11.6 percent, and other renewables (essentially wind) 1.4 percent. (See Figure 6-7.)

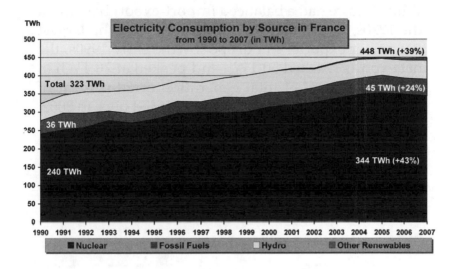

Sources: "*Résultats Techniques du Secteur Electrique en France 2007*", *Réseau de Transport d'Electricité, 2008;* "*2009 Statistiques de l'Energie Electrique en France*", *Réseau de Transport d'Electricité June 2008.*

## Figure 6-7. Evolution Of French Electricity Consumption, 1990-2007 (by Source, in kWh).

Instead of downsizing its nuclear plans in the 1980s, the public power company developed a very aggressive two front policy: long-term base load power export contracts and dumping of electricity into competitive markets like space heating and hot water heating. Foreign clients requested stiff conditions on supply guaranties. The French government did not hesitate to elevate the supply priority for foreign power customers to the level assigned to that of a French hospital. Foreign utilities in Belgium, Germany, Italy, Spain, Switzerland, and the UK were satisfied and agreed on large-scale long-term electricity purchase agreements.

While France had a balanced import-export balance in the 1970s, in the 1980s France turned into the largest net power exporter in Europe. By the early 1990s, the net exports reached 50 TWh and peaked at 70 TWh in 1995 (15 percent of total generation or 20 percent of national consumption), a value exceeded once in 2002, with an exceptional 76 TWh. Since then, there has been a clear tendency towards a decline of exports but an increase in imports. In 2009, France exported only 68.2 TWh, and the net exports plunged to 25.7 TWh, the lowest figure in over 20 years (see Figure 6-8).

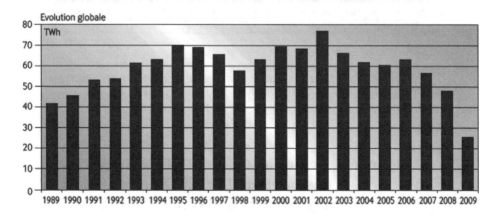

Source: *"2009 Statistiques de l'Energie Electrique en France"*, *Réseau de Transport d'Electricité*, June 2010, p. 11.

**Figure 6-8. French Net Power Exports, 1989-2009.**

A critical analysis of the impact of French electricity exports was carried out by the independent think tank INESTENE in 2002.[51] The report estimates that the power exports, analyzed over the years 1995 to 2001, were a major loss maker. According to INESTENE, the official revenues from exports did not cover the official nuclear generating costs. In addition, detailed modeling reveals that nuclear power produced less than three-quarters of the electricity that was exported—one-quarter being covered by coal fired power plants and around 3 percent by fuel oil—at much higher cost. In addition, INESTENE has made its own cost assessment factoring in the costs of electricity transport, the specific costs of the share of fossil fuel power, specific costs of nuclear power (research, fuel, investment, dismantling, insurance, and external costs). According to the INESTENE calculations annual losses vary between a minimum of €800 million and a maximum €6 billion.

In 2005, many of the long-term export contracts were not renewed. While long-term contracts represented about two-thirds of the total volume until 2005, they represented a little over one-third in 2006 and 2007. This means that the electricity market became much more volatile for France.

It is obvious that import and export electricity are not the same kind. France still has a huge overcapacity in base load power but increasingly lacks peak load capacity. Seasonal peak load exploded in the 1980s and 1990s, in particular, as a consequence of large-scale introduction of electric space heating. Not only did the daily load maximum more than double to almost 89 GW in early 2008, but also the difference between the lowest load day in summer and the highest load day in winter more than doubled to reach 57 GW by 2006. (See Figure 6-9.)

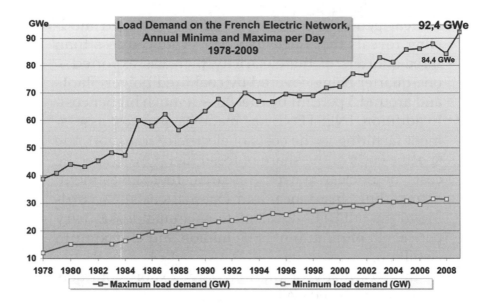

Source: Mycle Schneider, "*Le Nucléaire en France – Au-delà du mythe*", Greens-EFA Group, European Parliament, Brussels, Belgium, 2008.

**Figure 6-9. Seasonal Peak Load Explosion in France.**

Such a load curve is highly uneconomic because it requires a significant generating capacity for very short operational periods or the importation of peak load power at very high cost. Between 2005 and 2007, France imported an average of close to 30 TWh per year, of which 17.5 TWh came from Germany alone.[52] The exact costs of the power imports are not public. However, short-term peak load deliveries can be many times more expensive than base load exports in the framework of multi-annual agreements.

The peak load problem became so urgent that in 2006 EDF decided to restart 2,600 MW of mothballed oil-fired power plants, the oldest of which had started up in 1968 and added them to the thermal power plant fleet that had started up between 1950 and 1984. The new-old oil capacity can be compared to that of the new nuclear project at Flamanville with 1,600 MW which has been under construction since December 2007. Over the past few years, France generated twice as much electricity from oil-fired power plants as the UK and the situation is likely to become worse. At the same time, independent power producers have gained ground in France. The company POWEO alone aims at 3,400 MW of installed capacity by 2012, of which 600 MW is from renewable energy.[53] The rest of POW-EO's new capacity will be essentially natural gas peak load plants. The company calculated that the investment was worthwhile based on an operating time of 100 hours per year.[54]

Electric space heating is not only uneconomic, it is also an energetic absurdity and highly polluting. Instead of using primary energy directly (natural gas, oil, biomass, etc.), mainly fossil fuels, including coal in foreign countries, are burned in power plants. Around three-quarters of the energy is lost in the form of waste heat and distribution losses before the electricity is re-transformed into heat in homes. An assessment published by Gaz de France in 2007 puts the nuclear share of each additional kWh consumed by electric space heating as low as 10 percent. The high ratio of fossil fuels in the mix would lead to specific emissions of over 600 g of $CO_2$ per kWh, more than 10 times the official average emission per kWh. Even the French Secretary of State for Ecology, Nathalie Kosciusko-Morizet, calls the development of electric space heating an "error." She considers it "a French folly" to transform electricity into heat, and "even an aberration from a thermodynamic point of view."[55]

## WHAT IF . . .? OPTIMIZED EQUIPMENT AND ECONOMIC EVALUATION

The vastly over-dimensioned nuclear generating capacity clearly entailed the strategic choices to massively export power and to penetrate the market for heating. While EDF's commercial strategy on thermal uses did not have much success in the industry, up to 70 percent of new homes were equipped with electric space heating, and today over one-quarter of French homes are heated electrically.

An independent analysis published in 2006[56] looked into the question of what the French electricity generating system development would have looked like if it had been optimized and power exports had not been developed. A second analysis reviews what the generating park would have looked like if electric space heating had not been encouraged and massively implemented.[57]

The results show that, under optimized economic conditions, as of the early 1980s, the construction rhythm of nuclear plants would have slowed down significantly. No more than 33 GW of nuclear power from 36 plants would have been necessary compared to the 63 GW produced by the 58 nuclear plants that were built. In addition, the model suggested investing in new fossil fuel capacity starting in the early 1990s to cover the electric space heating needs.

The second scenario also assumed that the electric space heating, consuming close to 60 TWh annually by the end of the 1990s, was not developed. While the seasonal peak use implies a large share of non-nuclear components,[58] essentially fossil fuels, without construction of new coal-fired power plants, the scenario still would have resulted in less nuclear power. In other words, without power exports and electric space heating, an economically optimized French nuclear program would have been limited to less than 30 GW, the equivalent of the 34 x 900 MW reactors, the last of which was connected to the grid in 1987. The uneconomical decision to waste electricity in the form of heat[59] (space heating, hot water, and cooking) has resulted in French households having the highest consumption level in Europe since 1976. Today, per capita electricity consumption in France is some 25 percent higher than in Italy (Italy phased out nuclear energy after the Chernobyl accident in 1986) and 15 percent higher than the EU27 average.

The energy flow sheet in Figure 6-10 illustrates the phenomenal system losses of the power generating system. Only 27.5 percent of the energy contained in the primary resources injected (left side) is available in the form of final energy (right side).

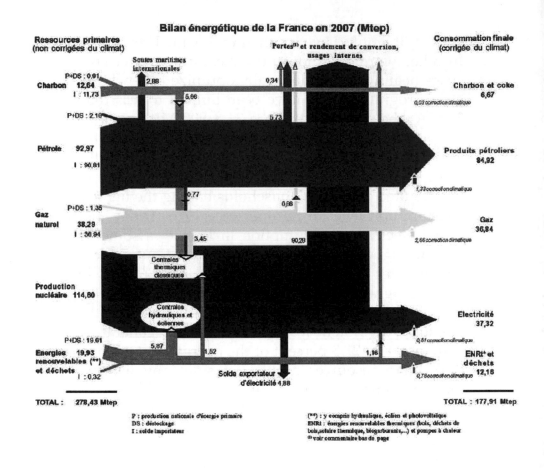

Figure 6-10. Energy Flow Sheet France (in Mtoe).

Source: *"L'énergie en France–Repères"*, Le Ministère de l'Ecologie, de l'Energie, du Développement Durable et de l'Aménagement du Territoire, 2008, available from *www.statistiques.equipement.gouv.fr/.../reperes_lenergie_en_France_cle1e446c.pdf*.

At the same time, the energy flow sheet illustrates why oil remains the main energy source in France as it is in most industrialized countries. In 2007, electricity represented 21 percent of the final energy consumed, of which nuclear power generated 77 percent. (See Figure 6-11.)

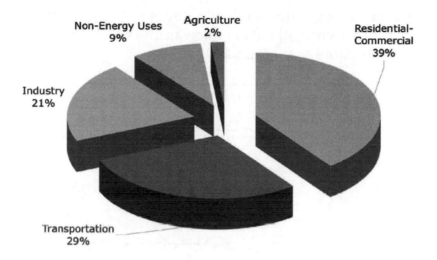

Non-Energy Uses
9%

Agriculture
2%

Residential-
Commercial
39%

Industry
21%

Transportation
29%

Source: *"Bilan énergétique de la France pour 2007"*, Direction
Générale de l'Energie et des Matières Premières (DGEMP), 2008,
available from *www.developpement-durable.gouv.fr/.../Bilan_energet-
ique_pour_l_annee_2007_cle2ba984.pdf.*

**Figure 6-11. Final Energy Consumption in France
in 2007 (By Sector, in Percent).**

## ENERGY INDEPENDENCE — FROM 50 PERCENT
## TO 8.5 PERCENT

It is remarkable to what extent the myth of energy
independence through nuclear power has survived
the last 35 years. One of the reasons is the artistic ma-
nipulation of basic data by the State administration
and the energy industry.

Figure 6-12 illustrates that, according to French
official accounting, primary electricity (essentially hydro and nuclear) has been around 100 percent import
independent before the first large scale nuclear program was launched in 1974. The graph suggests that

overall energy independence doubled between 1974 and 1990, when all but six nuclear units had come on-line, to stabilize at around 50 percent.

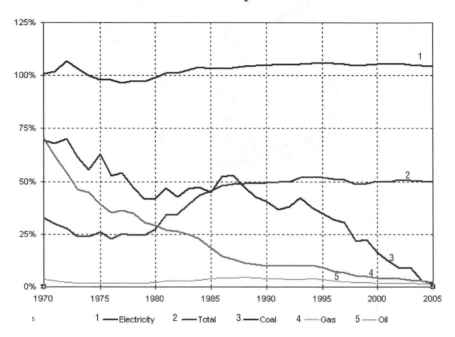

Sources : Ministry of Industry, 2008.[60]

**Figure 6-12. Energy Independence Trends
by Energy Source, 1970-2005.
(According to the French Ministry of Industry).**

However the figure hides a number of serious bi-ases that can be illustrated for the year 2007, when the French Ministry of Industry indicated energy inde-pendence as 50 percent:

a. Electricity exports should be excluded, because they do not influence energy independence. Thus 56.8 TWh, or 4.9 MTOE, should be substracted.

b. The auto-consumption of the nuclear sector is around 18 TWh or 1.6 MTOE (most of which is consumed by the uranium enrichment plant EURODIF alone) and should be deducted.

c. The degree of energy independence should be calculated on the final energy side. The two-thirds of energy wasted by nuclear plants (roughly equivalent to the oil consumption of the French automobile fleet) are incorporated in the ratio between energy produced and consumed nationally. The OECD's International Energy Agency equivalence introduces a lower ratio in final energy, accounting for the nuclear power contribution at 0.086 toe of final energy per MWh. If calculated on the final energy side, the independence level shrinks to less than 24 percent.

d. Finally, all primary nuclear resources, uranium, are imported. France stopped mining uranium in May 2001. While the production of energy from imported oil, gas, and coal is accounted for as imported energy, this logic is not applied to uranium. The argument is, on one hand, that the international conventions would consider uranium as primary material, not as an energy source and, on the other hand, that there are a number of politically stable, diversified sources for natural uranium that make its supply very secure. A third argument is that there is significant value added through transformation (conversion, enrichment). Those are valid arguments. However, they could also be applied to other energy sources and in particular to coal for diversity of supply and to oil for transformation (refining). If France did account for its uranium imports as such, the energy independence figure would obviously plunge.

e. However, some of the energy is generated by the reuse of plutonium and reprocessed uranium. In total, 22 French light water reactors are licensed to use plutonium fuel (MOX). About 100 tons of MOX, which generate 30-40 TWh of electricity, are used per year. Two reactors at the Cruas nuclear power plant use reprocessed uranium that generates 13 TWh. Plutonium and reprocessed uranium generate around 50 TWh, or hardly more than the equivalent of 10 percent of the final energy contribution of renewables. In total, the level of energy independence on the final energy side would be around 8.5 percent in 2007 (see Table 6-1).

|  | Mtoe | Level of Energy Independence |
|---|---|---|
| Nuclear Primary Energy Generation | 114.6 | 50.4% |
| + other Primary Energies (Renewables, etc.) | 21.8 | |
| (a) Electricity exports 56.8 TWh | - 4.9 | |
| (b) Nuclear auto-consumption ca. 18 TWh | - 1.6 | |
| **Primary Energy Generation/Independence** | **129.9** | **48.0%** |
| (c) Nuclear final energy contribution | 28.7 | |
| + Renewables* | 11.9 | |
| + Coal, oil, gas | 2.0 | |
| **Final Energy Generation/Independence I** | **42.6** | **23.9%** |
| (d) – Uranium imports | - 28.7 | |
| (e) + Plutonium & reprocessed uranium credit | + 1.3 | |
| **Final Energy Generation/Independence II** | **15.2** | **8.5%** |

*Respective national generation shares: renewables 97.7 percent; gas, 2.4 percent; coal, 1.2 percent; and oil, 1.2 percent.

Source: Mycle Schneider Consulting, 2008.

### Table 6-1. Adjusted Level of French Energy Independence in 2007.

## LOW ELECTRICITY PRICES — HIGH ENERGY BILLS

Since 1970, the French national energy bill has followed the price of oil. This has not changed much since the launch of the massive nuclear program in 1974. The bill increased with the oil embargoes in 1973 and 1979 and went down after the counter-oil shock in 1985. In 2008, France's energy bill reached almost €60 billion and was higher than ever before (Figures 6-13 and 14). The income from electricity exports has remained marginal in comparison.

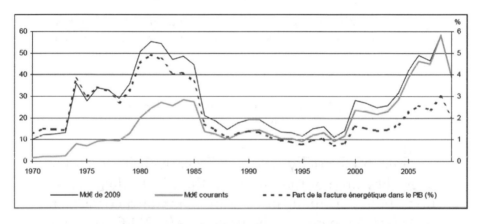

Source: "Bilan énergétique de la France pour 2009", Commissariat Général au Développement durable, June 2010, available from *www.developpement-durable.gouv.fr/Bilan-energetique-de-la-France,17214.html.*

**Figure 6-13. The French National Energy Bill, 1970-2009.**[61]

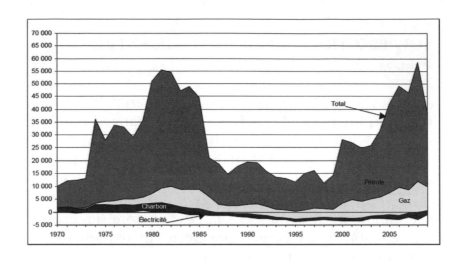

Source: "Bilan énergétique de la France pour 2009", Commissariat Général au Développement durable, June 2010, available from *www.developpement-durable.gouv.fr/Bilan-energetique-de-la-France,17214.html.*

**Figure 6-14. The French National Energy Bill 1970-2007.**[62]

French electricity prices are relatively low. However, they are not the lowest in the EU. In the case of household consumer prices, for a standard consumer of 3,500 kWh per year, the price was 0,1211 € per kWh at the beginning of 2007. That puts France at the 13th position of the 27 EU member countries, in the same range as Spain or the UK. The price comparison, according to Purchasing Power Standards (PPS), advances France to the third place behind Greece and Finland, but only marginally cheaper than the UK or Spain.

The average French industrial consumer with an annual consumption of 2,000 MWh paid 0.0587 € per kWh at the beginning of 2007, which puts France at the 6th position in the EU. The comparison, according to PPS, brings France to 4th place, behind Finland, Denmark, and Sweden.

## RESIDENTIAL SECTOR

France has a relatively large per capita electricity consumption. The low power prices played their role in that development, as can also be seen in countries like Finland and Sweden. However, the average consumption level of a French household is relatively meaningless since consumption is highly sensitive to whether or not a household is equipped with electric space and service water heating appliances. Consumption rises sharply in that case. Cheap electricity does not mean low energy bills.

French households have never spent as much on energy as they have in the past 3 years (see Figure 6-15). The relative share of the energy expenditures in the total household budgets has moved little since the late 1980s, and has remained between 5 and 6 percent. It followed the oil price shocks in 1973 and 1979, as well as the counter-oil shock in 1985 when slumping oil prices led to a drop by more than 2 percent in the relative burden of the energy bill.

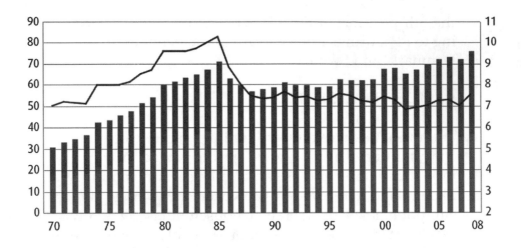

Source: *"L'énergie en France–Repères"*, Le Ministère de l'Ecologie, de l'Energie, du Développement Durable et de l'Aménagement du Territoire, 2008, available from *www.statistiques.equipement.gouv.fr/.../reperes_lenergie_en_France_cle1e446c.pdf.*

### Figure 6-15. Household Energy Consumption and Share in Total Household Consumption, 1970 to 2008.[63]

Understandably, high energy bills hit the poor significantly harder than the rich. In 2006, the 20 percent of French households with the highest income spent 6 percent of their income on energy, slightly less than in 2001, while the 20 percent with the lowest income had their share increased from 10 percent to 15 percent over the same period. The income of the rich grew faster than the energy prices, which was not the case of the low-income households. In the case of electricity alone, the difference between high and low income households even reaches a factor of three, with respectively 2 percent and 6 percent of the total budget.[64]

In spite of the significant share in the budget, poor people in France cannot cover their basic energy service needs. "The statistics of the National Housing Agency (ANAH) drawn from the housing inquiry are affirmative: three million French are cold in winter."[65] According to studies carried out by EDF, three million households, about 10 percent, are considered in a precarious energy situation, 500,000 households have access to the "tariff of primary necessity," and 300,000 households receive support from EDF's Solidarity Fund to cover their energy bills.[66] According to the Ministry of Economy, the total number of households eligible for the tariff of primary necessity, which was introduced in 2005, was two million as of August 2008, prior to the economic crisis. In August 2010, support levels were increased by 10 percent in each category of low income household and reduction levels are now 40 percent for an individual; 50 percent for an adult with one child, a couple without children or with one child, or for one adult with two children; and 60 percent for a couple with two or more children.

The impossibility of poor households paying their energy bills — and in particular their electricity bills in the case of electric space heating — also drains public social funds from the General Councils, the Family Social Assistance Fund (CAF), and others.[67] The number of demands for social assistance to pay energy bills increases by 15 percent every year.[68]

Depending on the source of the statistics, public expenditure from social funds for energy bills totals €150 million to over €200 million per year. Electric space heating represents half of the electricity consumption of the residential sector. In the private collective housing sector, the level of equipment with electric space heating even reaches 40 percent.[69] This is quite logical since the investment in electric systems is low compared with central heating systems, which is in the interest of the building owner, but the operating costs are the highest, which is at the expense of the inhabitants.

## INDUSTRIAL SECTOR

Concerning the French industrial sector, it has been continuously argued that the low cost of nuclear generated electricity was instrumental in supplying industry with low priced power and made it particularly competitive. At least two developments cast serious doubts on this line of reasoning.

1. The electricity consumption in the French industry has been decreasing significantly. Between 1996 and 2006, the industry decreased consumption at an average of 0.5 percent per year, following the general trend in the EU away from heavy towards service sector industries.[70] Between 2004 and 2009, the industry sector reduced consumption by almost 6 percent, heavy industry dropped even by 18 percent. Of course, this development is not policy driven but mainly due to the displacement of activities and the economic crisis.

2. France has constantly aggravated its foreign trade deficit, which it is estimated to have reached almost €40 billion in 2007. This result can be compared to the just under €200 billion trade surplus of Germany. The first half of 2008 brought a new record trade deficit for France with over €24 billion while Germany raised its surplus to over €103 billion.

The nuclear policy does not seem to influence either the capacity to keep electricity intensive industry in the country or the foreign trade balance. It should be recalled that Germany has a legally binding nuclear phase-out policy in place. Non-nuclear countries like Greece and Denmark also provide cheap power and keep the per capita consumption below the EU average.

## LIMITED RISK INSURANCES

France is a signatory of the 1960 Paris Convention and of the 1963 Brussels Convention on civil liability. The conventions allow signatory states to adapt the regulations according to their individual needs and wishes. One such condition is the capping of operator liabilities that has been reviewed and modified many times. In the case of France's EDF, the limit had been fixed at €91 million for any single incident on its installations, the lowest limit in Europe, according to a recent academic analysis:

> Of the total liability of €91 million today EDF covers only €31 million through insurance and the remaining €60 million through its own reserves. This is striking since we noticed that the price which EDF pays today for its insurance cover is the excessively high amount of €6.4 million per year. Compared to the objective value of the risk the premium paid for nuclear insurance to cover only the €31 million damages for the 58 French nuclear reactors is excessively high. Indeed, we estimate that the objective value of the risk per year to be around €17,980 million, whereas the insurance premium actually paid is €6.4 million a year. The value of the premium actually paid hence for a large extent does not correspond with the objective value of the risk.[71]

In case an event leads to costs beyond the operator's cap, the French state covers an additional €140 million, which can be increased by a further €150 million by the other contracting parties. In 2004, the obligatory liability cap for the operator under the Paris and Brussels conventions was raised to €700 million, the state intervention level to €500 million, and the contracting parties' additional coverage to €300 million (see Table 6-2).

| | Paris (1960) and Brussels Conventions (1963) caps for France, (1968) Law amended | Modification Protocol of the Paris and Brussels Conventions (2004) |
|---|---|---|
| Operator's liability cap | 91 | 700 |
| State invervention | 140 | 500 |
| Contracting Parties coverage | 150 | 300 |
| Total | 381 | 1,500 |

Source: Fiore and Faure, 2007.

**Table 6-2. Liability Coverage Caps Before and After The 2004 Convention Amendments (in Million euros).**[72]

A Chernobyl-size large-scale nuclear accident in France would most likely lead to hundreds of billions of euros of damage. Even after the increase by a factor of four of the total coverage, the total sum available of €1.5 billion remains small, less than half to be provided by the operator.

A study for the European Commission suggested that if EDF was required to fully insure its power plants with private insurance but using the current internationally agreed limits on liabilities, it would increase EDF's insurance premiums from €c0.0017/kWh, to

€c0.019/kWh, thus adding around 0.8 percent to the cost of generation. However, if there was no ceiling in place and an operator had to cover the full risk of a worst-case scenario accident, it would increase the insurance premiums to €c5.0/kWh, thus tripling the current total generating costs.[73]

## DECOMMISSIONING AND WASTE MANAGEMENT COST ASSESSMENT AND FUND MANAGEMENT[74]

The operation and decommissioning of nuclear power infrastructures leads to long-term liabilities. The sums involved are very significant. The French Court of Accounts has calculated back-end liabilities totalling €65 billion (undiscounted) for the three main French nuclear operators EDF, CEA, and AREVA as of the end of 2004.[75]

After national public and parliamentary debates, new legislation on nuclear waste research and management adopted on June 15, 2006, includes specific wording on the financing of decommissioning and waste management operations. Key articles of the new "Law on the Program Relative to the Sustainable Management of Radioactive Materials and Wastes" (hereafter New Waste Law)[76] includes the legal requirement to elaborate a National Plan for the Management of Radioactive Materials and Wastes and a National Inventory of Radioactive Materials and Wastes. Both are to be updated every 3 years. The National Radioactive Waste Management Agency (ANDRA) has to set up an internal restricted fund to finance the storage of long-lived high and medium level wastes. The fund will be fed by contributions from the nuclear operators under bilateral conventions. The nuclear operators have to set up internal restricted funds covered by dedicated assets managed under separate accountability. A National Financing Evaluation Commission of the Costs of Basic Nuclear Installations Dismantling and Spent Fuel and Radioactive Waste Management has been established that is comprised of representatives of the National Assembly and the Senate, as well as the Government and a number of experts that have to be independent of the nuclear operators and the energy industry.

While the new legal framework considerably changes the basis for the future availability of sufficient funding for nuclear decommissioning and waste management activities in France, a large number of uncertainties remain. These include:

a. The cost calculations underlying the provisions are nontransparent, and there is no public access to the data; the administrative authorities either have limited manpower[77] or are not consulted.[78] In the past, some cost calculations have proven wrong by an order of magnitude or more.

b. The spent fuel management policy choice has extreme impact on future costs. The final disposal cost estimates for long-lived high and intermediate level wastes vary by a factor of four or almost 45 billion euros between 13.5 and 58 billion. The current limitation of the reference scenario to the all-reprocessing option — evaluated as the cheapest geological disposal option[79] — has not been justified.

c. There is considerable opposition against the funding scheme adopted (internal restricted), which led the largest opposition group (Socialist Party) in the French National Assembly in a surprise move to abstain from voting for the New Waste Law. Two other parliamentary representations (Communist Party,[80] Green Party) voted against the law.

d. There is opposition against the current back-end strategy (reprocessing plus geological storage) from the civil society (nongovernmental organizations [NGOs], independent scientists, and consumer groups). The implementation of a second laboratory, legally required under 1991 legislation and firmly requested by trade unions and independent experts, has not been possible due to fierce local opposition. Policy changes in the future due to public pressure or legal claims are difficult to exclude.

e. The current cost estimates are based on the opening of a final geological disposal site for long-lived intermediate and high level wastes in 2020. After 6 years, the laboratory project at Bure was already more than 2 years behind schedule.

f. Safety analysis-based modifications of the technical specifications in waste conditioning, packaging, and storage can have significant impact on costs.[81]

g. The conditioning, sometimes reconditioning, and packaging of some waste categories (bitumen, graphite, and spent MOX) are still only in the development phase. Cost calculations necessarily have large uncertainties associated.

h. Following the shutdown over a 2-year period of a nuclear facility (for example, after an incident or an accident), the safety authorities can order the final closure and decommissioning of the facility. This could severely impact cost calculations and availability of backend funding.

i. Some materials currently not classified as waste might have to be managed as waste in the future (for example, reprocessed and depleted uranium, a portion of separated plutonium, spent plutonium fuels, and plutonium waste).

# BETWEEN PRODUCTIVITY AND NUCLEAR SAFETY — A FRAGILE BALANCE

The drop of the availability factor is an alarm signal for safety and is a wake-up call: Are we paying sufficient attention to staff competence as well as to maintenance quality and material aging?[82]

> Pierre Wiroth
> Inspector General for Nuclear Safety
> and Radiation Protection, EDF
> January 2008

The French nuclear power plants have a relatively low load factor. Historically, this has mainly been caused by the huge overcapacity. France is the only nuclear operator in the world that shuts down units on summer weekends because of lack of demand. Over 40 units are operated on a load following mode, which allows for the power output to be modified on the short term within a ca. 5 percent margin. Until the end of 2006, French reactors showed a cumulated lifetime energy availability of 77.3 percent. While the availability had increased since 2000 from 80.4 percent to 83.6 percent in 2006, in 2009 the load factor dropped, for the third year in a row, to reach 70.7 percent, an appalling result by international standards. The important difference between this latest deterioration and earlier developments is that the origin is clearly mainly of a technical nature, which raises the question whether EDF's reactor fleet would actually be able to reach availabilities in the order of over 90 percent like the best in the world.

EDF's own account of the dreadful 2009 nuclear performance reads like this:

Compared to 2008, the change in output of 28 TWh is a result of:

- social unrest that affected the campaign of unit shutdowns, and explains a loss of nuclear generation of about 17 TWh throughout the year. Unlike 2008, during which the generation was not impacted by social conflicts, these movements have caused extensions of unit shutdowns in 2009, mostly in the second half of the year, and generation losses due to lower output;

- several cases of incidental or extra shutdowns, mainly occurring in the second half of 2009 and particularly related to failure of equipment (steam generators, alternators, and transformers) the replacement of which was necessary and already programmed in part for 2010. Technical problems encountered this year on the steam generators, alternators, and transformers, respectively, had an impact on the Kd [availability factor] of 2.6 points, 1.7 points, and 0.5 points. These events led to a generation loss of about 6 TWh;

- various losses caused by environmental factors: the shutdown at the Blayais plant during the first quarter of 2009 due to the presence of vegetal waste and mud plugs in the Gironde after the Klaus storm and the lower output at the Cruas-Meysse plant following the scaling of cooling towers in 2009 led to a generation loss of about 3 TWh; various factors of lower impact: the greater use of the modulation of power in 2009 and the leap year effect in 2008 (which automatically led to the loss of a day of generation in 2009) resulted in a generation loss of about 2 TWh."[83]

EDF had estimated that it would take until 2010 to solve the technical problems of its steam generators.[84] The available industrial capacity limits the number of units to be subject to chemical cleaning to five or six per year. As of 2007, at least 15 of the 900 MW and

1300 MW units had already been identified and the safety authorities classified the issue as "generic."

An internal note of EDF's financial department directorate from the end of 2001 put the loss at €76 million per percent point. This figure has no doubt significantly increased with the climbing electricity prices over the last few years.

The steam generator plugging is only one of the latest of a list of serious generic problems that hit the French reactor fleet.[85] While there is no doubt that the high level of reactor standardization has multiple technical and economic advantages, it has also brought along the problem of systematic multiplication of problems into large parts of the reactor fleet.

The overall number of safety relevant events has increased steadily from 7.1 per reactor per year in 2000 to 10.93 in 2009. EDF used to stress that the number of more serious events was on the decline. This is no longer true. Between 2005 and 2009, the total number of events rated on the International Nuclear Event Scale (INES) increased from 759 to 795 (+5 percent). However, while the Level-0 events remained stable, the Level-1 events increased from 49 to 95 (+94 percent) over the same period.

EDF's Inspector General for Nuclear Safety and Radiation Protection stated in his report on the year 2007 that the new organization leading to a massive reduction of the costly stocks of replacement pieces for the nuclear power plants has led to a situation where "to dispose of them in time has become a major problem for the sites and between the sites."[86] The Inspector reports astounding cases where:

- pieces that were to be replaced are put back in place due to lack of spare parts;
- temporary advise is elaborated to compensate

over the long term for the absence of spare parts;

- plant outages are stretched due to delays in parts delivery and lead to costly time losses of subcontractors;
- unavailability of small consumables (screws, grease . . .) has become "a source of incomprehension and irritation."[87]

Two years later, the (new) Inspector General notes in the Annual Report for 2009 that "the timely availability of spare parts remains a problem today."[88]

Between 10,000 and 12,000 events are identified in EDF's plants every year, of which 600 to 800 are considered "significant events."[89] However, power reactors are only part of the French nuclear establishment. The summer of 2008 attracted great attention to facilities that had been previously unknown to the French public. A uranium spill was discovered in July 2008 at the clean-up company SOCATRI's[90] Tricastin site. Access to drinking water and any use of groundwater were banned in several surrounding municipalities. SOCATRI has issued a press release expressing its "regrets for the inconvenience generated by the incident and its media consequences."[91] Wine makers in the area have drawn the lessons of what is perceived as bad publicity, and the well known "Côteaux du Tricastin" will change its appellation as of the 2009 harvest.

The SOCATRI accident was revealed after the site had already exceeded regulatory annual emission limits the previous years. Carbon-14 emissions exceeded limits in 2006 and 2007 by a factor of 30 and 42, respectively. Tritium discharges were also exceeded by a factor of six and five, respectively. Rather than treating the problem at the source, SOCATRI requested — and obtained — increased discharge limits. In February 2008, the safety authorities granted the new annual limits: the carbon-14 limits (3,400 mega Becquerel) constitute an increase by a factor of 40 over the previous limit; the tritium limits (10 giga Becquerel) were boosted by a factor of 24.[92]

## EPR — EUROPEAN PROBLEM REACTOR?

> At the moment, AREVA has submitted half of the plans to us. Nuclear reactors are not built without plans, at least not in Finland.[93]

> Martin Landtman
> EPR Project Manager, TVO
> Olkiluoto, Finland, February 2008

### Olkiluoto-3, Finland.

In August 2005, construction of the first European Pressurized Water Reactor (EPR) began in Finland. The Olkiluoto-3 project has been plagued with difficulties since the first concrete was poured.

The utility TVO signed a turn-key contract with the Franco-German consortium Framatome-ANP, now AREVA NP (66 percent AREVA, 34 percent Siemens) to supply a 1,600 MW EPR. The Bavarian Landesbank — the Siemens headquarter is located in Bavaria — granted a loan of €1.95 billion, over 60 percent of the contract value, at a particularly preferential interest rate of 2.6 percent. The French public COFACE export credit agency covered an additional €720 million loan.

Five years after construction start the project is at least 3 1/2 years behind schedule and 90 percent over budget, the loss for the provider already provisioned stands at €2.7 billion as of July 2010. The reactor is now scheduled to come online in 2013. It remains unclear who will cover the additional cost. While TVO is insisting on the fix price conditions, AREVA has indicated that it will try to recover at least part of the additional costs.

In an unusually critical report the Finnish safety authorities, as early as 1 year after construction start, pinned down a number of reasons for the delays:

> The time and resources needed for the detailed design of the OL3 unit was clearly underestimated, when the overall schedule was agreed upon. . . . An additional problem arose from the fact that the supplier was not sufficiently familiar with the Finnish practises at the beginning of the project. . . . The major problems involve project management. . . . The power plant vendor has selected subcontractors with no prior experience in nuclear power plant construction to implement the project. These subcontractors have not received sufficient guidance and supervision to ensure smooth progress of their work. . . . As another example, the group monitored manufacturing of the reactor containment steel liner. The function of the steel liner is to ensure the leak-tightness and the containment and thus prevent any leaks of radioactive substances into the environment even in case of reactor damage. The selection and supervision of the liner manufacturer was left to the subcontractor who designed the liner and supplied it to FANP [AREVA NP]. The manufacturer had no earlier experience on manufacturing equipment for nuclear power plants. Requirements concerning quality and construction supervision were a surprise to the manufacturer. . . .[94]

On the attitude of AREVA NP as the vendor, the Finnish safety authorities note:

At this stage of construction there has already been many harmful changes in the vendor's site personnel and even the Site Manager has retired and [has been] replaced. This has made overall management, as well as detection and handling of problems difficult. . . . The incompetence in the constructor role becomes obvious in the preparations for concreting of the base slab. . . . The consortium has a habit of employing new people for problem solving, which seems to have resulted in even more confusion about responsibilities.[95]

Problems kept on coming. In an unprecedented move, three national nuclear safety authorities STUK (Finland), HSE (UK), and ASN (France) in October 2009 issued a joint statement raising concerns about the incompliance of the Control and Instrumentation (C&I) developed by AREVA-Siemens for the EPR:

Independence is important because, if a safety system provides protection against the failure of a control system, then they should not fail together. The EPR design, as originally proposed by the licensees and the manufacturer, AREVA, doesn't comply with the independence principle, as there is a very high degree of complex interconnectivity between the control and safety systems.[96]

As a consequence, the three safety authorities have asked AREVA "to make improvements to the initial EPR design."[97]

**Flamanville-3, France.**

On December 3, 2007, the first concrete was poured at the Flamanville-3 EPR project in France. The nuclear safety authorities carried out an inspection the same day. The inspectors note in their report that the quality control procedures for the base slab concrete are "unsatisfactory." Basic technical specifications and procedures have not been followed, including the concrete mixture, the input level, and the concrete test sample filing.[98]

A second inspection carried out on December 13, 2007, was aimed at the verification of the potential interaction of the building site with the operating two nuclear units. It revealed numerous cases of errors, violations of regulations, and lack of basic safety culture, including the erroneous assumption that the roof of the nuclear auxiliary building would be of reinforced concrete, a crane operator's access permit that had been expired for over a month, the total ignorance of the unit 2 operators about the potential impact of the building site (aside from the use of explosives for site preparation), and the lack of updating of the safety analysis of units 1 and 2.[99]

Several subsequent inspections revealed a number of additional anomalies that illustrated "a lack of rigor in the construction of the building site, difficulties in the management of external subcontractors and organizational deficiencies"; and on May 23, 2008, ASN ordered the cessation of concreting of all safety relevant parts of the plant.[100] Conditional restart of most of the concreting operations was granted on June 17, 2008, after EDF had submitted a plan to upgrade quality control and organization. By the end of September 2008, ASN still considered the organization "perfectible." ASN Inspectors had discovered that the documentation on welding "does not allow the justification of conformity with the referential."[101]

At the end of October 2008, the nuclear safety authorities identified quality control problems with the builder, AREVA. The Italian AREVA subcontractor Società delle Fucine did not apply the obligatory fabrication procedures. ASN gave AREVA 2 months to prove that the pieces forged by the Italian company are compatible with the required technical specifications. Otherwise the forgings will have to be redone.[102]

Finally, the problems were not just linked to the reactor project itself. The existing high power lines would not be sufficient to export the electricity from the new plant. An additional line is in the planning process, but the local population disapproved. On February 29, 2008, the citizens of Chèvreville, a 200-person village on the planned track of the power line, decided unanimously (90 percent participation in a local referendum) to boycott the municipal elections of March 9, 2008, in protest of the electricity grid extension.[103]

The two companies EDF and AREVA are in fierce competition to bring the first EPR online. EDF did not appreciate AREVA's offering the Olkiluoto project as a turnkey facility, because previously EDF was responsible for overall construction oversight, and AREVA's role remained limited to manufacturing. In an unprecedented move, in November 2008, EDF felt obliged to put out a press release claiming that the Flamanville project is still on schedule, thus providing a firm rebuttal of a statement by AREVA CEO Anne Lauvergeon who had stated in a radio interview that the project would be 1 year behind schedule.[104] In July 2010, EDF had to paddle back and officially admit that the Flamanville EPR project was not only 1 but 2 years behind schedule and at least €1.7 billion over budget. The plant, built at home by the largest and most experienced nuclear builder in the world is now anticipated to cost about €5 billion and is not expected to start operating before 2014.

This was not the only bad news that EDF-Group had to communicate to its majority shareholder, the French state. The company reported a decline of 22 percent in profit "mostly due to the recognition of a €1.1 billion provision for risks related to the Group's activities in the US."[105] In the meantime, AREVA experienced severe consequences due to the disastrous performance of its Olkiluoto project. In June 2010, the credit-rating agency Standard and Poor's downgraded the company from A to BBB+. "The downgrade reflects our view that AREVA's profitability will continue to be depressed over the next couple of years."[106] It is remarkable that the credit-rating agency justifies its decision in particular with the financial risks involved with the conflict between AREVA and EDF over the continued operation of the EURODIF (George-Besse-I) enrichment plant at Pierrelatte. While EDF considers it has access to plenty of enriched uranium at cheaper prices, AREVA badly needs EDF orders to operate the EURODIF plant until the follow-up plant George-Besse-II plant is to start up operating in 2013.

It could become worse for AREVA: "The ratings could come under pressure if the company were to experience further substantial delays in executing its asset disposal program or capital increase, particularly if this were to be combined with additional unforeseen expenditure and/or project cost overruns."[107]

## EROSION AND WORKFORCE CONCERNS

The French EPR project was not initiated because of the need to build new base-load power generation capacity. As demonstrated in previous chapters, there is still significant overcapacity available, and it would be economically inconceivable to build a

257

nuclear plant for power export only. Also, it is quite commonly agreed that the nuclear share has gone too high in France if compared with an ideal generating mix. Finally, the reactors are expected to operate for at least 40 years while the current average age of the French units is 25 years, the oldest having operated for 33 years. In other words, even at current consumption levels there is no need for replacement capacity until long after 2020.

The main reason why EDF is building the EPR is that there is widespread concern over the potentially devastating impact of the widening competence problem. In France, the situation is not much better than in other countries. There is a generational gap between the scientists, engineers, and technicians that have conceived, built, and operated the current generation of nuclear facilities and tomorrow's workforce needs. About 40 percent of EDF's current staff in reactor operation and maintenance will retire by 2015. In its Reference Report 2008, EDF states that "about half" of the operational and maintenance staff in production and engineering will retire between 2008 and 2015. EDF speaks clearly of an "unbalanced age structure" since more than 65 percent of the workforce is over 40, and introduces the problem of a lack of young skilled workers explicitly as a risk factor:

> The EDF group will do its utmost to recruit, retain, redeploy or renew these staff and skills in time and under satisfactory conditions. However, it cannot guarantee the measures adopted will always prove totally adequate, which may have an impact on its business and financial results.[108]

In 2008, the utility started hiring about 500 engineers annually for the nuclear sector alone. In mid May 2009, EDF advertised for positions for 50 operator-trainer engineers.[109] Where such a large number of experienced nuclear operators capable of training others could possibly come from remains a mystery. Reactor builder AREVA tried to hire 400 engineers in 2006 and another 750 in 2007. The level of success of the hiring efforts is not known. AREVA, like other nuclear companies, has formed partnerships with certain universities and engineering schools and "shepherds" students through their studies. AREVA's strategic marketing specialist Liz Smith explains that students can work effectively during studies and immediately upon graduation. "The strong bonds they form with AREVA during their studies increase loyalty to the company."[110] AREVA calls it "growing its own engineers," starting in middle and high school and through a "unique college program in order to meet tomorrow's demand for resources."[111]

It is obvious that the biggest share of the hired staff is not trained nuclear engineers or other nuclear scientists. The CEA affiliated national Institute for Nuclear Sciences and Techniques (INSTN) has only generated about 50 nuclear graduates per year. EDF has called upon the institute to double the number over the coming years.[112] As of 2009, the number of graduates reached 109, the highest number since 1970. This is an impressive surge. Other engineering schools generate a few dozen graduates more, but nothing anywhere in the range needed. Many of the graduates have gone into other professional sectors. The main operators have started a full-scale "seduction" campaign. In November 2007, the 20 nuclear engineering students at the Ecole des Mines in Nantes were picked up by an

EDF hostess who accompanied them to Paris where they were given a full day's program at EDF expense. AREVA did the same, combined with a visit to the La Hague reprocessing plant.

"Today competence renewal is the first management concern," stated EDF's Inspector General for Nuclear Safety in 2007. "At all levels management is concerned about immediate and future problems that the competence renewal brings along." The Inspector notes that "this concern is today generalized amongst all the nuclear stakeholders, suppliers, subcontractors, safety authorities, in France and elsewhere."[113]

The situation is aggravated by the fact that the "internal transfer market is closed," every site retaining its skilled workers in the "fear that they won't be replaced." Several nuclear power plant sites reported that in the course of the year "they did not get a single response to their vacancy announcements." The EDF Safety Inspector urged the reconstitution of "resource and time margins that don't exist any more within numerous departments and professions."[114] Three years later, the new EDF Inspector General notes little progress: "In conclusion, in spite of the continuous hiring, availability of the competences for today and tomorrow remains an issue of major concern for the managers."[115]

## CONCLUSIONS

There is no doubt that the French nuclear program represents a remarkable scientific, technological, and engineering performance. The implementation of a complex chain of facilities from uranium mining to waste disposal, from uranium conversion to reprocessing, from uranium enrichment to reactor operation over a period of 5 decades is the result of unrelenting persistence. The program has been designed, developed, and implemented under the guidance of a powerful technocrat elite, beyond governmental changes and outside parliamentary decisions and control. "A quoi ça sert ces débats parlementaires?" ("What are those parliamentary debates good for?") remarked Pierre Guillaumat in an interview in 1986.[116] The CEA's General Administrator in the 1950s, later minister of industry and defence and godfather of the Corps des Mines, the State elite that engineered the civil and military nuclear programs, did not believe in consulting public opinion. "No, I have never seen it except in Offenbach . . . . In hell there is public opinion, elsewhere I have never seen it."[117]

The autocratic decisionmaking process guaranteed the long-term implementation of the nuclear program in France. But the lack of democratic control mechanisms also led to a number of misconceptions, costly strategic errors, painful side effects, and a significant dependence on a single source of electricity.

A number of specific conditions have impacted on the costs of nuclear power in France and biased official cost figures. Those include:

- From the start, the French civil nuclear program has largely benefited from military developments and programs.
- Until recently, no general legislative act regulated the nuclear sector. This is certainly one of the reasons why licensing procedures were difficult to challenge in court, and thus did not lead to costly delays due to legal quarrels as in other countries.
- Access to information on nuclear issues has been restricted. It remains to be seen to what

extent the 2006 Nuclear Transparency and Safety Act provides a basis for change. So far, experience provides reasons to remain skeptical.

- The plutonium industry is a typical example of civil-military cross-subsidizing. It is also symptomatic of the incapacity of the establishment to adapt long-term strategies to changed realities. About 55 tons of French plutonium have been cumulated, as well as over 12,000 tons of spent fuel. In addition, over 30 tons of foreign plutonium are stored in France.
- International safeguards arrangements were designed in a way that leaves the use of dual use facilities and materials up to France, which drastically reduced operational costs of facilities that otherwise would have had to separate civil and military material flows.
- At least three-quarters of public research and development expenditures on energy between 1985 and 2001 went to nuclear fission. Little has changed since.
- Risk insurance levels have never reflected any realistic assessment of the potential consequences of a major accident. France has persistently practiced the lowest maximum liability limits in Europe.
- Decommissioning and waste management cost assessments leave a very large margin of uncertainties.
- New projects like the Franco-German EPR in Finland profit from very low interest, state guaranteed bank loans.

The massive development of the nuclear program in France was launched in 1974 as a response to the oil crisis in 1973. The record of the implementation of that program is far from convincing:

- The connection between oil and nuclear power is a widespread myth. In 1973 power generation represented less than 12 percent of the final oil consumption in France.
- In 2007, nuclear power provided 77 percent of the electricity but only 16 percent of final consumption energy in France. Almost three-quarters of the final energy used in France is provided by fossil fuels, close to half by oil. France has a higher per capita consumption of oil than neighboring nuclear phase-out countries Italy and Germany or the average EU27.
- The official energy independence level of 50 percent is highly biased. Disregarding electricity exports and auto-consumption and calculating independence levels on the final rather than on the primary energy side brings it down to 22 percent. Taking into account the fact that all of the consumed uranium is imported brings the French energy independence level down to 8.5 percent.
- Nuclear plants covered a large part of the artificially boosted increase in electricity consumption rather than substituted for other energy sources.
- The massive introduction of electric space heating (now >25 percent of homes) has led to an explosion of peak load, which, in turn, is leading to a highly uneconomic power consumption pattern. Electricity is also the most polluting heat form — much higher induced greenhouse gas emissions than gas or even oil heating — because of massive system losses.

- France's greenhouse gas emissions have practically stagnated since 1990. Provisional figures for 2007 suggest that $CO_2$ emissions were 10 percent higher than in 1995. Emissions are highly sensitive to the climate and to the technical availability of nuclear power plants.
- Expensive peak load power imports (virtually all on a short-term basis) are increasing fast and old oil-fired power plants (2,600 MW, one originally started up in 1968) are being reactivated.
- Electricity prices remain relatively low by EU standards. However, cheap electricity is not equal to low energy bills. In 2007, the French national energy bill reached the level of the early 1980s. Two million French households are eligible for primary necessity tariffs because they cannot pay their electricity bills.
- Assistance for the payment of electricity bills not only costs dozens of millions of euros to EDF, but also drains regional and national social funds. Cost estimates are as high as €150 million to over €200 million per year. The number of requests for assistance increases by 15 percent per year.
- Cheap power does not seem to lead to the anticipated and acclaimed industrial competitiveness. The French foreign trade deficit is estimated to have reached a record €40 billion in 2007 (to be compared to the €200 billion trade surplus of neighboring nuclear-phase-out country Germany). 2008 promises new records.

The current state of the nuclear program, new build projects, and the French promotion of nuclear technology around the world raise a number of questions:

- The high level of standardization provides the opportunity to effectively learn from experience. On the other hand, it increases significantly the risk of costly and potentially dangerous generic faults. The recently discovered steam generator plugging issue is only one of the latest in a series of safety relevant generic problems.

- The hunt for cost savings in the nuclear sector has led to side effects like a massive reduction in stocks of spare parts, which has led to supply bottleneck situations in various nuclear power plants.

- The maintenance of a high level of competence in the workforce has become the most urgent management issue for EDF. By 2015, about 40 percent of the nuclear operating and maintenance staff will be eligible for retirement. Already, several power plants sites are experiencing "no reply" situations as a result of vacancy announcements.

- After 3 years of construction, the new-build flagship project, the Franco-German AREVA-NP EPR in Olkiluoto, Finland, is over 2 years behind schedule and at least 50 percent or €1.5 billion over budget.

- The French nuclear safety authorities raised quality control issues within days of the start of the French EPR construction project at Flamanville and have not stopped raising concerns since.

- The promotion of nuclear technology and the transfer of know-how could increase the risk of nuclear weapons proliferation and inadequate safety conditions in new-comer countries.

Finally, the majority of French citizens, when polled, have been remarkably consistent and quite in line with the rest of the EU in their skepticism over the long-term nuclear power option. A 2007 study for the European Commission revealed that 59 percent of the French citizens polled were in favor of a reduction of nuclear power in the energy mix, while only 28 percent were in favor of increasing the nuclear share.

## ENDNOTES - CHAPTER 6

1. Roger Cohen, "Why American Needs Atomic Anne," *The New York Times*, January 23, 2008, available from *www.nytimes.com/2008/01/23/opinion/23iht-edcohen.3.9439528.html*.

2. John McCain, Remarks on U.S. Energy Policy delivered to the Center for Strategic and International Studies, Washington, DC: Ronald Reagan Building and International Trade Center, April 23, 2007.

3. Franco-Moroccan Economic Summit in Marrakech, October 24, 2008.

4. Peggy Hollinger and Ed Crooks, "Sarkozy makes nuclear energy plea," *Financial Times*, March 8, 2010, available from *www.ft.com/cms/s/0/fbd1c6f0-2aec-11df-886b-00144feabdc0,dwp_uuid=ea77f440-94f2-11df-af3b-00144feab49a.html#ixzz1810ZxIkE*.

5. Tribune du ministre des Affaires étrangères et européennes, M. Bernard Kouchner, dans le quotidian, "Réussir le passage à la nouvelle ère nucléaire," *Les Echos*, 29 avril 2008. (Minister of Foreign and European Affairs, Bernard Kouchner, *Les Echos*, "Making the transition to the new nuclear era," April 29, 2008.)

6. Law No. 46-628 dated April 8, 1946, on the nationalization of electricity and gas.

7. Caisse centrale des activités sociales des industries électriques et gazières or CCAS.

8. Marie-Christine Tabet, "CE d'EDF: la Cour des comptes demande des explications," *Le Figaro,* January 4, 2007. ("EDF's CE [Enterprise Committee]: the Court of Accounts requests explanations.)"

9. Including Iran that holds until today 10% of EURODIF via SOFIDIF (40% Iran, 60% AREVA) that is a 25% shareholder of EURODIF – VITALY FEDCHENKO Appendix 13C. Multilateral control of the nuclear fuel cycle SIPRI Yearbook 2006 available from *www.sipri.org/yearbook/2006/files/SIPRIYB0613c.pdf.*

10. Law No. 2006-686 dated June 13, 2006, relative to transparency and security in nuclear matters.

11. The author counted them at the time.

12. The Corps des Mines is fed historically by the top dozen graduates of the military Ecole Polytechnique plus, more recently top graduates from other elite engineering schools. In total, the annual admissions to the Corps des Mines is 20 or less. The cumulated number of living engineers of the Corps des Mines is approximately 700.

13. Exceptions confirm the rule: The current energy advisor to President Sarkozy is not from the Corps des Mines but from the Corps des Inspecteurs des Finances.

14. *IAEA Bulletin,* International Atomic Energy Agency, Autumn 1986.

15. Pierre Pellerin, who had been at the head of the SCPRI, since its creation in 1956, 30 years previously. Pellerin was indicted in 2006 for aggravated deceit.

16. *Ibid.*

17. Marc Léger and Laetitia Grammatico, "Nuclear Transparency and Safety Act: What Changes for French Nuclear Law?" *Nuclear Law Bulletin,* OECD-NEA, 2007.

18. HCTISN, "Avis sur le suivi radioécologique des eaux autour des installations nucléaires et sur la gestion des anciens sites d'entreposage de déchets radioactifs," 7 November 2008. (HCTISN, "Opinion on the radio-tracking water around nuclear facilities and the management of former storage sites for radioactive waste," November 7, 2008.)

19. BSNMS, —Rapport sur l'application des dispositions de la loi du 25 juillet 1980 sur la protection et le contrôle des matières nucléaire, Année 2007, HFDN, Ministère de l'Industrie, 2008. (BSNMS,-Report on the implementation of the provisions of the Act of July 25, 1980 on the protection and control of nuclear materials, Year 2007, HFDN, Ministry of Industry, 2008.)

20. Exposé des motifs du projet de loi n°1155, déposé le 6 novembre 1964, quoted according to Raymond Tourrain, —Rapport d'information sur l'état de la modernisation des forces nucléaires françaises, Assemblée Nationale, May 22, 1980. (Explanatory memorandum to the Bill No. 1155, filed November 6, 1964, quoted Tourraine According To Raymond, Information-Report on the status of French nuclear forces modernization, National Assembly, May 22, 1980.)

21. *Ibid.*

22. European Atomic Energy Community (EURATOM) Treaty, Article 2, Brussels, Belgium, July 1952.

23. BSCMNS, —Rapport sur l'application des dispositions de la loi du 25 juillet 1980 sur la protection et le contrôle des matières nucléaire, Année 1993, HFDN, Ministère de l'Industrie, 1994. English translation (BSCMNS,-Report on the implementation of the provisions of the Act of July 25, 1980 on the protection and control of nuclear materials, Year 1993 HFDN, Ministry of Industry, 1994.)

24. J. Régnier, COGEMA, in his presentation at the Fourth International Conference on Nuclear Fuel Reprocessing and Waste Management, London, UK, 24-28 April 1994 (RECOD '94 ).

25. IAEA, "The Safeguards Implementation Report," various years.

26. Thomas A. Shea, "IAEA Safeguards Implementation at Chemical Reprocessing Plants," IAEA, presentation to RECOD '94.

27. The text of the Agreement of 27 July 1978 between France, The European Atomic Energy Community and the International Atomic Energy Agency for the Application of Safeguards in France available from *www.iaea.org/Publications/Documents/Infcircs/Others/infcirc290.pdf Official IAEA English translation of the French original.*

28. *Ibid.*

29. BSCMNS, —Rapport sur l'application des dispositions de la loi du 25 juillet 1980 sur la protection et le contrôle des matières nucléaire, Année 1993, HFDN, Ministère de l'Industrie, 1994. English translation (BSCMNS,-Report on the implementation of the provisions of the Act of July 25, 1980 on the protection and control of nuclear materials, Year 1993 HFDN, Ministry of Industry, 1994.)

30. *Ibid.*

31. *Ibid.*

32. BSNMS, —Rapport sur l'application des dispositions de la loi du 25 juillet 1980 sur la protection et le contrôle des matières nucléaire, Année 2007, HFDN, Ministère de l'Industrie, 2008. (BSNMS,-Report on the implementation of the provisions of the Act of July 25, 1980 on the protection and control of nuclear materials, Year 2007, HFDN, Ministry of Industry, 2008.)

33. BSNMS, —Rapport sur l'application des dispositions de la loi du 25 juillet 1980 sur la protection et le contrôle des matières nucléaire, Année 2005, HFDN, Ministère de l'Industrie, 2006. (BSNMS,-Report on the implementation of the provisions of the Act of July 25, 1980 on the protection and control of nuclear materials, Year 2005, HFDN, Ministry of Industry, 2006.)

34. M. Méramédjian, then assistant director of the International Safeguards Application Division of the French Industry Ministry, personnel communication, August 4, 1995.

35. *Ibid.*

36. Mycle Schneider, Transnuklearaffäre – Über die Arbeit des Untersuchungsausschusses im EP, Regenbogenfraktion im EP (GRAEL), November 1988, p.61. ("Transnuclear Affair – About the Work of the Enquiry Committee in the European Parliament,"GRAEL Group [Green Alternative European Link] in the European Parliament.)

37. This is from notes the author obtained from research

38. This is from notes the author obtained from research

39. See Mathieu Pavageau, Mycle Schneider, "Japanese Plutonium and the French Nuclear Weapons Program," Paris, France: WISE-Paris, August 9, 1995.

40. Declan Butler and David Swinbanks, "Japanese plutonium suspected in French tests," *Nature*, August 31, 1995.

41. This section is based on the summary of Mycle Schneider, Yves Marignac, "Spent Nuclear Fuel Reprocessing in France," International Panel on Fissile Materials (IPFM), Princeton University, April 2008.

42. CGT-AREVA, "Négociations AREVA/EDF sur le retraitement recyclage des combustibles usés : le torchon brûle . . . ," Press Release, February 28, 2008.

43. K. R. Smith, A. P. Bexon, K. Sihra, J. R. Simmonds (HPA), J. Lochard, T. Schneider, C. Bataille (CEPN), "Guidance on the calculation, presentation and use of collective doses for routine discharges," UK Health Protection Agency/CEPN, Radiation Protection No. 144, commissioned by the European Commission, August 2006.

44. Mycle Schneider (Dir) *et al.*, "Research and Development on Nuclear Issues in France 1960-1997," Lohmar, Germany: commissioned by Energy Services, WISE-Paris, February 1998.

45. Oil consumption figures according to BP, Statistical World Review 2007, 2008; population figures according to DG TREN, EU Energy and Transport in Figures – Statistical Pocketbook 2007-2008, 2008.

46. Consumption of fossil fuels for power generation even increased by 24% between 1990 and 2007 (see Figure 6-7).

47. EEA, GHG Emissions and Trends – Profile France, 2008

48. Commissariat General du Développement Durable, "Bilan énergétique de la France pour 2009", June 2010

49. Ministère de l'Ecologie, "L'empreinte carbone de la demande finale intérieure de la France", August 2010

50. 34 x 900 MW, 20 x 1300 MW, 4 x 1500 MW.

51. Antoine Bonduelle, « Exportations de courant électrique : Qui perd, qui gagne ?" commissioned by Greenpeace France, IN-ESTENE, Paris, France, November 2002

52. The net imports from Germany over that period averaged 8 TWh!

53. POWEO, Press Release, January 31, 2007.

54. The year has 8,760 hours; 100 hours thus correspond to a load factor of hardly more than 1 percent. It is obvious that POWEO speculates on competing with EDF for higher load factors.

55. Jean-Michel Bezat, Kosciusko-Morizet "Notre objectif 2020 est realizable," *Le Monde,* October 1, 2008. (Kosciusko-Morizet "Our goal 2020 is achievable," *Le Monde,* October 1, 2008.)

56. Antoine Bonduelle, "La surcapacité nucléaire, quelle aurait pu être une stratégie d'équipement optimale", La Revue de l'Energie, N°569, p. 30-38, Paris, France, January-February 2006. (Antoine Bonduelle, "Nuclear-Overcapacity, what could be a strategy for optimal equipment?" *The Journal of Energy,* No. 569, p. 30-38, Paris, France, January-February 2006.)

57. The exercise has been carried out with the ELFIN (ELectricity FINancing) model, developed by the US EDF (Environmental Defense Fund).

58. The evaluation is complex and all the data is not on the table. See GDF, "Le chauffage au gaz naturel pour réduire les émissions de $CO_2$ ," July 2007.

59. Heat generation by electricity is necessarily significantly more wasteful than using the primary energy direct to generate heat (e.g., oil central heating and gas stove), because thermal power plants lose between 45 percent and 75 percent of the primary energy in the form of waste heat (unless they are combined heat and power plants).

60. Available from *www.industrie.gouv.fr/cgi-bin/industrie/ frame23e_loc.pl?bandeau=/energie/anglais/be_us.htm&gauche=/ energie/anglais/me_us.htm&droite=/energie/anglais/accueil.htm.*

61. Available from p. 4, *www.statistiques.equipement.gouv.fr/ IMG/pdf/reperes_lenergie_en_France_cle1e446c.pdf.*

62. *Ibid.*

63. *Ibid.*

64. ADEME, "Le poids des dépenses énergétiques dans le budget des ménages en France, ADEME & VOUS," *Stratégie & Etudes,* n°11, 3 April 2008. (ADEME, "The weight of energy expenditure in household budgets in France, ADEME & You," *Strategy & Studies,* No. 11, April 3, 2008.)

65. PUCA, ANAH, ADEME, "Appel à proposition de soutien à l'innovation PREBAT—Comité bâtiments existants—Réduction de la précarité énergétique," July 2007. (PUCA, ANAH, ADEME, "Call for proposals for supporting innovation PREBAT - Existing Buildings Committee - Reduction of Fuel Poverty," July 2007.)

66. *Ibid.*

67. PUCA.

68. ADEME, "Regards sur le Grenelle," September 2008. (ADEME, "Look the Grenelle," September 2008.)

69. Virginie Christel, et al., "Energie domestique — Des budgets sensibles aux prix des énergies importées," Ministère des Transports, de l'Equipement, du Tourisme et de la Mer, SESP-en bref, January 2006. (Virginia Christel, et al., "Energy home - price-sensitive budgets on energy imports," the Ministry of Transport, Infrastructure, Tourism, and Sea, SESP-Brief, January 2006.

70. EDF, "Statistiques de l'énergie électrique en France", June 2007. (EDF, "Statistics of electric power in France", June 2007.)

71. Karine Fiore and Michel Faure, "The civil liability of European nuclear operators: which coverage for the new 2004 Protocols? - Evidence from France," Maastricht, The Netherlands: Faculty of Law, Maastricht University, 2007.

72. Michael G. Faure and Karine Fiore, "An Economic Analysis of the Nuclear Liability Subsidy," *Pace Environmental Law (PELR) Review*, Vol. 26, No. 2, October 10, 2009, available from *ssrn.com/abstract=1503327*.

73. "Solutions for environment, economy and technology," Report for DG Environment, Environmentally harmful support measures in EU Member States, European Commission, January 2003, p. 132, quoted in Antony Froggatt and Simon Carroll, "Nuclear Third Party Insurance: The Nuclear Sector's 'Silent' Subsidy-State of Play and Opportunities in Europe," November 5, 2007.

74. This section is largely based on Mycle Schneider, "Country Report France," MSC, October 2006, contribution to Wolfgang IRREK (Dir.); Mycle Schneider et al., "Comparison among different decommissioning funds methodologies for nuclear installations," commissioned by the European Commission, coordinated by Wuppertal Institut, Wuppertal, Germany, March 2007, p. 170, available from *www.wupperinst.org/en/projects/project_details/index.html?&projekt_id=167&bid=137*.

75. The figure is even higher, almost €70 billion, if one does not take into account some overlap stemming from doubling pro-

visions due to joint operation by several operators of some facilities.

76. Loi de programme relatif à la gestion durable des matières et des déchets radioactifs, June 15, 2006.

77. For example, approximately one dedicated full-time staff person per key operator in the Industry Ministry.

78. The Nuclear Safety Authorities were not invited to join the Industry Ministry led Working Group that elaborated the reference cost scenario for geological disposal in 2005 (see DGEMP, 2005).

79. According to [DGEMP 2005] ; this is highly contradictory to a number of other national and international studies, for example, see CDP, 2000, and Girard, 2000.

80. Traditionally very pro-nuclear, the Communist Party spokesperson in her vote explanation has called the text "insufficient in research and financing."

81. The decrease of the admitted surface temperature of high-level waste from 150°C to 90°C multiplied the storage cost for this waste category by a factor of four.

82. Pierre Wiroth, "Rapport de l'Inspecteur Général pour la Sûreté Nucléaire et la Radioprotection," EDF, January 2008, p.6; ("Report of the General Inspector for Nuclear Safety and Radiation Protection").

83. The problem consists of extensive (up to 80%) plugging of the tube sheet penetrations. The phenomenon is estimated to evolve by 5% per year. The problem not only reduces the power output of the generator through the drop in heat exchange capacity, it also constitutes a safety problem because the tubes are affected by vibratory fatigue significantly faster. The phenomenon can thus lead to tube cracking within months (as happened at the Cruas power plant). EDF, "EDF 2009 Document de Référence," April 2010, available from *www.edf.com/fichiers/fckeditor/Commun/ Finance/Publications/Annee/2010/ddr/EDF_DDR2009_06_va.pdf.*

84. The most serious problem is probably the sump clogging issue that highlighted the fact that recirculation of primary cooling water would not have worked as anticipated in the case of a large loss of coolant accident. It was discovered in December 2003 and affected all of the 34 units of the 900 MW series. (see source in endnote 83)

85. *Ibid.*

86. Rapport de l'Inspecteur Général pour la Sûreté Nucléaire et la Radioprotection 2007," EDF, January 2008, ("Report of the Inspector General for Nuclear Safety and Radiation Protection 2007," EDF, January 2008).

87. *Ibid.*

88. For details, see Mycle Schneider et al., "Residual Risk – An Account of Events in Nuclear Power Plants Since the Chernobyl Accident in 1986", commissioned by MEP Rebecca Harms, May 1987, available from *www.greens-efa.org/cms/topics/rubrik/6/6659. energy@en.htm).*

89. *Ibid.*

90. SOCATRI is a subsidiary of the enrichment consortium EURODIF S.A.

91. SOCATRI, Press Release, July 11, 2008.

92. CRIIRAD, "Tricastin SOCATRI: ça continue !" 7 August 2008, ("Tricastin SOCATRI: it continues!" available from *www. criirad.org/actualites/dossiers-08/tricastin-juil08/socatri/carbone 14-aout08/criirad-cp8aout08.pdf.*

93. "Areva's annual results grow yet stronger," February 27, 2008, available from *www.world-nuclear-news.org/newsarticle. aspx?id=15934&terms=At%20the%20moment%2c%20AREVA%20 has%20submitted%20half%20of%20the%20plans%20to%20us.%20 Nuclear%20reactors%20are%20not%20built%20without%20 plans%2c%20at%20least%20not%20in%20Finland.*

94. STUK, Press Release, July 12, 2006, available from *www.*

*stuk.fi/stuk/tiedotteet/2006/en_GB/news_419/* ; STUK, "Management of Safety Requirements in Subcontracting During the Olkiluoto-3 Nuclear Power Plant Construction Phase," Investigation Report 1/06, translation dated September 1, 2006; full report is available from *www.stuk.fi/stuk/tiedotteet/2006/en_GB/news_419/_files/76545710906084186/default/investigation_report.pdf*.

95. *Ibid.*

96. Dave Clark, "French nuclear export drive tainted by safety fears," AFP, November 4, 2009.

97. ASN, "Joint Regulatory Position Statement on the EPR Pressurised Water Reactor," November 2, 2009.

98. ASN, DCN-Caen, Inspection letter to the Flamanville-3 project manager, January 25, 2008.

99. ASN, DCN-Caen, Inspection letter to the Flamanville director and the Flamanville-3 project manager, December 26, 2007.

100. ASN, Lettre d'information No. 2, June 2, 2008.

101. ASN, Letter to Director of Flamanville-3 construction project, September 30, 2008.

102. ASN, Note d'information, October 27, 2008.

103. AFP, "Une commune de la Manche vote le boycott des élections contre une ligne THT," La Romandie, March 6, 2008, ("A municipality of la Manche votes to boycott the elections in order to protest against a high power line").

104. EDF, Press Release, November 12, 2008.

105. *Ibid.*

106. "France-Based AREVA Downgraded To 'BBB+' On Continued Weakened Profitability; Outlook Stable" *Standard & Poor's,* June 28, 2010.

107. *Ibid.*

108. EDF-Group, "2008 Document de Référence – Leading the energy change", April 2009.

109. *Ibid.*

110. *Ibid.*

111. *Ibid.*

112. Mycle Schneider, Steve Thomas, Antony Froggatt, and Doug Koplow, "The World Nuclear Industry Status Report 2009," available from *www.nirs.org/neconomics/weltstatusbericht0908.pdf.*

113. This and following quotes from "Rapport de l'Inspecteur Général pour la Sûreté Nucléaire et la Radioprotection 2007," EDF, January 2008, ("Report of the Inspector General for Nuclear Safety and Radiation Protection 2007," EDF, January 2008).

114. *Ibid.*

115. "Rapport de l'Inspecteur Général pour la Sûreté Nucléaire et la Radioprotection 2009," EDF, 2010, ("Report of the Inspector General for Nuclear Safety and Radiation Protection 2009, EDF, 2010).

116. Interview with Georg Blume and the author, published in Damocles, No. 67, Autumn 1995.

117. Reference to Jacques Offenbach's Orpheus In The Underworld: "Hello, I'm Public Opinion."

# CHAPTER 7

# WHAT WILL BE REQUIRED OF THE BRITISH GOVERNMENT TO BUILD THE NEXT NUCLEAR POWER PLANT?

## Stephen Thomas

## INTRODUCTION

In May 2005, Tony Blair said, "Nuclear power is back on the agenda with a vengeance."[1] His chief scientific adviser and other government spokespeople suggested that up to 20 new nuclear units would be needed. This was taken by many, internationally, as a signal that the United Kingdom (UK) was about to launch an aggressive new program to build nuclear power stations. However, in evidence to a Parliamentary Select Committee, the Energy Minister, Malcolm Wicks said:

> It is not for government to say that we shall have X nuclear reactors and so on. Government will not be building nuclear reactors, will not say they want X number of nuclear reactors. I always thought myself that if at the moment one fifth of our electricity is from nuclear, if the market came forward with something to replicate that broadly in the future, from my own point of view it seems to me that would make a useful contribution to the mix. We are not going to do anything to facilitate that, nor this percentage nor that percentage.[2]

Subsequently, after a challenge by Greenpeace, the High Court found in February 2007 that the government's consultation process on nuclear power was inadequate and had to be repealed. "Mr. Justice Sullivan said that the consultation exercise was 'seriously flawed and that the process was manifestly inadequate and unfair' because insufficient information had been made available by the Government for consultees to make an 'intelligent response'."[3]

The government's Green Paper on energy published in May 2007 therefore made no specific commitments on nuclear power. However, one of Gordon Brown's first statements as the Prime Minister in June 2007 seemed to preempt the consultation. He told Parliament on July 4 that ". . . we have made the decision to continue with nuclear power, and . . . the security of our energy supply is best safeguarded by building a new generation of nuclear power stations."[4]

A new consultation was announced in May 2007 and was closed to submissions in October 2007. In January 2008, the government announced the result of the consultation, which again favoured new nuclear construction. The new White Paper stated: "[A]gainst the challenges of climate change and security of supply . . . the evidence in support of new nuclear power stations is compelling . . ."[5]

The commitment not to provide subsidies was reiterated: "It will be for energy companies to fund, develop and build new nuclear power stations in the UK, including meeting the full costs of decommissioning and their full share of waste management costs."[6] The utilities most likely to build nuclear plants, EDF and E.ON both supported the suggestion that subsidies would not be needed. Vincent de Rivaz, CEO of EDF Energy (UK) said: "We have made it clear we are not asking for subsidies, all costs will be borne by us."[7] While E.ON said in a press release: "It also believes that there is no requirement for either government subsidies or for a guaranteed long-term cost of carbon to make new nuclear power stations economic."[8]

This confidence is in contrast to the situation in the United States where the government has committed billions of dollars to subsidies for new nuclear plants and where industry has frequently stated that new nuclear plants without subsidies and guarantees would not be feasible. For example, in December 2007, Christopher Crane, President of Exelon Generation (one of the utilities that has stated an intention to build new nuclear plants), stated: "If the loan guarantee program is not in place by 2009, we will not go forward."[9]

This report examines whether the "free market" really will build new nuclear power plants in the UK without strong support from public funds other than a few enabling measures, for example, on licensing reactor designs, to include a review of what commitments the UK government has actually made. Additionally, this chapter looks at why owning and operating a nuclear power plant is so economically risky and what the specific risks are. It reviews the UK's track record with nuclear power, an important criterion used by the financial community to judge investment risk. This data show that nuclear power has a poor record in the UK. Financial costs borne by electricity consumers and taxpayers for this have been high, but, at least as important, the opportunity cost of placing resources in fruitless nuclear expansion programs has also been high. If these resources had gone into developing renewables and energy efficiency programs, the UK would be closer to making its electricity supply system sustainable.

This chapter also looks at experience in Finland and the United States with attempts to relaunch nuclear power programs. The order of the Olkiluoto 3 plant was portrayed by nuclear advocates as a demonstration that nuclear orders were possible in a liberalized electricity market without subsidy. However, closer examination of the terms of the deal show a number of apparent subsidies, while experience with the first 2 years of construction of the plant have been very poor, reinforcing how economically risky nuclear power plants are. The Bush administration tried to relaunch nuclear ordering, using federal subsidies to kick-start the process with a handful of new plants. Ordering was then expected to be self-sustaining. However, it became clear that the subsidy and guarantee program would have to be open-ended if nuclear orders beyond that of a handful of heavily subsidized units were to be placed.

This chapter reviews the claims that the use of financial instruments, such as bonds, could mean that the problems of dealing with economic risk could be overcome at low cost. Finally, a review of what guarantees and subsidies companies hoping to build nuclear power plants in Britain might seek and what the cost to the public might be is provided.

Issues of decommissioning and waste disposal have received a great deal of publicity in the UK. There have been extensive debates concerning the government's commitment to ensure that the full costs of decommissioning and a full share of waste management costs would be borne by energy companies and not subsidized by taxpayers. However, while the cost of these processes is high and very uncertain, the fact that they do not take place for up to a century or more after the plant is built means that, in any normal economic appraisal, these costs are discounted away. Ensuring that adequate financial arrangements are in place for decommissioning and waste disposal is therefore an important issue from a public policy point of view. However, the cost of these arrangements is not likely to be a major item for a company in its decision whether to build new nuclear plants. There is, therefore, only limited coverage of waste and decommissioning issues in this chapter.

## WHAT HAS THE GOVERNMENT PROMISED ON SUBSIDIES AND GUARANTEES?

The statement by Tony Blair indicating that nuclear power is once again a significant policy agenda issue caught the headlines, but it contains no specific promises. The government's more precise statements are less aggressive. The 2006 Review stated:

> Any new nuclear power stations would be proposed, developed, constructed and operated by the private sector, who would also meet full decommissioning costs and their full share of long-term waste management costs. The Government does not take a view on the future relative costs of different generating technologies. It is for the private sector to make these judgments, within the market framework established by government. The actual costs and economics of new nuclear will depend on, amongst other things, the contracts into which developers enter, and their cost of capital for financing the project.[10]

In evidence to the Trade and Industry Select Committee, Energy Minister Malcolm Wicks was more blunt.

> It is not for government to say that we shall have X nuclear reactors and so on. Government will not be building nuclear reactors, will not say they want X number of nuclear reactors. I always thought myself

that if at the moment one fifth of our electricity is from nuclear, if the market came forward with something to replicate that broadly in the future, from my own point of view it seems to me that would make a useful contribution to the mix. We are not going to do anything to facilitate that, nor this percentage nor that percentage.[11]

And in response to a question on subsidies: "Is that the Government's position? No direct subsidies and no indirect subsidies. Am I clear on that?" he said, "No cheques [checks] will be written, there will be no sweetheart deals." And:

No, there will not be any special fiscal arrangements for nuclear. It should not be a surprise, with respect, because we have said it very clearly in the Energy Review. You could pursue this if you wanted by saying that nuclear waste is quite a complex subject and we are going to look very carefully at that to make sure that the full costs of new nuclear waste are paid by the market.

The main concession was on licensing:

The idea of prelicensing is that you can say, here is a wind farm, here is a nuclear reactor, or a gas-powered station; let us prelicense it so that the regulators are satisfied that it is safe and all the other things as a piece of kit. Then the local inquiry can purely be about local issues rather than becoming a national or international occasion to reopen the whole debate about whether windmills or nuclear are desirable. That is what we are trying to do.[12]

## DECOMMISSIONING

The only apparent exception to the no-subsidies and guarantees rule concerns the arrangements for waste disposal. In February 2008, the government launched a Consultation on Funded Decommissioning Programme Guidance for New Nuclear Power Stations.[13] The title is misleading as the report covers waste disposal costs as well as decommissioning, and the proposals represent a significant departure from previous expectations relating to waste disposal as well as to decommissioning.

- On decommissioning, the government is proposing that companies would have to demonstrate detailed and costed plans for decommissioning, waste management, and disposal before they even begin construction of a nuclear power station;
- Set money aside into a secure and independent fund from day one of generating electricity; and,
- Have additional security in place to supplement the Fund should it be insufficient, for example, if the power station closes early.[14]

If these proposals are carried through with adequate measures to ensure that if estimated decommissioning costs increase, the companies will be required to make sufficient additional contributions to make up the shortfall, they appear a good base. However, closer examination reveals a number of issues not well-accounted for.

On timing, the proposals assume a plant will operate for 40 years. This will be followed by a 7-year defueling period. Stage 1 then follows, taking 5 years;

stage 2 is forecast to take a further 5 years; stage 3 is expected to take 10 years; and final site clearance is expected to take 6 years. So the elapsed time from plant closure to end of decommissioning is 37 years. This is a welcome shortening compared to the proposals for existing British plants, which is currently based on timescales of in excess of 100 years from plant closure to completion of decommissioning.

Little guidance is given on discounting. For example, no indication of the level of discount rate that can be assumed is shown, nor is it specified how long into the future liabilities can be discounted for. However, the area where it appears a guarantee is expected is in the following paragraph:

> We anticipate that operators will request that the Government provide them with a fixed unit price at the time they seek approval for their Funded Decommissioning Programme. This will occur alongside the regulators' licensing and permitting processes. At this time, the Secretary of State would use the cost modelling methodology it has developed, together with information from the NDA's parametric cost modelling work on the estimated costs of disposal facilities, to determine the fixed unit price, including the appropriate risk premium. The cost modelling methodology is described in greater detail at paragraphs 4.5.1 – 4.5.39 and further information on when we expect to be in a position to set a fixed unit price for operators is set out in the Roadmap paragraphs 2.25 – 2.32 and Table 2. To help future operators with their planning, the Government would expect to give operators a non-binding indicative price at an earlier date than when the Government would be willing to provide them with a final fixed unit price.[15]

This makes it clear that once the plant is ready to be built, the companies' contribution to the decommissioning fund would be capped, and, if costs increased beyond the level covered by the risk premium, taxpayers would have to foot the bill. Given the very rapid rate of escalation of decommissioning cost estimates in advance of the most challenging stage of decommissioning work actually being attempted, there must be a very large risk that the estimated cost will fall far short of the actual cost, even if the risk premium is included. This therefore represents a major taxpayer-funded cost guarantee.

## WASTE DISPOSAL

On waste disposal, the cost guarantees are much clearer. For low-level waste, no guarantees are involved. Operators will be expected to make their own arrangements for waste disposal and "will be required to meet these costs from operational expenditure for operational low level waste, and from the Fund for decommissioning low level waste."[16] While the cost of low-level waste disposal is far from stable, the process is technically well-established.

However, intermediate and high-level (spent fuel) waste is subject of a major cost guarantee backed by taxpayers (it is assumed that spent fuel will not be reprocessed). The consultation states: "The Government would expect to set a fixed unit price based on the operator's projected full share of waste disposal costs at the time when the approvals for the station are given, prior to construction of the station."[17]

Given that neither intermediate nor high-level waste disposal is established anywhere, the costs of such processes must be regarded as highly speculative. The government does try to provide evidence that the risk of cost escalation will be taken account of:

> In return for giving operators certainty over when they will transfer title to and liability for their waste and spent fuel to the Government, we will set the level of the risk premium to take account of the risk to the Government that the construction of disposal facilities is not complete by the date or dates specified in the agreed schedule. This risk premium will be built into the fixed unit price for the waste disposal service.[18]

However, as with decommissioning, such untried, technically challenging, and socially contentious processes must involve a huge degree of uncertainty.

## WHY IS NUCLEAR POWER ECONOMICALLY RISKY AND DOES THIS MATTER?

### Who Bears the Risk?

Any investment in a large new power station is economically risky because of the scale of the investment, the technologically challenging nature of power production, and the scope to choose options that turn out, for example because of movements in fossil fuel prices, to prove uneconomic. These risks were borne by consumers under the old model of organization of the electricity industry where electricity generation was a monopoly in a given territory. If the cost of a power plant was higher than forecast or it proved to be more expensive than the alternatives, the additional costs were paid by consumers. While this did expose consumers to investment risk, consumers were compensated because the cost of capital for new power stations was low since financiers could rely on the generation company recovering any costs it incurred from consumers.

One exception to this was in the United States where regulators could force generation companies to absorb some or all of the cost of investments if the regulator judged that the costs were excessive. In practice, this provision was not used until the mid-190s. Then, as nuclear plants began to come on line at prices far above their cost estimates, regulators began to disallow recovery of costs that they judged imprudent. Ordering nuclear power plants became a major economic risk for U.S. electric utilities, and ordering ceased in 1979, with all plants ordered after 1974 subsequently cancelled. Dozens of nuclear orders were cancelled to avoid exposure of utilities to this risk.

Elsewhere, developments in shifting investment risk from electricity consumers were not accomplished until the 1990s. One of the main motivations for the trend to reform and liberalize electricity industries was a desire to expose electricity generation companies to more investment risk, with the expectation that this would act as a financial discipline. If the company made a bad decision, the cost would be paid for from the profits of the company, not by consumers.

## Why Is Nuclear Power Particularly Economically Risky?

As argued above, any investment in a substantial power plant is a significant economic risk and, if the company building the plant bears the consequences of that risk the cost of capital will be much higher than under the old system. Financiers will fear that companies building new power plants could go bankrupt if the power plant cannot compete in the wholesale electricity market and will therefore charge a substantial risk premium on loans to build the plant to cover the risk that the loan will not be repaid if the company fails. Nuclear power is among the most capital intensive of power generation technologies with financial charges expected to account for more than half the total kilowatt (kWh) cost of generation. Therefore, making the electricity generation business a competitive one will inevitably disadvantage nuclear power compared to other less capital-intensive technologies.

However, nuclear power plants are far more economically risky than other types of power plant. This risk arises from a number of sources:

- Nuclear power plants are far more complex than most power plants, and there is far more potential for errors to be made in construction;
- Nuclear power plants are mostly constructed on-site, whereas other types of power plants can be assembled mainly in factories, where costs and quality are easier to control;
- Many of the costs arising from nuclear power generation are beyond the control of the companies, for example, projected costs for waste disposal and decommissioning have escalated sharply in the past couple of decades, while safety regulators may impose additional requirements on the company as a result of problems arising in other countries. For example, the Chernobyl and Three Mile Island accidents led to new regulatory requirements being imposed on plants where no problems had occurred.

However, the main perception of risk arises from the poor record of the nuclear industry in meeting its forecasts.

## WHO ARE THE INTERESTED PARTIES AND WHAT ARE THEIR MOTIVATIONS?

In a decision to order a nuclear power plant, there are three sets of interests directly involved: (1) the commercial companies selling and buying the plant, (2) the governments, and (3) the financial community, including financiers, credit rating agencies, and investment analysts. Each of these has a rather different perspective on the issue.

### The Companies.

The duty of a commercial company is to maximize the profits for its shareholders. In a perfect market, profits will be maximized by choosing the cheapest production technologies and maximizing internal efficiency. However, perfect markets do not exist and companies rely heavily on making strategic decisions, for example, to reduce their exposure to risk, or build their reputation, or build customer loyalty. Commercial companies discount future costs and benefits, so their outlook tends to be rather short-term, and financial consequences more than, say, 20 years in the future carry little weight. For plant vendors, some sales may have a particular value if, for example, they demonstrate a technology or open up a valuable new market. Companies may accept lower profits or even make a loss on a particular order if, in the long term, it strengthens the company's position. Generation companies will, all things being equal, look for the lowest cost technology if they are operating in a competitive market. However, if a technology needs protection from the market, it might be very attractive to a company if the government is prepared to provide that protection, for example, if the technology has particular environmental advantages.

**Government.**

Governments have a number of perspectives. They have a strategic duty to increase the competitiveness of the country's economy. Approaches vary widely on this from the interventionist approach, where they become involved in commercial deals, to the "hands-off" free market approach. Some governments see nuclear power as providing a cheap or at least a stable-price source of power that can largely be regarded as indigenous, and they therefore see it as their duty to promote nuclear power. Here, we will not debate whether these perceptions are valid, but most people would regard it as part of the government's responsibility to at least guide energy policy in a strategic direction. Even the most free-market of governments, such as those of Margaret Thatcher (UK) and Ronald Reagan (U.S.) have tried to promote nuclear power.

Governments of the home country of nuclear vendors may also try to promote reactor sales, for example, by providing loan guarantees or by making enabling political deals. The French government provided loan guarantees for the Olkiluoto Plant to promote the interests of the French vendor, Areva. The United States has recently concluded a bilateral agreement with India to allow U.S. companies to supply reactors and reactor technology to India. This breaks an international embargo on the supply of reactor technology and equipment going back more than 30 years resulting from India's nuclear weapons test in 1974.

**The Financial Community.**

In the past, while electricity was a monopoly industry, the financial community had a limited role in nuclear power investment decisions, at least for developed countries with stable economies. Full cost recovery from consumers was guaranteed, therefore the commercial risk attached to a nuclear power plant order was minimal. However, as was demonstrated by the collapse in 2002 of British Energy, owning and operating nuclear power plants is now a highly risky venture, and that risk is borne at least in part by the shareholders. Credit rating agencies will examine the investments and decisions of a company and use that information to assess their credit rating, which will, in turn, affect the cost of capital to that company. Financiers will assess the riskiness of a project and on that basis, as well as the general credit rating of the company, decide whether to lend money and at what rate. Investment analysts will look at the decisions of the companies assessing the likely profitability of the company. On that basis, they will decide whether to buy or sell shares, or recommend whether to buy or sell. Institutional investors have the power to force management changes if they are unhappy with the decisions being taken.

The decision to order a nuclear power plant is often seen as a two-way deal between the vendor and the utility, but the reality is that the third party of the deal is the financial community. If ordering a nuclear plant would adversely affect a company's share price or its credit rating, the company would have to think very hard before placing that order. The situation was summed up very neatly by Thomas Capps, CEO of a U.S. utility (Dominion) linked with a bid to build a nuclear plant under the Nuclear 2010 initiative: "We aren't going to build a nuclear plant anytime soon. Standard & Poor's and Moody's would have a heart attack. And my chief financial officer would, too."[19]

## WHAT ARE THE RISKS?

### Construction Cost and Time.

The usual rule-of-thumb for nuclear power is that about two-thirds of the generation cost is accounted for by fixed costs, that is, costs that will be incurred whether or not the plant is operated, and the rest by running costs. The main fixed costs are the cost of paying interest on the loans and repaying the capital, but the decommissioning cost is also included. In the United States, an assessment of 75 of the country's reactors showed predicted construction costs to have been $45 billion, but the actual costs were $145 billion.[20] In India, the country with the most recent and current construction experience, completion costs of the last 10 reactors have averaged at least 300 percent over budget.[21]

Over-runs in construction time also have high economic consequences. A delay in completing the plant will increase the interest that has to be paid on the loans needed to finance the plant. If the output of the plant is contracted to a customer, the plant owner might have to pay expensive compensation to the customer. The market value of a day's output of a 1,000 megawatt (MW) nuclear power plant could be around $1.5 million, so a year's delay could mean that $0.5 billion of power could have to be bought from the market. If supply is tight, the cost of buying this extra replacement power could be significantly more than the contract price.

## Operating Performance.

For a capital intensive technology like nuclear power, high utilization is of great importance so that the large fixed costs (repaying capital, paying interest, and paying for decommissioning) can be spread over as many saleable units of output as possible. In addition, nuclear power plants are physically inflexible, and it would not be wise to start up and shut down the plant or vary the output level more than is necessary. As a result, nuclear power plants are operated on "base-load," except in the very few countries (e.g., France) where the nuclear capacity represents such a high proportion of overall generating capacity that this is not possible. Even in France, the amount of load-following is small. A good measure of the reliability of the plant and how effective it is at producing saleable output is the capacity factor. The capacity factor is calculated as the output in a given period of time expressed as a percentage of the output that would have been produced if the unit had operated uninterrupted at its full design output level throughout the period concerned.[22]

Capacity factors of operating plants have been much poorer than forecast. The assumption by vendors and those promoting the technology has been that nuclear plants would be extremely reliable with the only interruptions to service being for maintenance and refueling, giving capacity factors of 85-95 percent. However, performance was poor, and around 1980, the average capacity factor for all plants worldwide was about 60 percent. To illustrate the impact on the economics of nuclear power, if we assume fixed costs represent two-thirds of the overall cost of power if the capacity factor is 90 percent, the overall cost would go up by a third if capacity factor was only 60 percent. To the extent that poor capacity factors are caused by equipment failures, the additional cost of maintenance and repair would further increase the unit cost of power. In a competitive market, a nuclear generator contracted to supply power that is unable to fulfill its commitment is likely to have to buy the "replacement" power for its customer, potentially at very high prices.

However, from the late 1980s onwards, the worldwide nuclear industry has made strenuous efforts to improve performance, and capacity factors now average more than 80 percent, for example, the United States now has an average of nearly 90 percent, compared to less than 60 percent in 1980, although the average lifetime capacity factor of America's nuclear power plants is still only 70 percent.

## Operating Costs.

Many people assume that nuclear power plants are essentially automatic machines requiring only the purchase of fuel and that they have very low running costs. The cost of fuel is relatively low and has been reasonably predictable. However, the assumption of low running costs was proved wrong in the late 1980s and early 1990s when a small number of U.S. nuclear power plants were retired because the cost of operating them (excluding repaying the fixed costs) was found to be greater than cost of building and operating a replacement gas-fired plant. It emerged that nonfuel operation and maintenance (O&M) costs were on average in excess of $22/MWh, while fuel costs were then more than $12/MWh.[23] Strenuous efforts were made to reduce nonfuel nuclear O&M costs, and by the mid

1990s, average nonfuel O&M costs had fallen to about $12.5/MWh and fuel costs to $4.5/MWh. However, it is important to note that these cost reductions were achieved mainly by improving the reliability of the plants rather than actually reducing costs. Many O&M costs are largely fixed — the cost of employing the staff and maintaining the plant — and vary little according to the level of output of the plant, so the more power that is produced, the lower the O&M cost per MWh. The threat of early closure on grounds of economics has now generally been lifted in the United States.

It is also worth noting that British Energy, which was essentially given its eight nuclear power plants when it was created in 1996, collapsed financially in 2002 because income from operation of the plants barely covered operating costs.

Fuel costs have fallen because the world uranium price has been low since the mid-1970s. U.S. fuel costs average about $5/MWh, but these are arguably artificially low because the U.S. Government assumes responsibility for disposal of spent fuel in return for a flat fee of $1/MWh. This is an arbitrary price set more than 2 decades ago and is not based on actual experience.

## Decommissioning and Waste Disposal Costs.

These costs are difficult to estimate because there is little experience with decommissioning commercial-scale plants, and the cost of disposal of waste (especially intermediate or long-lived waste) is uncertain. However, even schemes which provide a very high level of assurance that funds will be available when needed will not make a major difference to the overall economics. For example, if the owner was required to place the (discounted) sum forecast needed to carry out decommissioning at the start of the life of the plant, this would add only about 10 percent to the construction cost.

The problems come if the cost has been initially underestimated, the funds are lost or the company collapses before the plant completes its expected lifetime. All of these problems have been experienced in Britain. The expected decommissioning cost has gone up several-fold in real terms over the past couple of decades. In 1990, when the Central Electricity Generating Board (CEGB) was privatized, the accounting provisions made from contributions by consumers were not passed on to the successor company, Nuclear Electric. The subsidy that applied from 1990-96, described by Michael Heseltine[24] as being to "decommission old, unsafe nuclear plants" was, in fact, spent as cash flow by the company owning the plant, and the unspent portion has now been absorbed by the Treasury. The collapse of British Energy has meant that a significant proportion of their decommissioning costs will be paid by future taxpayers.[25]

**Insurance and Liability.**

There are two international legal instruments contributing to an international regime on nuclear liability: The International Atomic Energy Agency's Civil Liability for Nuclear Damage (1963 Vienna Convention) and the Organization for Economic Cooperation and Development's (OECD) Third Party Liability in the Field of Nuclear Energy (1960 Paris Convention) together with the linked Brussels Supplementary Convention of 1963. These conventions are linked by the Joint Protocol, adopted in 1988. The main purposes of the conventions are to:

1. Limit liability to a certain amount and limit the period for making claims;

2. Require insurance or other surety by operators;

3. Channel liability exclusively to the operator of the nuclear installation;

4. Impose strict liability on the nuclear operator, regardless of fault, but subject to exceptions (sometimes incorrectly referred to as absolute liability); and,

5. Grant exclusive jurisdiction to the courts of one country, normally the country in whose territory the incident occurs.

In 1997, a Protocol was adopted to amend the Vienna Convention, which entered into force in 2003, and in 2004, a Protocol was adopted on the Paris Conventions. These changed the definition of nuclear damage and changed the scope. For the Brussels Convention, new limits of liability were set as follows: operators (insured) €700 million; installation state (public funds) €500 million; and collective state contribution €300 million; a total liability of €1500 million. These new limits have to be ratified by all contracting parties and are currently not in force.

The scale of the costs caused by, for example, the Chernobyl disaster, which may be in the order of hundreds of billions of euros, means that conventional insurance coverage would probably not be available and, even if it was, its coverage might not be credible because a major accident would bankrupt the insurance companies.

It has been estimated that if Electricité de France (EDF), the main French electric utility, was required to fully insure its power plants with private insurance but using the current internationally agreed limit on liabilities of approximately €420 million, it would increase EDF's insurance premiums from €0.017/MWh, to €0.19/MWh, thus adding around 8 percent to the cost of generation. However, if there was no ceiling in place and an operator had to cover the full cost of a worst-case scenario accident, it would increase the insurance premiums to €5/MWh, thus increasing the cost of generation by around 300 percent.[26]

## A SHORT HISTORY OF THE BRITISH NUCLEAR POWER PROGRAM

The policy announced by the UK government in January 2008 will be the fifth attempt to relaunch the UK nuclear power program. The first generation nuclear power plants in Britain were of an indigenous design known as magnox. Between 1956 and 1971, 11 magnox stations were completed, but by the early 1960s, it was clear this design could never be economic. The magnox stations are generally portrayed as reliable workhorses and, if they had been followed by successful new designs, they would probably have been retired at the end of their 20-year design lifetime. However, the failure of subsequent programs has meant that they have operated long beyond their design lifetime, up to 40 years, and in 2008, two units remain in service. They have suffered corrosion problems, their reliability has been mediocre, and they represent a very expensive source of electricity.

In 1965, the UK government chose another British design to succeed the magnoxes, the advanced gas-cooled reactor (AGR), and five stations, each of about 1,200MW, were ordered. Instead of the 4 years forecast for building the plants, these took from 10-24 years from start of construction to commercial operation. None of the plants ever operated as designed, and all operate significantly below their design maximum output rating.

By 1970, the problems with the design were clear, and after a further 3 years of investigations, in 1973 the government chose another UK design, the Steam generating heavy water reactor (SGHWR). By 1977, the developers had to acknowledge that this design could not be built on a commercial scale and thus was abandoned. In 1977, the government adopted a dual reactor policy, two more AGRs were to be built, and the steps taken to be in a position to order a U.S. designed reactor, the pressurized water reactor (PWR) from Westinghouse. In 1979, the two AGR orders were confirmed, and the government announced a 10 reactor order program of Westinghouse PWRs, with the first order to be placed in 1981, and the others to follow at yearly intervals. By 1987, when the first order (Sizewell B) was actually placed, the program had been reduced to four units. At the same time, the government announced its intention to privatize the UK electricity industry and operate the generation sector as a competitive market.

In 1989, the government acknowledged that, after 2 years of effort, a plan could not be devised that would allow the privatization of the nuclear power plants. The main problem appeared to be the economic risk associated with building and operating the four new proposed PWRs (including Sizewell B). The operating plants were uneconomic, but their costs were largely known, and subsidies could be used to cover these. However, the risk of overrunning construction times and costs and poor operating reliability were much more open-ended for the new PWRs. The Energy Minister at the time claimed that "unprecedented guarantees were being sought. I am not willing to underwrite the private sector in this way. . . ."[27]

Studies preparing for privatization had revealed that, far from being cheap sources of generation, as had always been claimed, the operating cost alone of the magnox and the AGR stations was double that of the expected wholesale price of electricity. The construction cost, normally expected to account for more than two-thirds of the total generation cost from a nuclear power plant, had to be written off. The nuclear power plants were transferred to two new publicly owned companies, Nuclear Electric, Scottish Nuclear, and a huge consumer subsidy was introduced raising £1 billion per year simply to allow the companies to break even. The four unit PWR program was abandoned, although work was allowed to continue on Sizewell B, despite the clear evidence it would be hopelessly uneconomic.

In 1996, a year after the completion of Sizewell B, at a cost in excess of £3 billion when its costs were known and it seemed likely that its reliability would be reasonable, the government sold the seven AGRs and Sizewell B for about £1.7 billion to a new company, British Energy. The reliability of the AGRs had improved sufficiently that there appeared a reasonable expectation that their running costs could be met from the income from sales of electricity. While the company was required to make some contribution to the cost of decommissioning, the largest part of the cost was left to be met from British Energy's cash flow at the time decommissioning was carried out.

The privatization meant that eight nuclear stations, each of about 1200MW and paid for by consumers, were sold for about half the cost of building Sizewell B. Much of the cost of building Sizewell B was paid for from the £1 billion per year consumer subsidy, applied from 1990-96, money which consumers had been told would go to pay for decommissioning and waste disposal.

The UK wholesale electricity price remained unreasonably high from privatization in 1990 until 2001, when it fell sharply. British Energy quickly got into difficulties, and by September 2002, it had collapsed. The government eventually managed to force through a package of measures to save the company by assuming some of its liabilities and subsidizing its costs (for example on reprocessing), and the company was relaunched in 2005, with the government taking 64 percent share of ownership as the price for saving the company. In June 2007, the government sold 25 percent of the shares, and it expects to sell a further 10 percent, leaving the government with a 30 percent holdings. When the wholesale price of electricity falls again, the reliability of the plants deteriorates as they age and, as the impact of having to retire the oldest AGRs takes effect, it seems likely that the company will fail with taxpayers again having to take on the financial burdens it leaves behind.

Ironically, the main asset the company now has is ownership of the sites where many of the existing plants are. It is generally acknowledged that, in any new nuclear program in the UK, the first plants will be built on existing sites as it might be expected that public opposition here would be much less than at new sites. British Energy has neither the resources nor the credibility to build new nuclear plants, but it could earn significant income from the use of its existing sites, and a new nuclear program may be British Energy's best hope of survival, albeit essentially as a real estate company.

**Lessons.**

Even when the evidence is overwhelming that mistakes have been made, as happened with the AGR program, the SGHWR, and Sizewell B, the government will not abandon misconceived programs until long after they should have been cut and at great cost to taxpayers and electricity consumers. The failed attempts to relaunch nuclear power programs were based on hopelessly optimistic forecasts of construction costs and times, reliability, and operating cost.

However, the main outcome of this experience is the huge opportunity cost of these largely fruitless programs. They consumed the vast majority of government and electricity industry research and development (R&D) budgets, they dominated the attention of civil servants involved with the electricity industry, and they influenced UK industry to try to develop nuclear capabilities instead of more productive and profitable capabilities in renewable energy sources and energy efficient technologies.

## OLKILUOTO AND THE U.S. NUCLEAR POWER 2010 PROGRAM

**Olkiluoto.**

The Olkiluoto order is currently the only live new order in Western Europe or North America, and the first to be placed since the Civaux 2 order in France in 1993, which was coupled to the grid in 1999. It is the first plant of a new design, the European pressured water reactor (EPR), developed by the Franco-German company, Areva. The EPR is a 1,600MW pressurized water reactor (PWR) evolved from designs supplied by the two main owners of Areva NP, Areva (France) and Siemens (Germany). The customer is a company called Teollisuuden Voima Oy (TVO), owned by the large electric-intensive industries of Finland.

Olkiluoto is often portrayed as the exemplar of the capabilities of current designs. It is predicted to be cheaper to build and operate, and safer than its predecessors. It is also seen as a demonstration that nuclear power orders are feasible in liberalized electricity markets. Many commentators claimed that nuclear power orders were unfeasible in liberalized markets because consumers would no longer bear the full risk of building and operating new power plants. It is therefore important to examine the circumstances of the Olkiluoto order to see how far it really can be seen as a commercial order chosen in a free market and without subsidies and guarantees.

Before examining the specifics of the order, it is worth noting that Finland's experience with nuclear power has been much better than that of the UK. Finland ordered four relatively small nuclear power plants from 1971-75. Two of these at Loviisa (both 440MW net) used the first generation Russian design (VVER-440) but were upgraded to Western standards with the assistance of Siemens. The two at Olkiluoto (both 660MW net) use a Swedish BWR design similar to plants built in Sweden. The reliability of all four plants has been high, and, even today when reliability is much higher in the rest of the world than it was in the 1980s, all four Finnish units are in the top 20 percent in the league table of nuclear power plants ordered by lifetime capacity factor. So, the track record of Finland as a nuclear operator is better than that of the UK.

## Construction Cost and Time.

To reduce the risk to the buyer, Areva offered the plant under turnkey terms:

> It is a fixed price contract, with the consortium having total responsibility for plant equipment and buildings, construction of the entire plant up to and including commissioning (excluding excavation), licensability, schedule and performance. The overall project cost has been estimated by TVO at around €3bn.[28]

The turnkey terms fixed the price TVO would have to pay and allowed for fines to be levied on the contractors if the plant was late. The schedule allowed for a 48-month period from pouring of first concrete to first criticality.

From the start, the construction period has gone seriously wrong, so that after 18 months of construction in December 2006, the plant was 18 months behind schedule, and the vendor, Areva, was suffering severe losses.[29] This was not the result of a particular problem, but the result of a range of failures, including welding, delays in detailed designs, problems with concrete, and with the quality of some equipment. More generally, it seemed that none of the parties involved, including the vendor, the customer, or the safety regulator, had a clear enough understanding of the requirements that building a nuclear plant placed on them.

In December 2006, the French Ministry of Industry (the French government owns more than 90 percent of Areva) said that the losses to Areva had reached €700 million on a contract fixed at €3 billion. The turnkey contract should ensure that this cost escalation is not passed on to the customer, although the deal appeared to be under strain. Philippe Knoche, an Areva representative stated:

> Compensation principle. TVO did not accept this interpretation and the TVO project manager, Martin Landtman, when asked about Knoche's statement, said: 'I don't believe that Areva says this. The site is in the contractor's hands at the moment. Of course, in the end, TVO is responsible for what happens at the site. But the realisation of the project is Areva's responsibility.'[30]

Compensation for delays has already reached the limit of €300 million that would be payable for a delay of 18 months. The buyer will not receive compensation for further delays beyond those already incurred by September 2006.

Further problems were announced in August 2007, although these were not fully quantified in terms of delays to completion or additional costs. It was reported that the delays were partly due to problems meeting the requirement that the plant should be able to withstand an aircraft crashing into it, and partly because the volume of documentation required had been underestimated by the vendor.[31] One report stated that Areva NP was going to take an additional provision €500-700 million on top of the €700 million provision already made for losses.[32] In December 2007, Areva announced that the plant was not expected to be completed until summer 2011.

**Finance.**

The details of how the plant would be financed have not been published, but the European Renewable Energies Federation (EREF) and Greenpeace separately made complaints to the European Commission in December 2004 that they contravened European State aid regulations. The Commission did not begin to investigate the complaints until October 2006, and in September 2007, the Competition Commission dropped the case. According to EREF, the Bayerische Landesbank (owned by the state of Bavaria) led the syndicate (with Handelsbanken, Nordea, BNP Paribas, and J. P. Morgan) that provided the majority of the finance. It provided a loan of €1.95 billion, about 60 percent of the total cost at an interest rate of 2.6 percent. It is not clear if this is a real or a nominal rate. If it is a nominal rate, the real rate is effectively zero. Two export credit institutions are also involved: France's Coface, with a €610 million export credit guarantee covering Areva supplies, and the Swedish Export Agency SEK for €110 million.

**The Customer.**

The buyer, TVO, is an organization unique to Finland. PVO, the largest shareholder, holds 60 percent of TVO's shares. PVO is a not-for-profit company owned by Finnish electric-intensive industry that generates about 15 percent of Finland's electricity. Its shareholders are entitled to purchase electricity at cost in proportion to the size of their equity stakes. In return, they are obliged to pay fixed costs according to the percentage of their stakes and variable costs in proportion to the volume of electricity they consume. The other main shareholder in TVO is the largest Finnish electricity company, Fortum, with 25 percent of the shares. The majority of shares in Fortum are owned by the Finnish Government. This arrangement is effectively a life-of-plant contract for the output of Olkiluoto 3, at prices set to fully cover costs.

## Analysis of the Olkiluoto Experience.

Turnkey contracts have been few and far between in the history of nuclear power and have generally resulted in huge losses to the vendor. Nuclear power plants are immensely complex requiring a great deal of on-site work and input from a large number of organizations. It is therefore difficult for any one company to feel that they have sufficient control over the process so that they can guarantee the price to the customer. The most famous turnkey orders were the 12 placed in the United States in 1963-66.[33] The vendors lost huge amounts of money on these orders, but they achieved their objective. They convinced utilities that the vendors were confident of their designs and that buying a nuclear plant was no greater risk than buying a fossil fuel plant. Subsequent U.S. orders did not contain this protection for the buyer.

If the Olkiluoto order does accentuate the EPR technology, thereby opening the way for further orders, the losses incurred by Areva and Siemens might appear justifiable to their shareholders. However, experience has been so poor that, far from convincing new buyers, it might put them off, and potential buyers of the EPR in India and China are reported to be perturbed by the problems.[34] However, it seems unlikely that the owners of Areva could contemplate offering turnkey terms again until there is very clear evidence that the probability of cost and time overruns for an EPR had become extremely low.

The unique nature of the plant owner means that, far from competing in an open market, the owners have been able to insulate it very fully from the market by contracting for the lifetime's output of the plant at whatever cost is incurred. There is risk to the owners. The plant is likely to be at least 2 years behind in completion, and the owners will have to buy power from the market for that period, potentially at high prices. If the cost of power from Olkiluoto proves to be significantly higher than the wholesale market price, the owners will have to buy expensive power and, for the electric-intensive industry where the cost of power purchase could make up about half of the total costs, this could be catastrophic.

The European Commission has found that the finance did not involve unfair state aids. However, it is bizarre to find that loans to a prosperous Western European country have to be backed by export credit guarantees, and the cost of borrowing is blatantly far below commercial rates. The Olkiluoto order therefore does not provide any evidence that nuclear orders are feasible in a liberalized market without substantial public subsidies and guarantees. Experience so far reinforces the very high economic risks of cost and time over-runs involved in the construction of a nuclear power plant.

## THE U.S. NUCLEAR POWER 2010 PROGRAM

### The Program.

The Bush administration made a concerted effort to revive nuclear ordering with its 2002 Nuclear Power 2010 program. It has yet to achieve a new order. Under the program, the U.S. Department of Energy (DoE) expects to launch cooperative projects with the industry to:

> . . . obtain NRC approval of three sites to assure the availability of these potential locations for new nu-

clear power plants under the Early Site Permit (ESP) process . . . develop application preparation guidance for the combined Construction and Operating License (COL) and to resolve generic COL regulatory issues. (The COL process is a 'one-step' licensing process by which nuclear plant public health and safety concerns are resolved [prior to commencement of construction,] and before the NRC approves and issues a license to build and operate a new nuclear power plant.)[35]

A total of up to $450 million in grants is expected to be available for at least three projects. Two main organizations initially emerged to take advantage of these subsidies and have signed agreements with the DoE to develop COLs. Nustart, launched in 2004, was the first utility grouping to express an interest. It comprises a consortium of eight U.S. utilities including Constellation Energy, Entergy, Duke Power, Exelon, Florida Power & Light, Progress Energy, Southern Company, and the Tennessee Valley Authority (TVA), providing staff time, not cash).[36] EDF, and the vendors, Westinghouse and General Electric (GE) are members but have no voting rights.

This was followed up by the nuclear provisions of the U.S. Energy Policy Act of 2005 (EPACT 2005). The Bush program is best understood as an effort to reverse the power market lessons of the 1980s and 1990s. Since investors have proven unwilling to assume the risks of building new nuclear units, even after the improving of designs and the streamlining of the licensing process, EPACT 2005 reverts to the 1960s and 70s by reassigning risk back to those who are given no choice, this time the taxpayers instead of the customers.

The most important nuclear provisions of EPACT 2005 offered three types of support.[37] First, a limited number of new nuclear power plants can receive a

$18/MWh production tax credit for up to $125 million per 1000MW (or about 80 percent of what the plant could earn if it ran 100 percent of the time). The second benefit is a provision for federal loan guarantees covering up to 80 percent of the debt involved in the project (not the total cost). The third benefit provides up to $500 million in risk insurance for the first two units and $250 million for units 3-6. This insurance is to be paid if delays that are not the fault of the licensee slow the licensing process of the plant.[38]

By 2007, it was clear that the loan guarantees were not sufficient to reassure financiers. In April 2007 the U.S. Nuclear Energy Institute (NEI), the trade body for the nuclear industry, in a meeting with the U.S. Office of Management and Budget (OMB) lobbied for 100 percent debt coverage for up to 80 percent of the project cost. Subsequently, DoE proposed 90 percent of debt coverage by loan guarantees up to a maximum of 80 percent of total project cost, but this still did not satisfy the nuclear industry, which wanted guarantees for 100 percent of the debt. In August 2007, the OMB appeared to allow DoE the discretion to guarantee 100 percent of the debt.[39] In addition, it emerged that a provision in an Energy Bill passed by the Senate allowed a provision for up to $50 billion in loan guarantees for new nuclear power plants.[40] If we assume that a nuclear plant would cost $4 billion and that guarantees would apply up to the maximum 80 percent of project cost allowed, this would provide guarantees for at least 15 units.

**Analysis of the Program.**

The publicly-stated basis for the Finnish and UK nuclear programs was that nuclear orders did not need subsidies and guarantees, albeit the reality was very different for Finland. However, the basis of the U.S. program was that subsidies and guarantees for about four projects would be enough to kick-start ordering. The changes made in 2007 to the provisions mean that the support is much more extensive and open-ended than originally planned.

The provisions in Finland and the United States provide a good indicator of where UK companies wanting to build nuclear power plants will look for support. The largest elements are the loan guarantees and the market support. Comprehensive guarantees for the loans are vital because, as with Olkiluoto, this will dramatically reduce the cost of capital by shifting risk to taxpayers. Especially in regions where some form of wholesale electricity market exists, some form of price guarantee is necessary so that the nuclear plant is not exposed to the uncertainties of the market. The provisions on insuring against regulatory delays are also important, but their cost is significantly less.

Loan guarantees and regulatory insurance lower the price of nuclear power without lowering its cost, at least not for many years. This reduction occurs because some of the costs and risks are removed from the price charged to customers and onto the shoulders of taxpayers. For example, the production tax credit deprives the U.S. Treasury of funds that must be made up from other sources. Whether the benefit flows through to customers or is retained by investors will vary with the economic regulatory approach used but, either way, prices can be kept lower than would be the case if the credit did not exist. Similarly, the loan guarantees assure lenders that they will be repaid no matter what happens at the power plant. Essentially, their guaranteed loans are converted into government obligations. This lowers both the interest rate and the amount of more expensive equity capital that must be raised, as was the case for Olkiluoto in Finland.

Taken together and combined with other benefits recently conferred on the U.S. industry (such as the 20-year extension of the law limiting nuclear power plant exposure to liability for the costs of a serious accident), the benefits in the recent U.S. law have substantially increased the likelihood of a new U.S. nuclear power plant order in the next few years. Indeed, the incentives are structured to provide maximum benefit to plants ordered before the end of 2008.

During a conference in 2006, three U.S. electric utility CEOs made it clear that without the 2005 congressional action there was no possibility of nuclear orders, but even the extensive support now envisaged might not be sufficient to ensure new nuclear orders:

> [TXU CEO John Wilder] said there were now projects totaling about 26 gigawatts lining up for limited federal incentives, which could provide 'anywhere from a $2 per megawatt-hour advantage to a $20 per megawatt-hour advantage.' He said he didn't believe it would be known which companies would receive those benefits until about 2012. 'Quite frankly, that's all the difference between these projects working or not working.'[41]

NRG Energy President/CEO David Crane, also speaking on a September 26, 2006, conference panel with Wilder, said the measures in the Energy Policy Act of 2005 were key to his company's decision to pursue potential construction at South Texas Project. "I do think those are absolutely necessary to get nuclear plants under way," he said. "In fact, until I actually knew what they were, we would not have even contemplated it."[42]

Exelon Nuclear's President, Christopher Crane, said that the incentives were a key factor in his company's decision to prepare a COL. But other factors would influence whether Exelon commits to building a new reactor.[43]

## Can Use Of Financial Instruments Overcome the Problem of Risk?

Some commentators have suggested that the issues of economic risk can be dealt with by innovative use of insurance and financial instruments. David Newbery claimed that using these instruments, nuclear power plants could be built in the UK without use of government subsidies or other forms of government support.[44] Newbery's claim was based on the assumption that the main risk was market risk. Specifically identified are three risks: (1) with large amounts of intermittent renewables being built, at windy times, the energy spot price would occasionally crash; (2) the carbon price, set in the European Union (EU) Energy Trading Scheme (ETS) was uncertain; and, (3) in the future, the spot price of gas, which has a close relationship with the spot electricity price, was likely to be much less stable than it has been due to geo-political reasons.

Newbery proposed to deal with this risk by issuing bonds to small consumers so that the amount they paid for a specified amount of electricity was fixed. He gives an example under which a consumer would purchase a bond for £9 to buy 100kWh of electricity. If the retail price is higher than this, consumers would receive a larger dividend and, if it was lower, the dividend would also be lower, but consumers would have lower electricity bills. Given that a large proportion of small consumers do not understand how to switch electricity suppliers, much less understand the details of financial bonds, it seems highly unlikely consumers would see it as worthwhile to buy these bonds.

However, the main problem with Newbery's proposal is that he does not understand where the main issue of economic risk with nuclear power plants lies. Newbery says "suppose that construction, operating and regulatory risk can be insured, leaving only market price risk."[45] Why does Newbery assume that it will be cheaper to cover this risk through insurance rather than for it to be reflected in a high cost of capital? Insurers have access to the same information as financiers, and there is no reason to assume they will assess the risk differently.

Newbery assumes that any additional costs from the regulatory risk would be guaranteed by the government (taxpayers). This would be a subsidy and probably a rather large one. However, this figure is dwarfed by the risks arising from construction and operation. Olkiluoto, the Finnish nuclear power plant now under construction, was supposed to be the show-case for new nuclear technologies, but it is now 60 percent over budget (€1.5 billion) and 2.5 years late, with ample probability of additional cost and time overruns. If we assume the value of the output of a nuclear plant is €50/MWh, then the annual value of the output of a plant like Olkiluoto would be about €600 million, if it was reliable (achieving 90 percent of its maximum feasible output over the year). A nuclear company that cannot fulfill its contracts because the completion of the plant is late will have to buy replacement power from the market at the highest prices on offer. The delay of 2.5 years would result in losses of at least €1.5 billion from the energy not produced. An insurer that had covered Olkiluoto would therefore have to pay out €3 billion for cost and time overruns. What level of premium would be needed for an insurer to be willing to cover such a risk?

However, once the plant is complete, the technical risk does not end. Nuclear plants are not always reliable and, if we look at the four most recently completed plants in France, they averaged an availability of 45 percent in their first 4 years of operation. So if an insurer had insured these plants to operate at 90 percent availability, they would have had to pay out somewhere in the order of €4 billion, if we assume the replacement power could be bought at only average market price.

The reality is that using financial instruments cannot make risk disappear. Ultimately, the cost of bearing that risk has to be paid for and, in this case, it will be the public that pays for it, either taxpayers or electricity consumers.

## A UK PROGRAM

### Corporate Strategies.

On the basis of experience in Finland and the United States, it seems implausible that a nuclear power program can be launched in Britain without the support of public subsidies and guarantees. British Energy's financial collapse of 2002 probably means it is not plausible for it to pursue an application to build new nuclear power stations independently, although the sites it already owns mean that anyone hoping to build new nuclear capacity in the UK will probably have to involve British Energy. All of the six main UK electricity companies have expressed interest in par-

317

ticipating in plans to build new nuclear plants. EDF is usually seen as the most aggressive advocate of new nuclear capacity and has plans to build at least four new nuclear power plants (of the EPR design) in the UK. E.ON, also an experienced nuclear operator, is potentially an owner-operator, but has not yet specified the extent of its ambitions. RWE, like E.ON a German-based company with significant nuclear experience, has also stated its intention to invest in new nuclear capacity as an owner-operator. Centrica has said that it hopes to invest more than £3 billion (equivalent to one new unit) in new nuclear capacity in collaboration with other companies.[46] Scottish Power has not made a strong commitment to participating in new nuclear build, but Iberdrola, its Spanish owner, was reported to be in talks with British Energy in January about building a 1,600MW plant in the UK. In January 2008, EDF acknowledged that it was considering launching a takeover bid for Iberdrola and hence Scottish Power. Scottish power and Southern Energy have also held talks with British Energy about participating in new nuclear capacity.

This apparently united front in favor of nuclear seems hard to explain, given the implausibility of orders without subsidy and the government's apparently firm commitment not to provide subsidies and guarantees. However, while the companies are unwilling to use the words subsidies and guarantees, this appears to be due to a rather questionable view of what represents a subsidy or guarantee. For example, even the most aggressive of UK nuclear utilities, EDF, emphasized the need for some support: Plants could be built without subsidy "provided that there was agreement on the funding of decommissioning and waste disposal, a clear licensing and consent road map, and a credible carbon price."[47]

Unless the UK government is very naïve about the attractiveness of nuclear investment, or it does not actually expect any nuclear orders to be placed, there must be suspicions that the government and the companies are indulging in semantic distinctions about what constitutes a guarantee or a subsidy. The government expects it will take 7 years to pilot one or more designs through the expensive and time-consuming process of obtaining safety approval. This will be a major challenge for the nuclear safety body, the Nuclear Installations Inspectorate (NII), which is already understaffed and struggling to replace its aging workforce. There must be strong suspicions that, if after this effort no orders are forthcoming, the government of the day, by then with significant distance from today's government, will be tempted to introduce guarantees and subsidies. This will avoid the embarrassment of a UK government yet again diverting resources away from other energy options to a fruitless nuclear program.

On the face of it, utilities would seem to have no interest in building uneconomic facilities. However, for such utilities, nuclear orders would only be placed if there were clear provisions taking the plant out of the market. So, the more nuclear capacity a company owned, the less exposed to the market it would be. Companies cannot be held to statements by today's executives, so playing along with the government today simply puts them at the head of the queue for any subsidies that are made available. If the subsidies do not materialize or they are inadequate, the company can simply step out of the queue at no cost.

**Subsidies and Guarantees.**

Experience from Finland and the United States shows where these might be required.

## Decommissioning and Waste Disposal Cost.

As argued above, if decommissioning and waste disposal costs are accurately estimated from the start of operation, the delay from close of plant to completion of decommissioning and waste disposal is accurately forecast, provisions are invested securely and the rate of return the provisions can make is also accurately estimated, making provisions for decommissioning and waste disposal should not have a major impact on nuclear economics. Decommissioning and waste disposal take place so far in the future, the cost is effectively "discounted" away. However, if during the life of the plant, it emerges that the decommissioning and waste disposal costs have been underestimated, the provisions are lost or the return is less than expected, making up the additional money could be a major burden to the owner. Given the limited experience of decommissioning and waste disposal, and the rapid rise in decommissioning estimates, companies are likely, as noted by Vincent de Rivaz, to seek some cap on the contribution they have to make to pay for decommissioning and waste disposal.

The UK government quickly acceded to this pressure and, in a consultation published in February 2008, is offering to guarantee owners of nuclear power plants a fixed cost for decommissioning and waste disposal (intermediate- and high-level). However, the government is still claiming that subsidies and guarantees are not being offered:

> The Energy Bill and the guidance published today make clear that companies are liable by law to meet their full costs. 'Let me be clear—full means full.

Funds will be sufficient, secure and independent, it will be a criminal offence not to comply with the approved arrangements and we are taking powers to guard against unforeseen shortfalls.'[48]

It is not clear that such guarantees would have been needed, given that such comprehensive guarantees were not required in Finland and are not being discussed in the United States. Clearly, the companies will gratefully accept any additional guarantees they are offered but, given how far away these costs are, it seems unlikely that financiers would see them as a major risk.

In the United States, the government has taken title to spent fuel since 1978 and levies a fixed charge on utilities of only 0.1c/kWh for disposal of spent fuel. There is no "intermediate-level" U.S. category, and all waste that is not high-level is categorized as low-level. No cost guarantees exist for U.S. low-level waste.

Decommissioning funds are also not guaranteed in the United States. Costs estimates must be continually updated and, if a shortfall is anticipated, either because costs have escalated or the fund has not earned as much interest as expected, contributions must be increased.

### Construction Costs and Loan Guarantees.

The key to the Finnish order was the availability of a turnkey contract that seemed to place the risk of cost and time overruns on the vendor rather than the buyer. The UK would be a prestigious prize for any nuclear vendor, but at present it seems highly unlikely that any vendor could take the risk of offering any more than one unit on turnkey terms and probably then only if subsequent orders were committed and on less stringent terms to the vendor. Both the U.S. and Finnish programs have been based on loan guarantees paid for by the public, albeit in the Finnish case, the French and Swedish public. In addition, in some U.S. states, wholesale competition is being reined in and nuclear plants may be built under the traditional model of making them part of a regulated rate base. Under this, the company owning the plant would be guaranteed a fair rate of return on its investment. Publicly funded loan guarantees would appear to be essential if loans are to be offered at reasonable rates of interest unless nuclear plants are completely removed from the market.

**Market Guarantees.**

For the U.S. program, huge production tax credits are being offered that mean there is a high chance costs will be covered. For Olkiluoto, the plant's output is covered by an effective life-of-plant power purchase agreement at full cost recovery terms. Market guarantees would be likely to violate EU unfair state aids legislation, so some creative thinking, like a high guaranteed carbon price might be used to effectively provide support.

**Operating Costs and Reliability.**

The Finnish nuclear industry has always had a good record of reliably operating nuclear power plants, and the U.S. industry has turned around a very poor record of reliability over the past 15 years so that U.S. plants are now among the most reliable in the world. So both countries have a good track record of operation. However, while the UK nuclear industry has improved its performance since 1990, the reliability of its plants is probably worse than that of any other developed country. Whether investors would assume that the poor British record was not relevant, given that the operating companies would probably be French or German, remains to be seen. No vendor would guarantee the operating cost of a plant it sold nor would insurance coverage be available, so this is a risk it would be hard to assume.

### Regulatory Delays.

The Finnish regulator has been blamed, not necessarily justifiably, for some of the delays at Olkiluoto, and the U.S. program offers some insurance coverage against delays resulting from the regulatory process. Coverage might therefore be needed for UK plants, as envisaged by the EDF UK CEO, Vincent de Rivaz.

### Other Issues.

There are a number of other costs attached to building any new power plant that could be the subject of requests for subsidy. These include:

- Cost of connection to the transmission network. Particularly if the plant is built on a new site, or if it replaces a much smaller unit, there could be significant transmission reinforcement costs. National Grid Transco estimated that if all existing nuclear power stations were to be replaced, the cost of reinforcements to the transmission network would be £1.4 billion.[49]
- Spinning reserve costs. New nuclear power plants, especially if the EPR was chosen would represent the largest units in the system, up to

1,700MW. Spinning reserve is the amount of plant that must be kept in readiness for operation in case of the failure of the largest unit. PB Power noted that the current UK system is designed to allow the failure of two 660MW units. This was a standard derived in the 1970s when 660MW units were the largest units on the UK system. PB Power estimated that if an EPR was built with output of 1,580MW, an additional 260MW of spinning reserve would be needed at a cost, if supplied by a gas-fired plant of £1.3/MWh, or £2.1/MWh if supplied by a coal-fired plant. The EPR design is now likely to have a rating of about 1,700MW, so this cost may be an underestimate if 400MW of additional spinning reserve was needed. Of the other potential designs, the AP-1000 and the ACR-1000 would not need additional spinning reserve, while the ESBWR (1,520MW) would require about 200MW additional reserve.

## CONCLUSIONS

Politically, it seems that subsidies and guarantees are an anathema to a significant proportion of Members of Parliament. So if a new nuclear program is to go ahead in the UK, it has to be on the basis that no subsidies and guarantees will be given.

However, given the time-scale for new orders, which does not anticipate any orders being placed for 7 years or more, it is doubtful whether today's commitments from companies and government are worth anything. Energy market circumstances will change continuously for the next 7 years, and a commercial company operating in a competitive market will be able to claim that their commitment not to need subsidies had been overtaken by changes in energy markets. Equally, in 7 years the government will have little connection to today's government and will not feel bound by today's commitments. The suspicion must therefore be that statements by government and companies are only possible because those involved know they will not have to deliver on these commitments.

## ENDNOTES – CHAPTER 7

1. "Blair to push for new wave of nuclear construction in UK," Nucleonics Week, May 18, 2006.

2. Former UK Energy Minister Malcolm Wicks, "The Government's Energy Review," Testimony before the UK House of Commons, Trade and Industry Committee, London, UK, October 10, 2006, available from *www.publications.parliament.uk/pa/cm200506/cmselect/cmtrdind/uc1123-vii/uc112302.htm.*

3. Greenpeace Press Release, "Government's Nuclear Plans Declared Unlawful by High Court," February 15, 2007, available from *www.greenpeace.org.uk/media/press-releases/governments-nuclear-plans-declared-unlawful-by-high-court.*

4. G. Brown, Comments to parliament on July 4, 2007, available from *www.publications.parliament.uk/pa/cm200607/cmhansro/cmo70704/debtext/70704-0003.htm.*

5. UK Department for Business Enterprise and Regulatory Reform, "Meeting the Energy Challenge: A White Paper on Nuclear Power," Cm 7296, HMSO, 2008, p. 8.

6. *Ibid.*, p. 10.

7. "Going Nuclear," *Utility Week*, February 1, 2008.

8. E.ON Press Release, "E.ON welcomes new nuclear to UK power mix," E.ON Press Release, January 10, 2008; available from *www.pressreleases.eon-uk.com/blogs/eonukpressreleases/archive/2008/01/10/1165.aspx.*

9. "Loan Guarantees Tagged As Key For Nuclear Builds," Power, Finance and Risk, December 21, 2007.

10. UK Department of Trade and Industry, "The Energy Challenge: Energy Review Report,"' Cm 6887, HMSO, 2006, p. 113.

11. Former U.K. Energy Minister, Malcolm Wicks, "The Government's Energy Review."

12. *Ibid*.

13. Department for Business, Enterprise and Regulatory Reform, "Consultation on Funded Decommissioning Programme Guidance for New Nuclear Power Stations," February 22, 2008; available from *www.berr.gov.uk/files/file44486.pdf*.

14. *Ibid*.

15. *Ibid*., p. 16.

16. *Ibid*., p. 15.

17. *Ibid*., p. 104.

18. *Ibid*., p. 16.

19. M. Wald, "Interest in Reactors Builds, But Industry Is Still Cautious," *New York Times*, May 2, 2005.

20. "An analysis of Nuclear Power Plant Construction Costs," Department of Energy, Energy Information Administration, DOE/EIA-0411, 1986.

21. "Merrill Lynch Global Power and Gas Leaders Conference," *Nucleonics Week*, October 5, 2006.

22. *Ibid*.

23. David Newbery, "Reduce the risk of nuclear investment," *Financial Times*, January 9, 2008, p. 15.

24. *Ibid.*

25. Department for Business, Enterprise and Regulatory Reform Press release, "Clean Up Fund is Precondition For New Nuclear-Hutton," February 22, 2008, available from *www.gnn.gov.uk/ environment/fullDetail.asp?ReleaseID=354629&NewsAreaID=2&Na vigatedFromDepartment=True.*

26. "Environmentally harmful support measures in EU Member States," CE Solutions for Environment, Economy and Technology, Report for DG of the European Commission, January 2003, p. 132.

27. J. Wakeham, "House of Commons Debates. HC Debates," 1988/89, Vol. 159, November 9, 1989. "Areva-Siemens cannot accept 100 percent compensation responsibility, because the project is one of vast co-operation. The building site is joint, so we absolutely deny 100 percent."

28. Nuclear Power Progress; Site Work Underway on Finland's 1,600MWe EPR," *Modern Power Systems*, March 2004.

29. For a detailed review of the problems up to March 2007, see S. Thomas, P. Bradford, A. Froggatt, and D. Milborrow, "The Economics of Nuclear Power," Greenpeace International, 2007, available from *www.greenpeace.org/international/press/reports/the-economics-of-nuclear-power.*

30. Finnish Broadcasting Company TV News, January 30, 2007.

31. "Areva's nuclear delay threatens China contract," *Financial Times*, August 11, 2007, p. 19; and "Areva-Siemens consortium announces delay of Finnish nuclear reactor," *Datamonitor Newswire,* August 13, 2007.

32. "Areva to take 500-700 mln eur provisions for new Finnish reactor delay — report," *Thompson Financial News*, August 13, 2007.

33. S. Thomas, "The realities of nuclear power," Cambridge, UK: Cambridge University Press, 1988.

34. "Areva's nuclear delay threatens China contract," p. 19.

35. "Nuclear Power 2010," Department of Energy, available from *www.nuclear.energy.gov/np2010/activities.html*.

36. In December 2007, Constellation withdrew to pursue its own technology choice and was replaced by DTE Energy.

37. L. Parker and M. Holt, "Nuclear Power: Outlook for New U.S. Reactors' CRS Report for Congress," Order Code RL33442," 2007, available from *fas.org/sgp/crs/misc/RL33442.pdf*.

38. All three measures require implementing regulations, and the loan guarantees require an appropriation. So the actual scope and benefit of the subsidy is unclear.

39. "DOE Loan Proposal Seen As Likely Failure By Industry, Wall Street," *Energy Washington Week*, August 1, 2007.

40. "Senate bill could help finance nuclear plants," *International Herald Tribune*, August 1, 2007, p. 1.

41. "Merrill Lynch global power and gas leaders conference," *Nucleonics Week*, October 5, 2006.

42. *Ibid.*

43. *Ibid.*

44. David Newbery, "Reduce the risk of nuclear investment," *Financial Times*, January 9, 2008, p. 15.

45. *Ibid.*

46. "Centrica to invest GBP 3bn in nuclear," *The Express*, February 22, 2008.

47. *Ibid.*

48. Department for Business, Enterprise and Regulatory Reform Press release, "Clean Up Fund is Precondition For New Nuclear — Hutton," February 22, 2008, available from *www.gnn.gov*.

*uk/environment/fullDetail.asp?ReleaseID=354629&NewsAreaID=2&*
*NavigatedFromDepartment=True.*

49. "Grid 'will pay £1.4bn extra' for N-stations Transmission group must upgrade to cope with planned power plants," *Daily Telegraph,* July 13, 2006, p. 1.

# APPENDIX I

## DISCOUNTING, COST OF CAPITAL, AND REQUIRED RATE OF RETURN

A particularly difficult issue with nuclear economics is dealing with and putting on a common basis for comparison the streams of income and expenditure at different times in the life of a nuclear power plant. Under UK plans, the time from placing a reactor order to completion of decommissioning could span more than 200 years.

Conventionally, streams of income and expenditure incurring at different times are compared using discounted cash flow (DCF) methods. These are based on the intuitively reasonable proposition that income or expenditure incurred now should be weighted more heavily than income or expenditure earned in the future. For example, a liability that has to be discharged now will cost the full amount, but one that must be discharged in, say, 10 years can be met by investing a smaller sum and allowing the interest earned to make up the additional sum required. In a DCF analysis, all incomes and expenditures through time are brought to a common basis by "discounting." If an income of $100 is received in 1 year's time and the discount rate is 5 percent, the net present value of that income is $95.23—a sum of $95.23 would earn $4.77 in 1 year to make a total of $100. The discount rate is usually seen as the "opportunity cost" of the money, in other words, the rate of return (net of inflation) that would be earned if the sum of money was invested in an alternative use.

While this seems a reasonable process over periods of a decade or so and with relatively low discount rates, over long periods with high discount rates, the results of discounting can be very powerful, and the assumptions that are being made must be thought through. For example, if the discount rate is 15 percent, a cost incurred in 10 years of $100 would have a net present value of only $12.28. A cost incurred in 100 years, even if the discount rate was only 3 percent, would have a net present value of only $5.20, while at a discount rate of 15 percent, costs or benefits more than 15 years forward have a negligible value in an normal economic analysis (see Table 7-1).

| Discounting period (years) | 3% | 15% |
|---|---|---|
| 5 | 0.86 | 0.50 |
| 10 | 0.74 | 0.25 |
| 15 | 0.64 | 0.12 |
| 20 | 0.55 | 0.061 |
| 30 | 0.41 | 0.015 |
| 50 | 0.23 | 0.00092 |
| 10 | 0.052 | - |
| 150 | 0.012 | - |

Source: Author's calculations.

### Table 7-1. Impact of Discounting: Net Present Values.

If we apply this to nuclear plants operating in a competitive market where the cost of capital will be very high, this means that costs and benefits arising more than, say 10 years in the future, will have little weight in an evaluation of the economics of a nuclear power plant. Thus increasing the life of a plant from 30 years to 60 years will have little benefit, while refurbishment costs incurred after, say, 15 years will equally have little impact.

For decommissioning, for which under UK plans the most expensive stage, is not expected to be started until 135 years after plant closure, this means very large decommissioning costs will have little impact even with a very low discount rate consistent with investing funds in a very secure place with a low rate of return, such as 3 percent. If we assume a magnox plant will cost about $1.8 billion to decommission and the final stage accounts for 65 percent of the total (undiscounted) cost ($1.17 billion), a sum of only $28 million invested when the plant is closed will have grown sufficiently to pay for the final stage of decommissioning.

The implicit assumption with DCF methods is that the rate of return specified will be available for the entire period. Give that even government bonds, usually seen as the most secure form of investment, are only available for 30 years forward and that a period of 100 years of sustained economic growth is unprecedented in human history, this assumption seems difficult to justify. So, with nuclear power, there is the apparent paradox that, at the investment stage, a very high discount rate (or required rate of return) of 15 percent or more is likely to be applied to determine whether the investment will be profitable, while for decommissioning funds, a very low discount rate is applied to determine how much decommissioning funds can be expected to grow.

The key element resolving this paradox is risk. Nuclear power plant investment has always been risky because of the difficulty of controlling construction costs, the variability of performance, the risk of the impact of external events on operation, and the fact that many processes are yet to be fully proven (such as disposal of high level waste and decommissioning). In a competitive environment, there are additional risks because of the rigidity of the cost structure. Most of the costs will be incurred whether or not the plant is operated. Thus while nuclear plants will do well when the wholesale price is high (as was the case with British Energy from 1996-99), they will do poorly when the wholesale price is low (2000-02). The fact that the plant has made good profits for a decade will not protect it from bankruptcy in the bad years, and financiers will therefore see investment in nuclear power as extremely risky and will apply a very high interest rate reflecting the risk that the money loaned could easily be lost.

# CHAPTER 8

# A CASE STUDY OF SUBSIDIES TO CALVERT CLIFFS

## Doug Koplow

## OVERVIEW

Sharply rising energy prices in 2007 and the first part of 2008, growing concerns over climate change, and geopolitical instability in major fossil fuel producing regions of the world have focused increasing attention on energy security and supply diversification. The nuclear industry was well-positioned to enter this fray. Capacity factors at existing reactors have been slowly climbing. A series of massive capital write-downs at these reactors over the past 2 decades meant that much of the cost to build the facilities had already been dumped onto taxpayers and ratepayers. Industry boosters have highlighted low operating costs only, as if capital costs do not exist. Finally, nuclear's role as a baseload generating source with relatively low carbon emissions has been transformed by well-funded and well-staffed industry trade associations into claims that their resource was the only viable "carbon-free" resource available to meet our growing energy demand.

Many countries, including the United States, have bought these arguments virtually whole cloth. Despite cost projections running into the hundreds of billions of dollars for the nuclear solution, the hope of a clean, domestic, low carbon nuclear future has been subjected to little critical review. This is unfortunate. While we do face very real energy security and climate change challenges, transforming our economy will require thousands of small actions and a heightened level of market transparency and accountability.

The economics of nuclear power are far from transparent. The technology is riddled with complex public subsidies to new reactors that are both opaque and quite difficult to value. Industry sound bites mask key information so that public subsidies for the sector will likely exceed the private capital put at risk, hardly a formula for sound financial decisionmaking. These taxpayer "investments" are really highly concentrated, politically-targeted bets on a narrow set of technologies and management teams.

Choosing who to subsidize with billions in public largesse does not encourage the rational, technical evaluations needed to maximize success rates. Instead, the recipients of this support are at least as likely to be determined based on their political connections and the sophistication of their lobbying as they are on the large scale market viability of their approach.

A case study of the proposed new reactor at Calvert Cliffs in Lusby, Maryland, provides a useful window into the dynamics and implications of federal nuclear policy today. The analysis demonstrates not only that the taxpayer ends up as the largest de facto investor in this project, but also that, while we bear most of the downside risk, we share little of the upside benefits should the plant ultimately be successful. The data also highlight that despite nuclear's relatively low carbon footprint, the cost per unit of greenhouse gas avoided is far more expensive than many other alternatives.

This chapter begins with some historical context on the role of government subsidies for nuclear power in the United States. It then shifts to the specific case of Calvert Cliffs, including the venture structure, projected costs, and acknowledged or embedded subsidies. The final sections of the chapter evaluate the cost-efficiency of a nuclear power option to address energy security and global warming concerns.

## NUCLEAR VIABILITY: RELIANT ON SUBSIDIES FOR MORE THAN A HALF-CENTURY

Despite industry efforts to frame nuclear energy as the cheapest option, the reality is that nuclear power's very survival has required large and continuous government support. The industry routinely argues that subsidies are transitional, needed only for a short time to gain operational experience with new reactor designs. After these "first of a kind" costs have been amortized, the argument goes, the industry will be self-reliant.

All sorts of industries are challenged by the need to invest in continuous technical improvements to remain competitive. Unlike most industries that rely on private capital for this need, the nuclear power sector has been making the transitional support argument since the earliest civilian reactors. A 1954 advertisement from the General Electric civilian reactor program notes this clearly:

> We already know the kinds of plants which will be feasible, how they will operate, and we can estimate what their expenses will be. In 5 years—certainly within 10—a number of them will be operating at about the same cost as those using coal. They will be privately financed, built without government subsidy.[1]

Clearly, 5 or 10 years were not enough. If fact, more than 50 years later, almost identical claims are still being made by the industry. Yet, in the intervening half-century of "transitional" support, the federal government has provided a growing array of subsidies to bolster nearly every step in the nuclear fuel cycle. Some of these programs have fed the industry for virtually its entire existence.

Of greatest importance to nuclear viability have been the subsidies that effectively socialize the most intractable risks of nuclear energy: damages from accidents (capped via the Price-Anderson Act first passed in 1957) and management of extremely long-lived radioactive wastes (where the federal government has guaranteed ultimate responsibility for management in return for a small variable surcharge per unit of power sold).

Uranium enrichment services are another example, because the complexity and scale of operations early in the industry's evolution would have made them cost-prohibitive. In the United States, these facilities were historically government-owned and remain so in a number of other countries. U.S. enrichment operations were privatized in 1998, though not before providing decades of large subsidies to civilian reactor customers. The U.S. Enrichment Corporation (USEC), as the privatized organization is known, inherited key assets of its public predecessor, while leaving the cleanup of contaminated sites a taxpayer liability.

Not every energy technology has these types of impairments. As a result, the more the federal government does to shift the costs and risks of dealing with these issues away from investors, the more harm is done to the competitive position of alternative energy resources.

Subsidies for capital formation have also been extremely important to nuclear energy, because the resource is perhaps the most sensitive of all energy technologies to the cost of capital. Large, complex plants that take many years to build carry inherent risks of significant shifts in market conditions before the plants come on line. Their technical rigidity precludes mid-course corrections (other than delay or abandonment), yet their scale requires high capacity utilization for them to be efficient. Invested funds can be tied up for years, accruing substantial financing costs as well. Finally, the need to pre-sell power via advance power contracts, while mitigating the market risk upon completion, also opens the facility to large financial obligations to meet these contracts via power purchases if the start of operations of the nuclear plant is delayed.

A common theme in government support for the sector has been to bring down capital and financing costs, either through direct subsidies (accelerated depreciation and various tax credits) or by shifting risks to rate payers (such as by including project and interest costs in a regulated utility rate base during the period of construction). These interventions are sold as being low- or no-cost to the government. The idea that providing large amounts of credits and guarantees is somehow costless to the provider is pure fantasy, as the recent financial meltdown so clearly illustrates.

Other subsidies involved support for uranium mining and stockpiling; a half-century of government-financed research and development into reactor technologies, waste management and cleanup, and enrichment technologies; and special tax breaks for plant decommissioning.

Although there is no comprehensive record of historical subsidies for nuclear power since inception, a review of a number of studies that have been done over the years demonstrates government's central role in the sector's market viability. Table 8-1 illustrates that subsidies were generally equal to one-third or more of the value of the power produced. While such levels of support may not be surprising for very new industries with little installed base, to see subsidy levels so high over the course of 5 decades is quite striking.

| Period of Analysis | Federal Subsidy, $Billions | | Subsidy, cents/ kWh | | Avg Subsidy as % of Industrial Price | Analysis | Notes |
|---|---|---|---|---|---|---|---|
| | Low | High | Low | High | | | |
| 2008 | - | - | 5.0 | 8.3 | 113-189% | Koplow/Earth Track calcula- tions - subsidies to a new reactor | Share of national average wholesale rates, 2002-06 |
| 1947-99 | 178.0 | - | 1.5 | - | NA | Goldberg/Re- newable Energy Porfolio Project (2000) | P-A not estimated. |
| 1968-90 | 122.3 | - | 2.3 | - | 33% | Komanoff/ Greenpeace (1992) | P-A not estimated. |
| 1950-90 | 142.4 | - | 2.6 | - | NA | Komanoff/ Greenpeace (1992) | |
| 1989 | 7.6 | 16.2 | 1.4 | 3.1 | 32% | Koplow/Alliance to Save Energy (1993) | |
| 1985 | 26.8 | - | 7.0 | - | 83% | Heede, Morgan, Ridley/Center for Renewable Resources (1985) | P-A not estimated. |
| 1981 | - | - | 5.9 | 12.3 | 105% | Chapman et al./ US EPA (1981) | Tax expenditures only. |
| 1950-79 | - | - | 4.1 | 6.0 | NA | Bowring/En- ergy Information Administration (1980) | Tax and credit subsidies not estimated. |

Source: Koplow (2009).

## Table 8-1. Subsidizing Plant Construction and Operation (In 2007 Dollars).

# VENTURE OVERVIEW OF CALVERT CLIFFS 3

Calvert Cliffs, located in Lusby, MD, already serves as host to two existing nuclear reactors with a total capacity of 1,700 megawatts (MW). These units came online in 1975 and 1977.[2] The reactors are owned by Constellation Energy Group (CEG), a holding company formed in 1999 from the holdings of the Baltimore Gas and Electric Company.

No new nuclear reactors have been built in the United States for decades. Although the industry likes to blame regulatory bureaucracy for the problem, others point out that the majority of reactors were cancelled after license approval on economic grounds.[3] It is clear, however, that constructing a new reactor is a far more complicated financial undertaking than buying and operating an existing one.

The corporate structure set up to build new reactors at Calvert Cliffs provides important insights into political and economic strategies Constellation is using to manage risk and boost returns to shareholders. These are important complementary strategies for obtaining very large government subsidies. Constellation and other nuclear firms face successive challenges. After fighting a vigorous political battle to create a new wave of large subsidies that shift risk of new construction away from investors, the firms must now manage the deployment of those subsidies to ensure they support their specific projects. With many of the most lucrative subsidies time- or capacity-limited, Constellation must work to extend expiring policies, and to capture available subsidies instead of having them flow to rivals.

## CORPORATE STRUCTURE

The new reactors (Constellation discusses just Calvert Cliffs 3 at the site, but it is clear their plans include a number of additional reactors around the country) are to be developed and built by a new corporate joint venture. Though complicated, getting a picture of the corporate structure is important for providing context to the new reactor plan. Three significant findings are evident. First, the firm has adopted a joint venture approach to building new reactors in order to spread risks. This is a logical structure, one that has been adopted by all of the new-build nuclear projects underway. Second, the corporate structure remains in flux, having already been through a series of important modifications despite the young age of the venture. These shifts are likely to continue in response to significant changes in market conditions or public policy circumstances. Third, the growing role of foreign governments in the U.S. nuclear rennaissance can be seen clearly through the evolution in Constellation's deal structure. This involvement certainly weakens claims that nuclear power boosts domestic energy security.

Also of note is the highly compartmentalized corporate structure adopted for this venture. This compartmentalization may give Constellation greater flexibility to modify parts of their venture as conditions change. A more important goal, however, is probably to control financial and operating risks by isolating the parent firms from the liabilities associated with the new nuclear venture as much as possible. Though this insulation may be good for Constellation shareholders, it may be very bad from the perspective of the taxpayer or surrounding community — the groups who will suffer if the venture does not go as planned.

The last wave of U.S. reactor construction resulted in massive capital write-offs. Similarly, poor incentive structures within the mortgage and commercial debt were significant factors in the growing losses, and resultant taxpayer bailouts, of financial firms. These examples should underscore how important proper risk management and incentive alignment is with these new-build scenarios. Unfortunately, public policy seems to be moving in the opposite direction with more subsidy programs with complex and opaque rules. Lost in the press to move ahead with new reactors is the fact that proper review and challenge of these programs is most critical at their inception, before taxpayers become contractually obligated to back tens of billions of dollars in new reactor investments.

## NUCLEAR KEY PARTNERS

### UniStar.

The first formulation of Constellation's joint venture was UniStar Nuclear LLC, launched in 2005. This entity was a partnership between Constellation and Areva. Constellation is the largest seller of wholesale and retail electricity, "[l]arger than maybe the next three competitors combined," according to Joe Turnage, a Senior Vice President in Constellation's Generating Group.[4] The involvement of Areva brought in both the French and German governments. Areva NP, the division of Areva slated to produce the Evolutionary Power Reactor (EPR) to be used at Calvert Cliffs was formed in 2001 by the combination of Siemens (roughly 30 percent owned by the German government) and Framatome (owned by the French government). The role of the German government is diminishing as that of France increases. Siemens announced in January 2009 that it would divest its interest in Areva NP, selling its interest to the Areva parent company, Areva S.A.[5] Areva S.A. is approximately 80 percent owned by the French government.[6]

July 2007 brought significant changes to UniStar Nuclear with the formation of a similarly-named new partnership, UniStar Nuclear Energy LLC (UNE). UNE is owned by Constellation Energy and Electricite de France (EdF), and absorbed the earlier partnership. While the French government was already involved with Calvert Cliffs 3 through Areva S.A., EdF is also 85 percent owned by the French government.[7] This investment also gave EdF roughly 9.5 percent of Constellation Energy, UNE's parent. In December of 2008, EdF significantly upped its ownership of Constellation's nuclear venture, with an additional investment of $4.5 billion.[8]

Foreign ownership brings with it some interesting challenges. Calvert Cliffs will be owned and operated by a firm that has substantial involvement by the French government. The provider of critical heavy forgings will also be non-U.S. — either French or Japanese. Enrichment services, as well, are increasingly being supplied by non-U.S. firms — though USEC remains a U.S. competitor.

One obvious challenge with this situation is its legality. Section 103(d) of the Atomic Energy Act states that "No license may be issued to an alien or any corporation or other entity if the Commission knows or has reason to believe it is owned, controlled, or dominated by an alien, a foreign corporation, or a foreign government."[9] An earlier *New York Times* article noted that the purpose of this clause was related to nuclear security, but that relative to U.S. firms "EdF's expertise in power plant construction is far more current."[10] FERC's decision to accept the EdF purchase demonstrates their belief that Section 103(d) does not apply in this circumstance, though subsequent challenge seems likely.[11]

Precedent also matters here. Would the involvement of countries such as China or Russia be subjected to greater constraints and review under Section 103(d) than that given to French involvement? Will the rapid acceptance of French government involvement in Calvert Cliffs 3 make it more difficult to argue national security concerns about foreign ownership in other circumstances?

Another important part of the venture structure has been the use of contractual relationships with key suppliers outside of joint venture partners. These include Bechtel (architect, engineer, and builder for the new plants), Accenture (plant-related information technology systems), and Alstom (nuclear turbine generators). Accenture's contractual involvement with the plant is interesting, as the firm recently conducted a global survey of public attitudes to nuclear power that "found that, overall, sentiment has swung in favor of nuclear energy."[12] Poll takers normally do not have a financial stake in the outcome of the polls. (See Figure 8-1.)

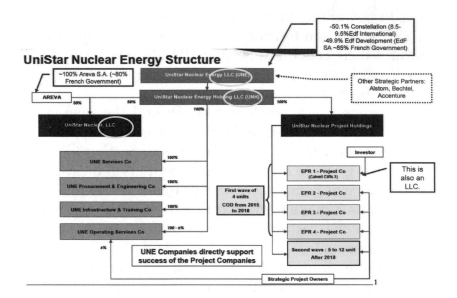

**UniStar Nuclear Energy Structure**

Sources: Basic chart developed by Turnage/Constellation, 2008. Additional material is from Mariotte, 2008; EdF, 2009; Gil/Reuters, 2009; and FERC, 2009.

**Figure 8-1. Unistar Nuclear Energy Structure.**

## Venture Strategy.

UniStar's venture strategy can be discerned in part from corporate statements and publications, and in part from looking at the decisions they have made thus far. The discussion below addresses both their market strategy and a closely related political strategy.

## Market Strategy.

The French model of nuclear plant deployment seems to provide the core framework in UniStar's strategy: market standardization, close integration with the political system, and achievement of economies of scale. Of particular import is:

- First Mover Advantage. UNE has worked to move early as a new U.S. reactor builder in order to secure critical inputs. Constrained material inputs, such as heavy forgings, are central to this effort, because delivery delays can ripple forward to delay a plant opening and can be extremely costly. However, the first mover advantage is perhaps even more important with respect to securing access to key government subsidies such as loan guarantees, construction delay insurance, and production tax credits that are (at least for now) limited to the first handful of reactors. UNE was the first firm to submit combined operating license (COL) paperwork and to establish contracts on key heavy forgings. They also moved quickly with early standardization of their reactor design, based on a model already being deployed in Europe. This strategy seems to be working: Calvert Cliffs 3 has made the Department of Energy's (DoE) short-list of five projects to receive highly lucrative loan guarantees. Two or three of these projects will be funded under current budget authority.[13]

- Economies of Scale. Reactor standardization is an oft-listed success factor in the French nuclear power program, and is being replicated in the UniStar venture. Other elements of achieving economies of scale include:
  - Adopting a single design and licensing process to roll out at multiple sites, including provision of contract licensing for other firms using the same Areva reactor.

— Adopting a single design and licensing process to roll out at multiple sites, including provision of contract licensing for other firms using the same Areva reactor.
— Working with large partner firms with deep pockets and staying power in the market.
— Establishing long-term, stable relationships with a small handful of well connected partner organizations (specifically Alstom, Bechtel, Accenture, and Areva). This allows learning to spread through their broader supply chain. Partner firms are also more willing to incur high initial fixed costs if they are confident they will not be cut out of future developments.

UniStar notes a few related marketplace goals as well. These include achieving a predictable construction and maintenance schedule, streamlined and efficient operations at a high capacity factor, and reduced costs. All of these goals are logical objectives, though it is hard to guess whether they can be met once construction begins. Despite a highly favorable regulatory environment, two current projects for similar Areva reactors in Olkiluoto, Finland, and Flamanville, France, are both way over budget. These delays are indicative of the challenges UNE will likely face in the United States even once they are over licensing hurdles.

**Political Strategy.**

Nuclear power has always relied heavily on political support to make it viable. The new reactors will be no different. Some of UNE's market strategies have ancillary benefits in the political arena. These are supplemented by a variety of direct political initiatives to reshape the political terrain for Calvert Cliffs 3 to one more favorable to the firm.

- Powerful Partners. Members of the UNE venture, as well as the core set of subcontractors, are all large, politically savvy firms with long experience in working with governments to achieve their market goals. Both Areva S.A. and EdF are heavily government owned. The French government also owned more than 26 percent of Alstom through June of 2006, when its stake was sold to Bouygues SA, a firm with close connections with French President Nicholas Sarkozy.[14] Nuclear energy is viewed as a strategic industry of France, so aligning the French government with a new reactor at Calvert Cliffs is likely to have significant financial and political dividends for the project sponsor. The firm hopes the French government will provide a project guarantee for 30 percent of the project cost, with an additional 50 percent guarantee from the U.S. Government. Bechtel is a major player in large U.S. construction projects, well versed in the politics these projects often entail. Finally, reactor partner Areva-NP has used Japan Steel Works to produce heavy forgings for similar reactors it is building in France and Finland,[15] and it will benefit from Japanese government support should a similar supply arrangement be used in the United States.
- Suppressing Political Challenge. Past U.S. reactor construction was often heavily contested in the courts and the government. Delays of any

sort on a large project can be expensive. They are particularly troublesome if substantial investments have already been made on which interest is accruing, or if the delays boost the risk of missing power supply guarantees. Delays can also increase the overall market risks of the projects, since much can change in the demand pattern and pricing for electricity over a span of a couple of years. UniStar has deployed a number of strategies to minimize the likelihood of their business decisions being challenged:

— Collocating new reactors with old ones. Locating new reactors on the same site as old reactors reduces siting battles as well as allowing the new reactor to share some preexisting ancillary infrastructure investment.

— Lobbying. Constellation spent $100,000 in the first half of 2007 "to lobby the federal government on the issue [of loan guarantees], disclosure forms show."[16] Constellation's total spending on lobbying increased six-fold between 2006 and 2008, to nearly $3 million.[17]

— Reduce public oversight of environmental impacts. James Curtiss, a director of Constellation and head of the law firm Winston & Strawn's energy practice, worked with the NRC to change the definition of construction such that, according to an NRC official Andrew Kugler, it would exclude from NRC oversight "probably 90 percent of the true environmental impacts of construction."[18]

— Reduce or eliminate public input to licensing. Intervenors must be granted standing to have their opinions heard in a licensing case. Historically, proximity to a reactor was sufficient since such parties would be harmed in the case of an accident. UniStar has advocated in NRC filings to replace this with a probabilistic assessment of risk based on modeling of the core damage frequency.[19] Intervenors argue that the new standard would rely on modeling by the applicant, and, if upheld, "no intervention—and thus no meaningful public involvement in the NRC's reactor licensing process—would be possible for any reactor design that could claim similar low risks."[20]

- Balance Promotion of Reactor as Both "New and Innovative" and "Tested and Low Risk." UniStar faces a challenge in its reactor designs due to conflicting pressures. To be eligible for the most lucrative federal subsidies, the reactor design must be new and innovative. Yet, investors rationally worry that very new technologies have much greater risks of poor performance and cost over-runs. As a result of trying to meet both of these objectives, UniStar's promotional materials tend to be somewhat schizophrenic, describing the reactor as "advanced" and "state of the art" as well as "evolutionary" and employing "technologies that have been licensed in the United States for more than 40 years."[21]
- Publicize Jobs Creation. All big industrial projects use local job creation as a selling point to garner community support of their project. Calvert Cliffs 3 is no exception. Constellation notes that the project will provide approximately

4,000 jobs during peak construction, and boost permanent jobs within Calvert County by about 360.[22] While some new jobs will be created, the exact numbers are always tough to benchmark. As of August 2006, Constellation was the fourth largest employer in the Tri-county area of Southern Maryland, with 1,143 jobs.[23] Employment levels in 2004 were flagged at a similar 1,140.[24] The County's "Brief Economic Facts, 2006-2007" notes only 800 jobs for Constellation at the Calvert Cliffs Site.[25] The cause of the discrepancy could be measurement error, job shifts, or reduced need for labor—the source does not say. However, it is useful to note that the difference between the two values is almost exactly the number of new permanent jobs the firm says will be created by Calvert Cliffs 3.[26]

## SUBSIDIES ARE CENTRAL TO VIABILITY OF CALVERT CLIFFS 3

Public subsidies have always been a central plank of UniStar's new reactor development program, something the firm has been quite up-front about. Questioning before the California Energy Commission in June of 2007 is a good example:

Associate Member Geesman: "And just to revisit the cap[ital] question again. Your business model is premised on receiving the federal loan guarantee for each of your four projects. Is that correct?"

Dr. Turnage: "That is correct."[27]

Around the same time, Constellation's Co-CEO noted to the *New York Times* that "Without loan guarantees we will not build nuclear reactors."[28] UniStar's President George Vanderheyden notes that "Everywhere else in the world where entities are pursuing advanced new nuclear plants, it is all governments. Only here in the U.S. do we try to make private companies build these plants."[29]

Nuclear power benefits from more than 20 subsidies, most of which are applicable to the Calvert Cliffs 3 project. These programs, listed in Table 8-2, support all key cost elements in the nuclear fuel cycle, from research and development to plant construction and operations, through closure and post-closure issues. The structure and value of some of these subsidies on plant economics are discussed in the subsequent sections. Many of these subsidy values shown are based on UniStar's own estimates.

| | Revelance to Calvert Cliffs 3 | Anticipated Subsidy Magnitude |
|---|---|---|
| **Subsidies to Capital Costs** <br> **Cost of Funds** | | |
| Federal loan guarantees/Clean Energy Bank | Eligible | Very large |
| Advantaged credit, foreign banks | Eligible | Large |
| Ratebasing of work-in-process | Merchant plant; not relevant. | Very large for eligible facilities |
| Regulatory risk delay insurance | Eligible | Medium |
| **Cost of Capital Goods** | | |
| Accelerated depreciation | Automatic | Large |
| Research and development | Pro-rata beneficiary | Low to Medium |
| **Output based subsidies** | | |
| Production tax credit | Eligible | Large |
| Market Price support | | |
| Renewable portfolio standard | Included in OH; under consideration elsewhere, but not yet in MD. | Potentially large for eligible facilities |
| **Subsidies to Operating Costs** <br> **Fuel and Enrichment** | | |
| Cap on liability: fuel cycle, transport, contractors. | Pro-rata beneficiary | Moderate |
| Excess of percentage over cost depletion for uranium | Pro-rata beneficiary | Low |
| HEU dilution programs | Pro-rata beneficiary | Unknown |
| Enrichment D&D: LT funding shortfall | Pro-rata beneficiary | Low |
| Virtually free patenting of federal hardrock mining claims (including uranium) | Pro-rata beneficiary | Low |
| No royalty payments on uranium extracted from federal lands | Pro-rata beneficiary | Low |
| Inadequate bonding for uranium mine sites | Pro-rata beneficiary | Low |
| **Insurance** | | |
| Cap on liability: reactor accidents | Automatic | Large |
| **Regulatory oversight** | | |
| Incomplete recovery of NRC oversight costs. | Pro-rata beneficiary | Low; most costs now covered. |

## Table 8-2. A Compendium of Government Subsidies for Nuclear Power.

| | Revelance to Calvert Cliffs 3 | Anticipated Subsidy Magnitude |
|---|---|---|
| **Taxes** | | |
| Calvert County, MD property tax abatement | Specific to plant | Relatively small |
| Depreciated value rather than as-sessed value as MD tax base | Automatic | Relatively small |
| **Security** | | |
| Low design basis threat for reactors | Plant designed for higher than standard | Unknown |
| Ancillary costs to prevent proliferation | Pro-rata beneficiary | Unknown |
| **Emissions and waste management** | | |
| Windfall $CO_2$ credits from grandfather-ing based on energy output. | Depends on $CO_2$ control regime. | Potentially Large |
| Inadequacy of waste disposal fee - spent fuel | Pro-rata beneficiary | Low-Moderate |
| Payments for late delivery of disposal services | Not relevant since new reactor not covered by old agreement. | Litigation likely to result in vey high federal payments. |
| **Subsidies to Closure/Post-Closure** | | |
| Decommissioning trusts: preferential tax rates, special transfers; under accrual. | Only preferential tax rates would be relevant for a new reactor. | Relatively small |

## Table 8-2. A Compendium of Government Subsidies for Nuclear Power. (cont.)

## FEDERAL LOAN GUARANTEES

Capital markets provide funds to finance new investments. The most common forms of capital are equity and debt. With equity, an investor owns a slice of the firm, and the value of that ownership interest varies with the fortunes of the company. Debt is a contract in which the lender provides cash to a borrower in return for a set of pre-defined payments of the amount lent, plus interest. Because the return to investors through equity (via dividends or a growing value for the shares owned) is not contractually guaranteed, investors normally require a higher return on

355

equity than on debt. For both classes of instruments, the higher the perceived risk of the venture, the higher the rate of return investors will demand.

An important distinction must be made between the risk level of the firm versus the risk level of the project. Firm-level information on the cost of capital is often used as a benchmark for the financing assumptions for a new nuclear power plant. Large coal projects may be used as proxies as well. In both cases, costs are tweaked upward slightly to allow for the greater uncertainty of nuclear. This approach tends to understate the appropriate return targets for the nuclear project because nuclear power is considered a much higher financial risk than either the firm or alternative large power plant proxies.

The perception of greater risk is well placed, derived in large part from the actual historical performance of the industry. Historical cost overruns for the construction of the existing fleet of reactors topped $300 billion (in 2006 dollars); and sunk costs in reactor projects that were abandoned prior to completion added another $40 to $50 billion.[30] Another roughly $100 billion (in 2007 dollars) was deemed uneconomic at the time the electric industry was deregulated and was shifted to ratepayers as "stranded costs."[31]

The historical performance of these investments was, in large part, driven by market characteristics and risks that remain concerns today. The very large scale of reactors, their high fixed costs, and their long construction period create significant investment risks associated with misestimating what the market will look like when the plant construction finally enters production. The financial penalties from being wrong are quite large, as even with good market conditions, the economics of reactors require that they operate at a high capacity utilization to be profitable.

Absent federal intervention, the risk profile of new plants suggests that debt providers would require a high share of equity in the plant. They would also require returns on both debt and equity that would be too high for the energy produced to compete in the marketplace. While the industry views this as a negative outcome, it is actually a core function of capital markets, and quite a useful role for society. By requiring higher returns on higher risk ventures, capital markets provide strong incentives to find smaller scale or more rapidly deployable solutions that pose lower financial and market risks, yet still address the problem (e.g., creating more electricity) in comparable ways.

In this case, however, the federal government has on offer large loan guarantees. For eligible nuclear reactors or enrichment facilities, the high risk of default is shifted from their investors to taxpayers. The sums are significant: $20.5 billion has thus far been authorized for the nuclear sector, all but $2.0 billion earmarked for reactors. The industry is pushing for much higher levels, approaching $100 billion. Much of the debate has focused on the high default risk of the federal guarantees. These are real: both the Congressional Budget Office and the Government Accountability Office expect 50 percent of the loans to default.[32]

Often overlooked is the fact that the guarantees have tremendous value regardless of the default. There are two main reasons for this. First, they allow the plants to use a much higher share of debt (which is lower cost) than would otherwise be possible. The guarantees under present law will cover a project structure up to 80 percent debt. Second, the guarantees bring down the cost of that debt dramatically, since investors care only about the federal government's risk of default (close to zero) rather than the chance that the nuclear reactor will go bust.

Together these factors greatly reduce the cost of financing a new nuclear plant. UniStar estimates the program will save them 3.7 cents per kilowatt hour (kWh) on a levelized cost basis, a cost reduction of nearly 40 percent.[33] As shown in Table 8-3, this translates to nearly $500 million per year in savings per reactor. The authorizing statute allows the guarantees to stay out for a maximum of 30 years—which a rational owner will do since the cost of funds is so low. This translates to a public investment of nearly $13 billion for a single nuclear reactor, an astonishing amount of public support for a single private facility.

| Value of Energy Subsidies to a UniStar EPR Nuclear Reactor | | |
|---|---|---|
| | Value | Source/Notes |
| **1) Constellation Energy Core Inputs (embedded in levelized cost estimates in Turnage, 2008b)** | | |
| Reactor size (MW) | 1,600 | (1) |
| Overnight cost (2007$/kW) | 3,500 | |
| Reactor delivery date | 2,016 | |
| Capacity Factor (avg). | 0.953 | |
| ROE | 0.15 | |
| D/E with guarantees | 80/20 | |
| D/E no guarantees | 50/50 | (1) |
| Duration of debt | 30 | |
| | | |
| **2) UniStar estimated savings from LG/MWhr (2007$/MWhr)** | | |
| Base case break-even | 57 | (1) |
| Break-even, no loan guarantees | 94 | (2) |
| Incremental savings from LG | 37 | |
| | | |
| **3) Convert MWhr values to annual savings** | | |
| MWh/year | 13,357,248 | (3) |
| LG savings/year ($millions) | 494 | |
| Duration of loan guarantee | 30 | |
| PV of savings from LG ($millions) | 14,827 | (4) |
| | | |
| **Sources and Notes** | | |
| (1) | Joe Turnage, "New Nuclear Development: Part of the Strategy for a Lower Carbon Energy Future," International Trade Administration, Nuclear Energy Summit, October 8, 2008, pp. 24, 25. | |
| (2) | Note that this still includes other subsidies | |
| (3) | Hours per year x capacity factor | |
| (4) | Because the cost scenarios represent levelized costs, converting to a PV does not require discounting, as doing so would simply be reinflating the values already in their cost model over the operating life, then discounting them back to 2007 dollars with the same discount rate. | |

## Table 8-3. Value of Energy Subsidies to a UniStar EPR Nuclear Reactor.

These savings are not "free" money, as the industry likes to portray them. Quite the contrary: the savings to a specific industrial facility arise because their business risk is being moved from the investors who will profit from the new reactor to generally taxpayers. It is clearly a good deal for the nuclear industry; far less clear is how the taxpayer is benefiting.

## PRODUCTION TAX CREDITS

The Energy Policy Act of 2005 introduced a 1.8 cent/kWh production tax credit for new nuclear power plants. The nuclear production tax credit (PTC) is limited in two ways. First, no single plant can claim more than $125 million per year in credits; or claim the credit for more than 8 years. Second, current statutes stipulate that a maximum of 6,000 megawatts (MW) of capacity will be able to claim the credit.

Although the Department of Energy has discretion in how the eligible capacity to receive the PTC is allocated across projects, it is reasonable to assume that each plant will get a smaller share of the total available subsidy the larger the number of new reactors that get built. UniStar's early cost estimates assumed they would get full access to the PTC; newer cost estimates assume they will get half of what they are eligible for. Politically, however, the energy-related PTCs are frequently tinkered with. Thus, it is plausible that if many plants are queued up to be built, Congress would simply increase the allowable number of credits. Senator Lisa Murkowski (R-Alaska), for example, has proposed doubling the cap to 12,000 MW.[34] EIA projections assume 8,000 MW of capacity will ultimately tap into the credit,[35] indicative of this possibility.

## ADDITIONAL SUBSIDIES ASSUMED PART OF UNISTAR'S BASELINE COSTS

While UniStar's cost models do explicitly include the federal loan guarantees and production tax credits, the $57/MWH levelized cost base case scenario also includes many other subsidies that help keep costs down. Were these subsidies to be removed, the delivered cost of power would rise even further. Due to the large number of subsidies for the nuclear fuel cycle (see Table 8-2), the following discussion only addresses a handful that are considered to be significant.

### Accelerated Depreciation.

Normal accounting rules allow capital investments to be deducted from taxable income over the service life of the investment. When deductions are accelerated, corporations receive higher than normal deductions in the early years of the investment. Funds that would have otherwise gone to the taxing authority are retained as additional cash within the firm, and can be used for other purposes. The provision acts as an interest-free loan. Towards the end of the asset life, the allowed deductions actually go below the baseline (since total deductions are capped at 100 percent of the investment), accelerated depreciation still provides net subsidies on a present-value basis.

The larger the investment, and the more rapid the write-off relative to actual service life, the larger the subsidy will be. Nuclear reactors, which can last 40-60 years, can be written off from taxes entirely in only 15 years. This generates a reduction in levelized power costs of roughly 0.3 to 0.6 cents per kWh. Price escalation in plant costs suggest the actual levelized value of accelerated depreciation may end up higher than this figure.

**Accident Liability.**

Most industrial enterprises face accident risks. What makes nuclear energy different is the potential for much larger scale damage through the release and dispersal of high-level, long-lived, radioactivity. Thankfully, the probability of a major accident at a U.S. reactor is very low. However, the potential damages should one occur would be extremely high.

The Price-Anderson Act (P-A), first passed in 1957 and renewed multiple times since then, caps the liability of a reactor owner for damages they cause to people and property outside their plant walls in the case of an accident. Under P-A in its current form, a primary tier of insurance (presently $300 million per reactor) must be purchased by the firms. A secondary level of insurance has been created through retrospective pooling of payments from all reactors should an accident at any single reactor exceed the available primary coverage. This second tier coverage provides, in aggregate, more than $10 billion nationwide.

As shown in Table 8-4, while the size of the total pool seems large, it is not. Payment of retrospective premiums is capped at $15 million per reactor per year, resulting in a delay of more than 6 years from the accident until final payment. Converting the P-A pool to a present value is appropriate given the long payment period, and the fact that most of the damage is caused immediately upon the accident. On a present value basis, the pool coverage is about 30 percent lower — roughly $7.7 billion. This level of damages is exceeded on a routine basis in storm events such as hurricanes.

|  | Nominal | Present Value | Notes |
|---|---|---|---|
| **Total payments from Calvert 3 to offsite parties** | | | |
| Primary Insurance, $mils | $300.0 | $300.0 | (1) |
| Retrospective premiums, $mils | $100.6 | $66.5 | (2), (3) |
| *Total liability for Calvert 3* | $400.6 | $366.5 | |
| **Additional resources from other reactors** | | | |
| Retrospective premiums, $mils | $10,462.4 | $6,920.7 | |
| **Total available to offsite parties** | $10,863.0 | $7,653.8 | |
| **Adequacy of Coverage** | | | |
| Balt/WDC MSA 2000 Population, millions | | 7.6 | (4) |
| Total insurance available, $/person | | $1,007 | (5) |
| Calvert 3 coverage, $/person | | $48 | (5) |

**Notes and Sources:**

(1) P-A coverage requirements were last revised in the Energy Policy Act of 2005.

(2) Retrospective premiums are capped at $15 million/year, so each reactor will need more than 6 years of payments to fully pay in their amount due. Calculations assume 105 reactors, 104 currently in operation plus Calvert Cliffs 3. Statutory retrospective premia of $95.8m per reactor can have a 5 percent surcharge levied, upping the total to $100.6m/reactor. Multiyear payments have been discounted at 12 percent real.

(3) This reflects UniStar financing assumptions of 50 percent debt at 12 percent and 50 percent equity at 18 percent, less 3 percent assumed inflation rate.

(4) "Ranking Tables for Metropolitan Areas: 1990 and 2000," U.S. Bureau of the Census, April 2, 2001, available from *www.census.gov/population/www/cen2000/briefs/phc-t3/tables/tab03.txt*.

(5) Aggregate coverage available per person before P-A cap reached; and Calvert Cliffs 3 portion of that coverage per person in the surrounding region.

# Table 8-4. Insurance Coverage for an Accident at Calvert Cliffs 3.

In addition, the pool of coverage has grown much more slowly than the population density surrounding the plants, the value of real estate and infrastructure in the potentially affected areas, or court recognition (via jury awards) of ancillary damages in accidents, such as environmental damages and lost wages for injured workers. In the case of Calvert Cliffs 3, total coverage in the related Baltimore/Washington combined statistical area barely tops $1,000 per person before the private coverage maxes out. This small amount would need to cover loss of property as well as morbidity or mortality from an accident. The portion paid by Calvert Cliffs 3 directly to cover the off-site accident risk from its own operations (Tier 1 coverage plus its share of Tier 2) would be less than $50 per person affected.

While the original plan on P-A was for it to last roughly 10 years—at which point private insurance would be available, primary coverage levels have increased little on a real basis. Industry continues to claim that accident coverage remains highly constrained, and that increased requirements for them to internalize the accident risks from their operations would be unworkable. Surprisingly, however, there seem to be fewer constraints on the policies the utilities want to protect themselves from risk rather than third parties.

For example, Calvert Cliff 3's Tier 1 and 2 responsibilities under P-A force them to cover damages only up to $370 million present value. In contrast, based on a review of financial filings with the Security and Exchange Commission, Constellation Energy's insurance coverage at existing locations indicate that they would carry more than ten times as much insurance cover ($4.2 billion) for damage to their own property and interruption of service.

Any time there are statutory caps on liability below reasonably expected damages, a subsidy is conferred on the recipient industry. Quantitatively, this subsidy is equal to the premiums that would be required to purchase full coverage, less the premiums actually paid for the partial coverage under P-A. Valuing this amount is not easy, since it requires some data on the probability distribution of both accidents and damages. The subsidy estimates shown in Table 8-5 are based on work by Heyes.[36] They should be viewed as indicative rather than precise, as even he believes additional work is needed to develop more accurate values.[37]

| | Low | High | Notes |
|---|---|---|---|
| | *Cents per kWh* | | |
| I. Private investment in Calvert Cliffs 3 | | | |
| Base case of Calvert Cliffs | 5.7 | 5.7 | Constellation estimate, Oct. 2008 |
| | - | - | |
| II. Public investment in Calvert Cliffs 3 | | | |
| *A. Selected EPACT subsidies* | | | |
| Production tax credits | 0.5 | 0.5 | Constellation estimate assuming 50% access to PTCs |
| Loan Guarantees, 100% of debt | 3.7 | 3.7 | Constellation estimate, Oct. 2008 |
| Industry total estimated cost | 9.9 | 9.9 | |
| | | | |
| *B. Additional subsidies ignored in Constellation models* | | | |
| Accelerated depreciation | 0.3 | 0.6 | 15 yr 150% DB vs. service life. |
| Price-Anderson cap on reactors | 0.5 | 2.5 | Based on Heyes (2002); values uncertain. |
| Waste fund short-fall | - | 0.2 | Based on Rothwell (2005). |
| Calvert Co. property tax abatement | 0.0 | 0.0 | $20m/year, but not visible on a per kWh basis. |
| Reduced cost of capital from de-lay insurance, first two reactors | - | 0.8 | High estimate based on Bradford (2007). |
| *Add-in missing subsidies* | 0.8 | 4.1 | |
| | | | |
| III. Total cost of nuclear power | | | |
| Public subsidy | 5.0 | 8.3 | |
| Public/private share | 87% | 145% | |
| Subsidy/avg. wholesale rates, 2002-06 | 113% | 189% | |
| Full cost of power | 10.7 | 14.0 | |

**Table 8-5. Public Subsidies to Calvert Cliffs 3 Approach Private Capital at Risk and Exceed Value of Power Produced.**

More recent estimates contained within a CBO report estimate the subsidy value of P-A caps at less than $600,000 per reactor year. This estimate is not considered realistic, and therefore not included.[38] As there is not much resolution on the origin of this value, it is difficult to pinpoint the drivers behind such a low number. However, politically it would be highly unlikely for the industry to fight so fiercely for more than half a century to retain this subsidy if the value to them really was so insignificant.

One common issue with these lower estimates is that they estimate subsidy costs for a handful of scenarios, rather than for a much bigger universe of accident scenarios. For example, the probability of an accident with damages in excess of $12 billion may be low, but if one sums the probability of an accident for the entire range of $300 million through tens of billions, the numbers turn out larger. It is not clear whether this specific limitation applied to the CBO work or not.

**Management of Long-Lived Nuclear Waste.**

High level radioactive waste must be isolated and managed for thousands of years. At any point during this period, accident or theft can happen, bringing with it potential liabilities to the waste generator and site manager, should they still be in operation. A suitable waste repository is quite difficult to site and build, and faces severe risks for cost escalation.

The combination of technical complexity and difficult, though long-lived, risk exposure is not one that investors or owners like very much. These factors could well have made civilian nuclear power uninvestible. Even if the waste management concerns did not block investment entirely, it is clear that they would have further worsened the already challenging economics of nuclear power.

Federal subsidies have solved this problem for the industry. First, the government stepped in and agreed to take on full ownership of the waste from the plant owners, eliminating uncertain and very long-term liabilities. Given the technical risks and political concerns related to a high level repository, the government's contractual obligations were very poorly structured, containing no risk sharing on delays. Second, the fact that the government agreed to take on this liability in return for a small fee per kWh that is passed through to consumers is also quite important. In so doing, a very large and uncertain fixed cost has been shifted to a very small and predictable variable cost. Both of these factors generate subsidies: the former through reduced sector risk, bringing down cost of capital; and in the latter, if the federal collections underestimate the funds that will ultimately be needed.

This structure has turned out quite badly for the taxpayer. The federal government has been unable to meet its promised deadlines, and therefore has been subjected to breach of contract litigation by the industry and has lost. Payments are already going to utilities to cover on-site storage, and are expected to escalate sharply over time. The tax liabilities have a present value according to the U.S. Department of Energy of at least $7 billion,[39] and ranging as high as $80 billion.[40] For a new reactor, economist Geoffrey Rothwell estimates per kWh surcharges would need to be at least 0.2 c/kWh higher to cover waste disposal costs taken on by the government.[41]

### Calvert County Property Tax Abatement.

In an effort to increase the chances of getting a new reactor at Calvert Cliffs, the Calvert County Board of Commissioners approved a 50 percent reduction in property taxes over the first 15 years of plant operations. This is expected to save the company $20 million per year. The company currently pays $15.5 million in annual property taxes.[42] While too small to even register on a per kWh basis, this is a very sizeable subsidy for a county-level government to offer. The property tax abatement to the new reactor is equivalent to roughly 7 percent of the County's 2009 budget of $296 million, and larger than their entire annual debt service.[43]

## INTEGRATING UNISTAR COST ESTIMATES AND ADDITIONAL SUBSIDY DATA

In an effort to sell the idea of a new reactor at Calvert Cliffs and to educate people about what such an effort would entail, Constellation staff have provided many briefings over the past 4 years on the venture. Most of them have been conducted by Joe Turnage, a Senior Vice President in the Constellation's Generation Group. His presentations provide a valuable resource for understanding the economics of the new reactors based on an industry view of the market and their cost of capital. One can also see how core assumptions have changed over time as market realities demonstrated problems with original assumptions. This section reviews specific information on the value of subsidies to Calvert Cliffs, then provides some additional contextual information on the subsidy value of federal loan guarantees.

## Value of Government Subsidies Clear From Constellation Cost Models.

Running through the results from Constellation's own cost models (the models themselves are not public) clearly illustrates why the firm has focused so heavily on government support. The models calculate the levelized cost per MWh of delivered energy from a new reactor, based on the firm's internal assumptions regarding financing, cost of capital and equipment, and operating parameters. Levelized costs represent the average price they expect to be able to deliver electricity to the wholesale market during the life of the plant and pay back their full investment, including financing costs.

As of October 2008, Turnage projected the break-even price for their firm at $57/MWH.[44] In an earlier presentation, he noted that "at $80/MWH, these plants would not likely be built."[45] Higher delivered costs increase the risk that when the plant finally comes on line, its cost structure will be too high to enable Uni-Star to recoup its investment and earn a profit.

Interestingly, without the government subsidies, UniStar's own models illustrate there is no way they would be competitive. Without loan guarantees for all of the project debt (assumed at 80 percent of the project cost), the levelized cost from Calvert Cliffs 3 would spike from $57/MWH up to $94/MWH. This scenario appears still to assume that the plant would receive lucrative production tax credits worth roughly $5/MWH; the price of power without either of these two programs would be almost $100/MWH.[46]

As a frame of reference, U.S. average wholesale power prices in 2007 — a time of surging energy prices — were roughly \$57/MWH.[47] UniStar's new reactor would just about have broken even, assuming everything on construction and operation went according to plan. The average wholesale electricity price for the U.S. during the 2001-2007 period was only \$47/MWH.

Table 8-5 provides a more detailed summary of the public and private costs associated with the Calvert Cliffs 3 reactor. Some key conclusions:

- Full Levilized Cost of Power Is Not Competitive Based on Unistar's Own Data. The largest cost elements (net of subsidies levelized cost of new EPR reactor, production tax credits, and loan guarantees) take Turnage's own inputs as given. These factors alone, which put the levelized cost of nuclear power at \$99/MWH, render the resource uncompetitive.
- Public Sector Investment Nearly Equal To, or Larger Than, Private Capital Put at Risk. Under the high cost estimate, the public sector investment in Calvert Cliffs 3 is nearly 150% that put in by the plant owners themselves. Should the investment pay off, the public sector would have no direct stake in the venture's profits.
- Subsidies Are Worth More Than the Power. The concept of "value added" measures how much more a product is worth than the sum cost of its inputs. Striking in Table 8-5 is the fact that Calvert Cliffs appears to be a value subtracting enterprise, where input costs are actually worth more than the power one gets out at the other end. Subsidies are 113 to nearly 190 percent of the wholesale value of power, even assuming in the low estimate that there are no subsidies to waste management or from delay insurance. A 5- year average was used to prevent single-year

price fluctuations from skewing the results. Value-subtracting businesses do not normally survive in market economies because investors bleed cash. With nuclear power, public subsidies drive this anomaly.

As discussed below, however, some of Turnage's assumptions are not realistic; and their "net-of-subsidies" values still include some important subsidies to nuclear power. Correcting these assumptions can be expected to further worsen the economics of the proposed Calvert Cliffs reactor.

- Levelized Cost of Reactor Likely too Low. Turnage assumes an overnight capital cost of a new reactor at $3,500/kW of capacity. The overnight value estimates the cost if the plant could be built in one day; "all-in" costs reflect the need to finance the plant, as well as incur costs to integrate it with the grid.
  - The value used in the Turnage cost models is well below the $5,746/kW overnight cost for this same reactor estimated by the Congressional Research Service.[48] This shift alone would bring the levelized cost well above $72/kW, even with loan guarantees and the PTC.
  - Turnage's estimate assumes equity providers would want a return on equity of 15 percent (down from 18 percent in earlier iterations). As noted above, however, investment hurdle rates are driven by the riskiness of the project, not the firm. A new-build nuclear reactor is viewed as quite high risk, and would therefore require a higher-than normal return on equity in order for investments to

proceed. It is useful to note that the return on equity for Exelon Corporation, a large U.S. utility with many nuclear reactors for which it did not bear the construction risks, has averaged more than 20 percent over the trailing 5 years.[49] It is hard to imagine investors accepting a lower return for a higher risk project in the case of UniStar. Thus, without federal guarantees on the debt, the cost of equity should be expected to rise well above Turnage's 15 percent target. There is much higher to go: risks commensurate with early stage venture capital can have hurdle rates of 30 percent or higher.

- "Stress" Cases also understate likely reactor costs. To evaluate how well the venture would succeed if certain conditions were worse than expected, Turnage estimated levelized costs assuming no federal guarantees were available; and that the lifetime capacity factor dropped from 95.3 percent to 85 percent.
    - Under a merchant model, Turnage assumes UniStar could still finance 50 percent of the venture with debt, at a 12 percent interest rate. Yet, Constellation's 5-year debt-to-total capital ratio has averaged only slightly above 50 percent for existing facilities for the 5 years prior to October 2007.[50] Higher-risk new projects would be expected to have higher equity requirements than the existing plant fleet in a merchant environment.
    - The non-partisan Keystone study of nuclear economics, issued in June 2007, estimated equity ratios even for non-nuclear merchant

plants would need to be at 65-70 percent.[51] However, the recent collapse of credit markets suggests even higher equity ratios might be needed.

— Jim Harding, a main author of the Keystone report, also views lifetime capacity factors for new plants deploying new technologies at 75-85 percent.[52]

## EVALUATING THE SOCIAL BENEFITS OF CALVERT CLIFFS 3

In return for billions of dollars in subsidies for Calvert Cliffs 3, the taxpayer is expected to get two main social benefits: energy security and reduced emissions of greenhouse gases. Both of these claims begin to erode under closer scrutiny.

On the energy security front, proponents argue the nuclear power can reduce or replace our reliance on oil imports from unstable regions. This line of reasoning has a number of weaknesses. First, it will be many years before electricity and oil are substitutes, and electrical power on the grid provided by nuclear power stations would be able to fuel our transport fleet. At present, these two markets are almost unrelated. Even hybrid vehicles, which do rely on some electrical motive energy, get that energy from onboard combustion of fossil fuels, not from the grid. Second, nuclear power is an increasingly international venture, with key components produced abroad. Key plant components such as heavy forgings are a good example, and are not made in the United States. Enrichment services and uranium are also provided by international markets with some U.S. presence, but also with heavy reliance on foreign firms and mines.

Finally, there is the link between reactors and terror risks. This can arise through attacks on plants, or through the linkage between the civilian power sector and weapons proliferation. With respect to the former, the NRC ruled unanimously in January 2007 that nuclear plants don't need to protect themselves against attacks using airplanes.[53] However, Constellation has said their design basis is harder, and could withstand a direct hit from a civilian or military jet aircraft. With regards to proliferation, it is unlikely that a single new reactor at Calvert Cliffs will have a material impact on proliferation risks. However, reactor construction on the order projected to mitigate any sizeable portion of global GHG emissions clearly would.

The climate change picture is even more interesting, as this has been a major push behind public subsidy to new reactors. While it is true that nuclear power does have quite low emissions of greenhouse gases (GHGs) per unit of energy produced, those figures are not zero. In addition, the economic costs for the reactors are quite high once both the public and the private investments are taken into account. As a result, the cost per unit of $CO_2$ equivalent removed through the nuclear fuel cycle turns out to be significantly higher than many other options with shorter implementation periods and much lower market and financial risks.

Figure 8-2 illustrates this graphically, integrating data on the marginal cost of abatement from evaluative work done by McKinsey with estimates of subsidies to new build nuclear reactors done by Earth Track. As can be seen, the lower cost options tend to be in improved efficiency, systems management, and land use modifications. Subsidies alone to nuclear power exceed the costs of many of these other alternatives and greatly exceed the market value of the off-sets on the carbon market. Many scientists believe we have a limited window to address climate change concerns, and it is quite important that our investments into GHG reduction are done efficiently, targeting the lowest cost, lowest risk options first.

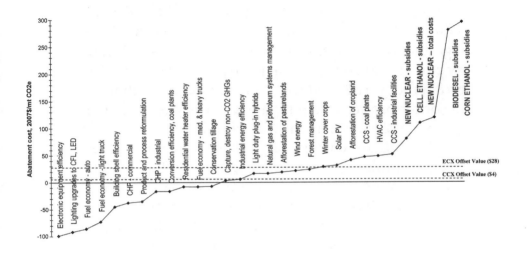

Sources:

*Abatement technologies*: McKinsey & Company (2007), mid-range case.

*Offset prices*: Average of contract values from CCX (2008-10) and ECX (2008-12).

*Subsidy data*: Earth Track, Inc. Chart prepared by Earth Track, Inc. for Greenpeace Solutions, 2008.

### Figure 8-2. Subsidizing Nuclear Energy Is an Expensive Way to Address Climate Change.

## CONCLUSIONS

Calvert Cliffs 3 is one of a number of projects around the world to restart nuclear energy through the construction of many new reactors. A close review of the corporate structure and public support to this initiative indicates that much of the financial and operating risks are being shifted from investors to the taxpayer and the surrounding population. Smaller scale, emerging power sources are also likely to be hurt in two ways. First, subsidies will enable uneconomic reactors to be built. Second, even if the massive capital investments in the reactors are lost entirely due to bankruptcy or restructuring, the reactors would continue to function. Their low operating costs would squeeze the margins of many alternative resources that had not been so heavily subsidized.

Once subsidies are added to private investment costs at the reactors, Calvert Cliffs 3 would not be commercially competitive. Public subsidies alone are likely to exceed the value of the power that the facility produces. Public investment is nearly equal to (low estimate), or greatly exceeds (high estimate), the private investment into the new plant. Nonetheless, the taxpayer will not share in the "upside," should the plant be financially successful.

Even from a greenhouse gas perspective, nuclear power is an expensive solution. Once reductions are normalized to the cost per metric ton of $CO_2$ equivalent reduced, it is evident that there are a variety of other technologies and options that are far less expensive, as well as having lower financial risks, smaller unit sizes, and more rapid deployment schedules. The availability of these other options can be seen by how much lower the market value of carbon offsets is relative to the cost of abatement via the nuclear fuel cycle.

While the United States faces real energy security and climate change challenges, this does not mean that earmarking tens or hundreds of billions of dollars in subsidies to the nuclear sector is a worthwhile or effective strategy. Any subsidies that are to be deployed to reach these policy end-goals should be competitively tendered, forcing nuclear to compete on an efficiency basis with alternative energy pathways.

The federal government's foray into large scale subsidization of energy credit, both through loan guarantees and more recently, via clean energy "banks" is particularly worrying. There is little evidence that the federal government has the technical skills to manage programs on these scales, or the ability to shelter decisions from being politicized. Oversight structures and the alignment of incentives to increase the chance of project success are both lacking. Once these deals are approved, there will be little that can be done in terms of mid-course corrections to reduce the size of taxpayer losses or the competitive impediments that widescale subsidization of large, baseload nuclear capacity will create for smaller-scale alternatives.

## ENDNOTES - CHAPTER 8

1. General Electric Advertisement, *National Geographic Magazine*, January 1954.

2. Linda Vasallo, "Calvert County, MD, Board of County Commissioners Support Expansion at Calvert Cliffs Nuclear Power Plant," 2007.

3. Peter Bradford, "Recent Developments Affecting State Regulation of Nuclear Power," *Regulatory Assistance Project Issues Letter*, July 2008.

4. California Energy Commission (CEC), Transcript to the Committee Workshop before the California Energy Resources Conservation and Development Corporation in the matter of "Preparation of the 2007 Integrated Energy Policy Report," Sacramento, CA, June 28, 2007, p. 273.

5. Siemens, "Siemens to divest its state in Areva NP joint venture," Press Release, January 26, 2009.

6. "French Utility Areva Q4 Sales Up, Eyes More Growth," *Reuters*, January 29, 2009.

7. Electricite de France (EdF), "Shareholding structure at December 31st, 2008," available from *www.shareholders.edf.com*.

8. U.S. Federal Energy Regulatory Commision (FERC), "FERC Approves EDF Purchase of Constellation Energy Stake," Press Release, February 19, 2009.

9. Michael Mariotte, "Joint Intervenors Reply to NRC Staff's Answer to Petition to Intervene and Applicants' Answer to Petition to Intervene," *Nuclear Information and Resource Service*, 2008, p. 20.

10. Matthew Wald, "French-American Venture Plans New Reactors in the U.S.," *The New York Times*, July 21, 2007.

11. Mariotte, p. 20.

12. Accenture, "Consumers Warm to Nuclear Power in Fight Against Fossil Fuel Dependency, Accenture Survey Finds," Corporate Press Release, March 17, 2009.

13. Eileen O'Grady, "Five U.S. nuclear plants make DoE loan short-list," Update 1, *Reuters*, February 18, 2009.

14. Alstom, "Ownership of Alstom Shares," *Alstom-Annual Report 2007/08*, p. 223; Mycle Schneider, Mycle Schneider Consulting, e-mail communication with Doug Koplow, March 19, 2009.

15. Steve Kidd, "New nuclear build-sufficient supply capacity?" *Nuclear Engineering International*, March 3, 2009.

16. Paul Adams, "Economics of nuclear power are rethought: Loan guarantees could transform energy industry," *Baltimore Sun*, September 4, 2007.

17. "Annual Lobbying for Constellation Energy," compiled by the Center for Responsive Politics, available from *www.OpenSecrets.org*.

18. Elliot Blair Smith, "Nuclear Utilities Redefine One Word to Bulldoze for New Plants," *Bloomberg*, September 27, 2007.

19. UniStar, "Calvert Cliffs 3 Nuclear Power Plant: Powering the Mid-Atlantic's Economic Future," 2008; available from *www.unistarnuclear.com/projects/cc3_econ.html*.

20. Mariotte, p. 8.

21. "French Utility Areva Q4 Sales Up, Eyes More Growth," *Reuters*, January 29, 2009; UniStar EPR, UniStar Nuclear Energy, "The U.S. Evolutionary Power Reactor," September 15, 2005, available from *www.unistarnuclear.com/press/091505-press.html*.

22. UniStar, "Calvert Cliffs 3 Nuclear Power Plant."

23. Tri-County Council, *Growth, Southern Maryland*, August 2006, p. 4.

24. *Ibid.*

25. Calvert County, MD, "Brief Economic Facts, 2006-2007," 2007.

26. MD DBED, Maryland Department of Business & Economic Development, "Calvert County, Maryland: Brief Economic Facts," 2007, available from *www.ChoseMaryland.org*.

27. CEC, Transcript to the Committee Workshop, p. 302.

28. Wald.

29. Peter Behr, "Nuclear Power: A key energy industry nervously awaits its 'rebirth'," *E&E*, April 27, 2009.

30. David Schlissel, Michael Mullett, and Robert Alvarez, *Nuclear Loan Guarantees: Another Taxpayer Bailout Ahead?*, Union of

Concerned Scientists, March 2009, p. 11.

31. Christopher Seiple, "Stranded Investment: The Other Side of the Story," *Public Utilities Fortnightly*, March 15 1997.

32. U.S. Congressional Budget Office (CBO), "Cost Estimate for S. 14, Energy Policy Act of 2003," May 7, 2003. U.S. Government Accountability Office (GAO), "Department of Energy: New Loan Guarantee Program Should Complete Activities Necessary for Effective and Accountable Program Management," *GAO-08-750*, Washington, DC, July 2008.

33. Joe Turnage, "New Nuclear Development: Part of the Strategy for a Lower Carbon Energy Future," presentation at the International Trade Administration Nuclear Energy Summit, Washington, DC, October 8, 2008, pp. 24-25.

34. Katherine Ling, "Nuclear Power: Senate GOP to offer plan for industry incentives, reprocessing," *E&E*, May 5, 2009.

35. Mark Holt, "Nuclear Energy Policy," *CRS Report to Congress*, updated September 2, 2008, p. 5.

36. Anthony Heyes, "Determining the Price of Price-Anderson," *Regulation*, Winter 2002-03.

37. Anthony Heyes, E-mail correspondence with Doug Koplow, Earth Track, October 27, 2005.

38. Justin Falk, *Nuclear Power's Role in Generating Electricity*, Washington, DC: U.S. Congressional Budget Office (CBO), May 2008, p. 29.

39. Kim Cawley, "The Federal Government's Liabilities Under the Nuclear Waste Policy Act, " Testimony before the Committee on Budget, U.S. House of Representatives, Washington, DC: U.S. Congress, October 4, 2007.

40. Dave Berlin, "Nuclear Waste Storage," *Citizens Against Government Waste*, September 23, 2004.

41. Geoffrey Rothwell, E-mail correspondence with Doug Ko-

plow, Earth Track, November 1, 2005.

42. Jamie Smith Hopkins and Paul Adams, "Calvert County solicits reactor," *Baltimore Sun*, August 9, 2006.

43. "Budget Summary: FY2009 Commissioners Report," Calvert County, MD, 2009.

44. Turnage, "New Nuclear Development," p. 25.

45. Joe Turnage, "A Strategic Analysis of the Investment Opportunity for Advanced Nuclear Generation," presentation at the MIT American Nuclear Society Seminar Series, Cambridge, MA, March 12, 3007, p. 48.

46. Turnage, "New Nuclear Development."

47. United States Energy Information Administration, "Table 2. Average Wholesale Price by NERC region, 2001-2007," based on Form EIA-861, Annual Electric Power Industry Report, available from *www.eia.doe.gov/cneaf/electricity/wholesale/wholesalet2.xls*.

48. Stan Kaplen, *Power Plants: Characteristics and Costs*, Washington, DC: Congressional Research Service, November 13, 2008.

49. "Management Effectiveness Ratios, Exelon Corp.," *Thomson Reuters*, March 19, 2009.

50. "New Nuclear Generation in the United States: Keeping Options Open vs. Addressing and Inevitable Necessity," *Moody's Corporate Finance*, October 2007.

51. *Nuclear Power Joint Fact-Finding*, Washington, DC: Keystone Center, June 2007.

52. Jim Harding, *Economics of New Nuclear Power and Proliferation Risks in a Carbon-Constrained World*, prepared for the Nuclear Proliferation Education Project, June 2007.

53. Steve Mufson, "Nuclear Agency: Air Defenses Impractical," *Washington Post*, January 30, 2007.

# CHAPTER 9

## NUCLEAR POWER IN SAUDI ARABIA, EGYPT, AND TURKEY: HOW COST EFFECTIVE?

### Peter Tynan
### John Stephenson

## INTRODUCTION

The interest in nuclear energy in the Middle East and North Africa has become widespread in recent years. Although most attention has been focused on the progress of Iran in its nuclear program, six other countries in the region have signed agreements to proceed with nuclear power development and another 10 have expressed interest or conducted studies related to nuclear power.[1] (See Figure 9-1.)

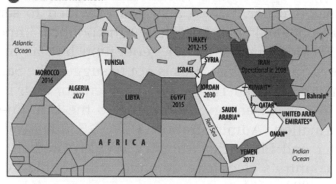

Source: *Christian Science Monitor*, November 1, 2007.

**Figure 9-1. Interest in Nuclear Power in the Middle East and North Africa.**

Speculation on the motivation for this interest in nuclear energy incorporates both political and economic rationales. Politically, the arguments focus on the regional rivals of those states now seeking nuclear power. Observers have argued that Shi'ite Iran's nuclear push has instigated the growing interest among Sunni states.[2] As Iran's nuclear program strengthens despite international pressures, Sunni interest in nuclear alternatives may concurrently intensify. Other political arguments point to Israel as a chief reason for further Middle Eastern nuclear development,[3] with its lack of participation in nonproliferation treaties.

In addition to these political rationales for nuclear development, official sources more often focus on economic arguments for nuclear power. These include dwindling oil reserves, a lack of natural resources, or lucrative export opportunities when natural gas prices are high. Officials further emphasize the growth of many Middle Eastern states, such as the United Arab Emirates (UAE), which "argues that it needs nuclear energy to satisfy soaring demand for power and desalinated water."[4] Even analyses provided to senior U.S. policymakers acknowledge the political impetus for seeking nuclear power generation but, as in the case of Saudi Arabia, "this is not to suggest the Saudis do not have an energy-based argument for their interest in nuclear energy. According to the U.S. Energy Information Administration (EIA), Saudi Arabia's Water and Electricity Ministry (WEM) predicts that the country's electricity demand will double by the years 2023–25."[5]

This analysis focuses on evaluating the economic and resource arguments for the development of nuclear power that are oft cited with three case studies:

Saudi Arabia, Egypt, and Turkey. These three case studies were selected for having unique characteristics, but also being representative of other countries with interest in nuclear power. Saudi Arabia has large fossil fuel reserves which form the base for its economy; maintains a strong sovereign credit rating; and has fast rising electricity demand which is partly driven by desalination needs. Comparable countries to Saudi Arabia include Libya and the UAE. Egypt is more comparable to Algeria, Morocco, Tunisia, and Yemen, having some domestic fossil fuel reserves but a poor credit rating. Finally, Turkey represents a fairly unique case study with no domestic fossil fuel reserve and a strong economy based on non-oil/gas sectors. By reviewing and analyzing three very different case studies, this analysis seeks to shed light on the broader applicability of nuclear power in the region.

In evaluating the economic and resources arguments for nuclear power, this analytical framework takes into account four key components to determine the best energy sources for meeting a country's future energy gap: (1) total potential capacity, (2) relative cost, (3) energy security and location of sources, and (4) environmental issues. See Figure 9-2.

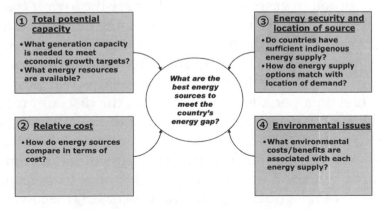

Figure 9-2. Analytical Framework.

## Total Potential Capacity.

This analysis seeks to answer two key questions, namely: (1) what generation capacity is needed to meet economic growth targets; and (2) what energy resources are available? These factors vary widely by country. In most cases, a combination of energy sources, rather than a single technology, would be required to meet future demand growth and to cover both the peak and base load demands.

## Relative Cost.

The development of generation capacity needs to take into account the relative costs across a variety of generation options. While relative cost should not be the only selection criteria due to the risk mitigation benefits of having a more balanced national energy portfolio, they play a crucial role in determining the commercial viability and development of generation capacity. Though country specific-factors can impact relative cost comparisons, it is useful to understand structural factors that affect relative cost.

Although gas, coal, and nuclear are the lowest cost options according to global surveys, the discount rate plays a critical role in determining overall cost effectiveness. With the large upfront capital costs of nuclear generation development and the relatively low cost of fuel on a per kilowatt hour basis, the discount rate plays a critical role in determining the relative costs across these options. At a 5 percent discount rate, the levelized cost of nuclear is $29/megawatt hour, compared with $47 for natural gas. But at a 10 percent discount rate, nuclear generation costs $43/megawatt hour, compared with $51 for natural gas (see Figures 9-3 and 9-4). The implication is that in countries with higher cost of capital, as is the case with most developing countries, the cost advantage of nuclear power declines. Discount rates influence every infrastructure project and may range from 8-15 percent. Furthermore, moving beyond the high-level global survey, it is important to broaden the factors being considered.

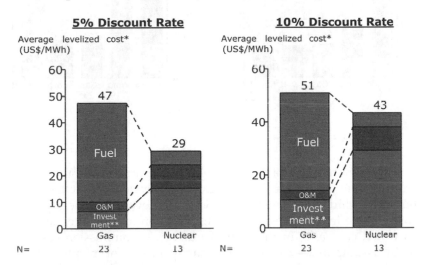

Notes and Sources:

*Average of data from survey of new facilities in 21 countries, mostly OECD, but also include four developing countries. Levelized generation cost include initial investment cost, operation and maintenance cost, fuel cost, and in the case of nuclear; main assumptions—85 percent capacity factor for plants; 40-year lifetime for coal and nuclear plants; for other plants, lifetime come from country level responses; fuel price projection based on each country's models.

**Investment cost for nuclear power includes decommission cost. Source: "Projected Cost of Generating Electricity 2005 Update"—NEA/IEA.

**Figure 9-3. Cost Comparison of Natural Gas and Nuclear Generation.**

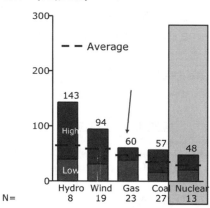
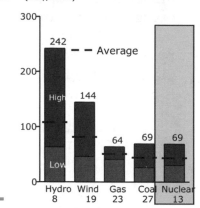

Notes and Sources:

*Data from survey of new facilities in 21 countries, mostly Organization for Economic Cooperation and Development (OECD) but also include four developing countries. Levelized generation cost include initial investment cost, operation and maintenance cost, fuel cost, and, in the case of nuclear, decommission cost; main assumptions—85 percent capacity factor for plants; 40-year lifetime for coal and nuclear plants; for other plants, lifetime comes from country level responses; fuel price projection based on each country's models. "Projected Cost of Generating Electricity 2005 Update"—Nuclear Energy Agency (NEA)/IEA.

**Figure 9-4. Global Survey of Comparative Costs.**

The global survey of cost comparisons ignores a number of factors, and in doing so, underestimates the relative cost of nuclear generation. First, the 45 global benchmarks of nuclear generation costs reflect vendor numbers which many experts believe to be too low.[6] Second, nuclear power plants generate significant amounts of electricity in a more centralized manner, requiring extensive transmission and distribution networks. Not only does this require additional investment, but also results in higher amounts of energy loss, as even best-in-class networks incur a ~20 percent system loss of electricity. Third, cost items which are not directly borne by the power plant operator, such as the cost of a robust government regulatory body, the cost and time to build up human capital to operate the plants, and the cost of insurance and loan guarantees, are much higher for nuclear power plants versus other resources and are not included in the calculations. Taken together, real cost of nuclear power plants may be significantly higher than the costs reflected in the IEA survey. While some analysts estimate low costs for new plants, such as the EIA, University of Chicago, and vendors with estimates of $1,500-$2,100 per kilowatt, other analysts, such as Keystone Center, Standard & Poor's, and Moody's, estimate a much higher range from $3,600-$6,000 per kilowatt.[7]

Also, a number of cost saving potentials exist for renewable sources such as hydro (particularly small hydro projects) and wind that are not taken into account here. First, renewable power generation can be decentralized and local, providing savings in transmission infrastructure. Second, most of the renewable technologies are still in the process of development, with cost decreasing over time due to both increase in scale and advancement in technology (see Figure 9-5). Third, in a potential regime where there is a carbon tax, renewable sources have the advantage of zero emission. Lastly, even though per unit cost of electricity from renewable sources may be higher, renewable projects require much less upfront capital due to smaller sizes of the generators and can be brought online incrementally. This is an advantage, particularly if the cost of capital is high, and when energy demand growth in a particular area is uncertain and thus a more modular expansion of generation capacity is desirable.

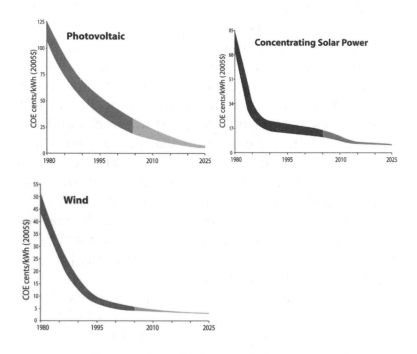

**Figure 9-5. Cost Curve For Solar Photovoltaic, Concentrating Solar Power, and Wind, 1980-2025.**[8]

**Energy Security and Location of Source.**

This component seeks to address the questions about: (1) whether countries have sufficient indigenous energy supply sources; and (2) how energy supply options match with local demand. Where countries depend on imports to satisfy energy demand, developing indigenous sources of power, such as nuclear or renewable energy, makes more sense. In cases where the natural endowment of energy sources is located far from the population centers that require that energy, alternatives also have to be considered.

**Environmental Issues.**

Finally, environmental issues are assessed to determine what the environmental costs and benefits are with each of the supply options. Hydrocarbon based resources emit a variety of pollutants, including $SO_2$ that contributes to acid rain and $CO_2$ which impacts global climate change. However, within the hydrocarbon family, there is a large difference in environmental impact, with natural gas being much cleaner than oil or coal. Though nuclear power does not emit $CO_2$ or $SO_2$, it does pose potential threats to the environment in the form of accidents and waste disposal. Renewable sources tend to have the best environmental record, although large hydropower projects can also adversely affect nearby ecosystems. See Figure 9-6.

gCO2 emitted / MJ

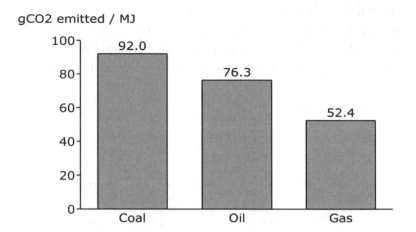

Source: Intergovernmental Panel on Climate Change, "Working Group III Fourth Assessment Report," Chapter 4: Energy Supply.

**Figure 9-6. Comparison of $CO_2$ Emissions Across Fossil Fuel Resources.**

**Viability.**

An evaluation of governance indicators also highlights important points about the ease (and difficulties) of implementing a policy to develop nuclear power. The successful development and operation of nuclear plants in countries such as France, Great Britain, the United States, and Finland have all occurred in countries with relatively strong governance and regulatory effectiveness. But despite strong governance, many projects in France, Great Britain, the United States, and Finland have not been completed on time; have unclear commercial viability; have required significant government subsidies and bailouts; or have encountered other problems. This suggests that even with relatively high marks in terms of government effectiveness, regulatory quality, and the control of corruption (see Figure 9-7), nuclear power development is inherently difficult to coordinate between the private and public sectors.

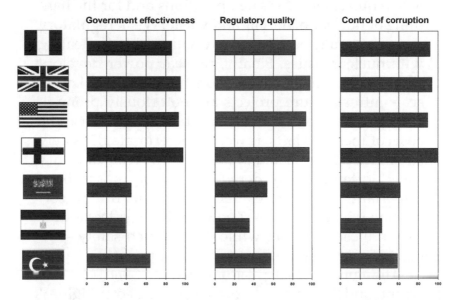

Source: *World Bank Governance Indicators*, 2007, available from *info.worldbank.org/governance/wgi/index.asp*.

**Figure 9-7. World Bank Governance Indicator Comparison.**

By comparison, Saudi Arabia, Egypt, and Turkey have considerably lower scores across government effectiveness, regulatory quality, and control of corruption, as rated by the World Bank governance indicators, which serve as a general measure of governance.

Each of these dimensions could have considerable implications on the likelihood and effectiveness of nuclear development in these countries. Given the strong role of government in the subsidization of nuclear power in other countries, weaker government effectiveness could reduce the viability of even subsidized projects. Nuclear power also has significant regulatory requirements for safe operations and for the handling of nuclear materials and waste. Weak regulatory systems would be particularly vulnerable in dealing with nuclear issues. Finally, nuclear power development requires huge upfront construction costs. Large-scale infrastructure projects are notoriously prone to corruption which can both substantially increase the costs of these projects as well as deteriorate the safety standards of completed projects. These real "ancillary costs," often not considered or factored in, can have important consequences. Increased costs due to corruption will make nuclear power relatively less attractive in these countries while deteriorated safety standards could have catastrophic consequences.

The specific energy situations for Saudi Arabia, Egypt, and Turkey will now be addressed in light of the issue of whether nuclear power is necessary to meet their electricity generation needs.

## SAUDI ARABIA

### Total Potential Capacity.

Energy demand in Saudi Arabia is expected to grow relatively rapidly in the next 25 years. Based on estimates of 3.5 percent gross domestic product (GDP) growth, it is expected that Saudi Arabia's generation capacity will need to grow at 4.4 percent, necessitating an addition of 66.1 gigawatts of ᵍ  ᵃtion capacity by 2030.[9] Although current use is rou     53 percent residential and 20 percent industrial, gro  ᵗ in electricity demand is expected to be largely driven by requirements of new desalination plants. Of the current 24 gigawatts (GW) planned for development in Saudi Arabia, 15GW are estimated to be for desalination needs.[10] See Figure 9-8.

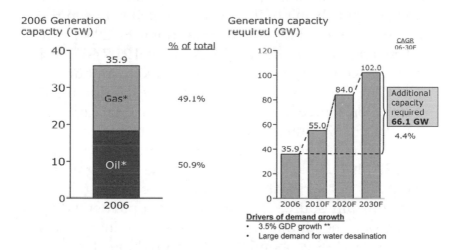

Notes and Sources:

*Gas and Oil generation capacity split based on 2005 proportions.

**GDP growth rate from 2003-2030. *Saudi Arabia Electricity and Cogeneration Regulatory Authority Annual Report 2006; International Energy Agency Statistics; World Energy Outlook 2005,* IEA.

### Figure 9-8. Saudi Arabia's Current and Projected Generating Capacity.

Saudi Arabia's electricity generation capacity currently relies on fossil fuel resources despite having considerable renewable energy resources. With 35.9

GW of total generation capacity, 50.9 percent is generated using oil and 49.1 percent uses natural gas.[11] The solar and wind resources available to Saudi Arabia remain largely untapped, with no signal from the national government to promote significant investment.

The development model for the power sector in Saudi Arabia is to shift more generation to gas powered turbines, especially in times of high oil prices, and use oil resources to fill the gap between supply and demand. In line with this strategy, almost all new power plants being built now can switch between the two fuels. According to IEA projections, by 2030, new oil generation capacity would likely total 22.7 GW and new gas generation capacity would likely amount to 43.4 GW. This would make natural gas the dominant generation source, with 60 percent of the total capacity by 2030.[12] This strategy for developing new electricity generation has been emphasized by industry experts at ARAMCO, who have said that: "The Saudi government has not seriously considered electricity generation from any source other than gas, supplemented by crude oil. The abundance of the resource just means that there's less economic need for anything else."[13]

**Relative Cost.**

For natural gas generation, the primary cost component is the fuel, so the relative economics of generation options in Saudi Arabia must incorporate significantly lower natural gas prices. While natural gas prices in Western Europe, North America, and East Asia fluctuate between $6 and $8 per million British thermal units (BTUs), the costs are considerably lower in gas-producing countries like Saudi Arabia, where the cost is roughly $0.8 per million BTUs (see Figure

9-9). As a result, the relative costs between natural gas and nuclear generation are vastly different in Saudi Arabia, with nuclear costing $43 per megawatt hour and natural gas generation costing less than $20 per megawatt hour.

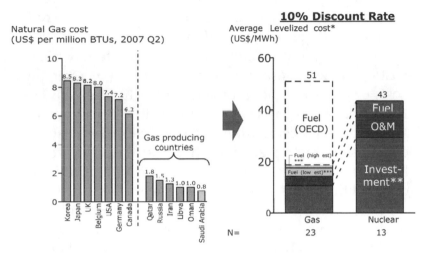

Notes:
*Average of data from survey of new facilities in 21 countries, mostly OECD, but also include 4 developing countries. Levelized generation cost include initial investment cost, operation and maintenance cost, fuel cost, and in the case of nuclear; main assumptions—85 percent capacity factor for plants, 40-year lifetime for coal and nuclear plants, for other plants lifetime come from country level responses, fuel price projection based on each country's models.

**Investment cost for nuclear power includes decommission cost.

***calculated as Saudi's gas price as a percent of the lowest and highest price in the OECD gas price data available.

Source: American Chemistry Council; "Projected Cost of Generating Electricity 2005 Update," NEA/IEA.

**Figure 9-9. Natural Gas Costs and Resulting Cost Comparison of Natural Gas and Nuclear Generation in Saudi Arabia.**

The question then arises about the opportunity cost of natural gas and the ability of Saudi Arabia to export it lucratively. Although more detailed analysis could yield the exact threshold whereby the opportunity cost is too great and natural gas should be exported instead of used for electricity generation, a rough approximation suggests that threshold has not yet occurred. First, natural gas prices would have to be high enough for the revenue generated from the sale to cover the cost of nuclear power in Saudi Arabia. To cover the costs of nuclear generation, current natural gas prices would have to increase by 18 percent to break even.[14] However, this estimation assumes that the costs of developing nuclear generation in Saudi Arabia would be the same as the world average, despite additional construction costs and a weaker government and regulatory effectiveness. Taking nuclear costs from more roughly comparable countries, the costs for nuclear could be as much as $58 per megawatt hour,[15] thereby requiring a roughly 41 percent increase in natural gas prices before meeting the threshold whereby exportation of natural gas makes sense in Saudi Arabia. An 18 percent increase in natural gas prices in Western Europe, North America, and East Asia would mean a cost per million BTUs of $9, and a 41 percent increase would mean a cost of approximately $11 per million BTUs. For nuclear development and exportation of natural gas to make sense, prices would need to remain high. Recent decreases in price suggest that this threshold is even further away.

Second, an assessment should take into account the additional costs incurred by building infrastructure for natural gas exportation. This exportation could occur either via pipelines (which would require regional agreements between Saudi Arabia and its neighbors) or via liquefied natural gas tankers. Either case would require significant investments in infrastructure would put the threshold for lucrative exportation of natural gas even higher.

**Energy Security and Location of Energy Sources.**

Saudi Arabia has abundant reserve of natural gas to provide for its growing demand for electricity generation. Current reserves stand at ~240 trillion cubic feet, with a 2005 annual production level of 2.87 trillion cubic feet.[16] There have been recent shortages of natural gas in the domestic market, leading to incidents of electricity blackouts. The supply crunch occurred due to traditional practices of only producing associated gas (gas co-produced from oil wells). Recent change in government policy to limit gas flaring and encourage production from independent gas field should address the supply shortage.[17] Realizing that natural gas is important for the domestic electricity market, the government has stepped up exploration projects with the hopes of adding another 50 trillion cubic feet to reserves by 2016. It has also for now earmarked all natural gas for domestic consumption.

Natural gas supply within Saudi Arabia is transported by the Master Gas System. The Master Gas System came online in 1982 predominantly to transport associated gas from the Ghawar field, and since then has expanded to transport gas from a variety of fields to industrial centers at Yanbu and along the Red Sea coast.[18] Saudi ARAMCO intends to build a further 3,000 km of gas pipeline by 2006 to expand the Master Gas System.[19] Continued investment will ensure meeting the demand of the rapidly expanding gas power sector. (See Figure 9-10.)

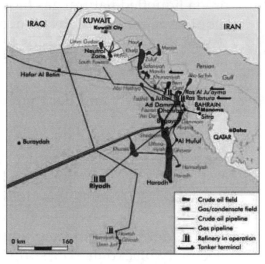

Source: *World Energy Outlook 2005*, International Energy Agency.

**Figure 9-10. Natural Gas and Oil Infrastructure in Saudi Arabia.**

**Environmental Impacts.**

Environmental improvements from electricity production could result from proper development of Saudi Arabia's solar and wind resources. By contributing nearly half of the carbon emissions of coal-generation, the use of natural gas for electricity generation helps to keep Saudi Arabia's per capita emissions relatively low. However, the environmental impact of Saudi Arabia's power sector could be further improved should the government move away from a policy of neglecting the country's renewable resources.

## Conclusions.

While Saudi Arabia will see an increase in the need for electricity, it has focused on developing natural gas generation to meet this growing need. In the foreseeable future, natural gas provides a cost competitive generation option vis-à-vis nuclear. The opportunity cost of natural gas exportation may not be high enough to induce Saudi Arabia to sell its natural gas and build nuclear power instead. Furthermore, Saudi Arabia has invested in its Master Gas System to leverage its natural gas resources for electricity generation. This strategy will likely have a positive environmental impact, though increased focus on renewable energy opportunities in terms of solar power and wind power could further improve Saudi Arabia's environmental impact.

## EGYPT

### Total Potential Capacity.

In 2006, Egypt had 20.5 gigawatts of electricity generation capacity.[20] The residential and industrial sector dominate the consumption of electricity at 36.5 percent and 35.2 percent, respectively; however, electricity use from these two sectors have also grown the slowest in the past 5 years at 7.1 percent and 6.5 percent per annum, respectively.[21] The government and public sectors make up 12.4 percent of electricity consumption, and have grown at 8.5 percent per annum between 2001-02 and 2005-06.[22] The remaining 16.9 percent of electricity consumption is split almost evenly between commercial, agriculture, and other sectors, with the commercial sector growing the fastest at 12.9 percent per annum for the same time period.[23]

The IEA projects that required electricity generation capacity in Egypt will grow more slowly in the next 25 years than it has in the past 30 years, at an annual rate of 2.4 percent between 2006 and 2030.[24] The drivers behind this slowing growth of electricity demand includes a slowing economy which is projected at 3.6 percent GDP growth over the same time period due to an aging population, as well as a maturing economy that tends to be less energy intensive.[25] With this projection, Egypt would need 15.5 gigawatts of additional generation capacity by 2030. (See Figure 9-11.)

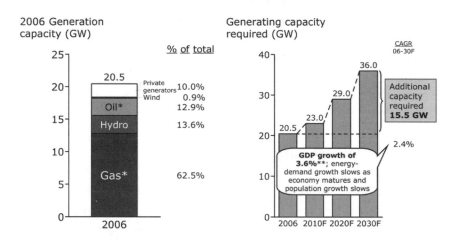

Notes and Sources:

*Gas and Oil generation capacity estimated based on total "thermal" generation capacity in 2006 and average ratio of gas to oil generation capacity in 2003 and 2010F.

**GDP growth rate from 2003-2030; *Egyptian Electric Holding Company Annual Report 2005-2006*; Demand projection from *World Energy Outlook 2005 – the Middle East and North Africa Insights*, IEA.

**Figure 9-11. Egypt's Current and Projected Generating Capacity.**

The importance of natural gas in Egyptian electricity generation grew dramatically during the late 1990s, following heavy foreign investment in the Egyptian gas sector and dwindling oil production.[26] In 2006, 62.5 percent of the electricity generation comes from gas-fired turbines owned by the Egyptian Electricity Holding Company, 13.6 percent from hydropower plants, and 12.9 percent from oil.[27] Renewable sources other than hydropower make up an insignificant amount of generation. In addition, approximately 10 percent of the electricity generated comes from private operators licensed by the government, a vast majority of whom use gas-fired turbines.[28] In the next 25 years, natural gas powered plants will likely continue to play an increasing role in Egyptian electricity generation, with the share of oil powered plants declining slightly, and other power sources such as hydro, wind, and other renewable energy contributing only very slightly to the 15.5 GW additional generation capacity required by 2030.[29]

### Natural Gas.

Natural gas will most likely be the greatest contributor to new electricity generation until 2030, making up more than 90 percent of the additional capacity required. Egypt has made a few recent major discoveries and has seen its gas reserves increase dramatically,[30] currently at 68.5 trillion cubic feet.[31] Significant investment has increased Egypt's production capacity to roughly 5 billion cubic feet per day in 2007. Even though Egypt is at the same time building up its capacity to export its natural gas, the government is deeply concerned about maintaining an adequate supply for domestic use. To this end, it has limited gas reserves available for export, 25 percent, down a third from previous reserves.[32] Most of the planned buildup in electricity generation has focused on gas. Already, there are enough confirmed projects to add 5.4 gigawatts of gas capacity to Egypt.[33] An additional 9.2 gigawatts of gas-fired capacity is expected to meet 2030 demand.[34]

## Oil.

As Egypt's oil reserves dwindle and production drops, the use of oil for electricity generation will continue to decrease. Over the next 20 years, there are no plans to construct anymore power plants fueled by oil. In fact, one or two oil powered plants are likely to be retired, leading to the shrinking of absolute capacity from oil powered generation.[35]

## Hydro.

Development of new hydro resources will most likely be limited. Egypt's largest hydro resources have already been exploited in the large Aswan Dam projects, leaving comparatively smaller opportunities. A few smaller projects, however, are already in the plans. Four hydro power units in Naga Hammadi are due to be completed by May 2008, with combined 640 megawatts of capacity; a project in Kanater Delta in Damietta to be completed in 2010, with 130 megawatts of capacity; and another at New Asiut Barrage scheduled for completion by 2014, with 320 megawatts of capacity.[36]

## Renewables.

The Egyptian government has not made any serious efforts to invest in renewable energy development. There are a few isolated projects in the plans, including a wind power project in the Suez financed by the Netherlands with ~60 megawatts generating capacity, and a combined solar/gas power project at Kureimat subsidized by the Global Environment Facility that will have ~31 megawatts of solar capacity.[37] Taken together, and without further government policies supporting aggressive development, renewable sources will only contribute marginally to the additional generating capacity required by 2030.

### Nuclear.

At the encouragement of President Mubarak to prioritize nuclear energy development for Egypt, the Egyptian Ministry of Electricity and Energy has officially authorized the construction of three nuclear power plants in al Dab'ah region in Egypt's northwest with a total generating capacity of 1.8 gigawatts.[38] According to the plan, the first plant will begin operation in 2015-16, with the other two scheduled for completion by 2017-18 and 2019-20, respectively.[39] If successful, these three nuclear plants will contribute to ~5 percent of Egypt's electricity generating capacity by 2030.

### Relative Cost.

As with the case for Saudi Arabia, the cost of gas generation is significantly lower than the world average, given much lower cost for fuel in a gas producing country. The average cost for gas producing countries is approximately $1.2 per million BTUs, compared to the range of $6 to $8 faced by import markets (see Figure 9-12). As a result, the relative costs between natural gas and nuclear generation are vastly different in

Egypt, with nuclear costing $43 per megawatt hour and natural gas generation costing less than $22 per megawatt hour.

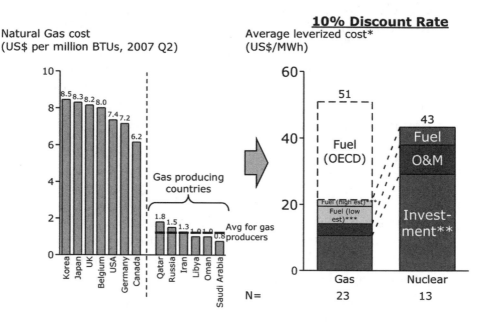

Notes and Sources:

*Average of data from survey of new facilities in 21 countries, mostly OECD but also include 4 developing countries. Leverized generation cost includes initial investment cost, operation and maintenance cost, fuel cost, and in the case of nuclear; main assumptions—85 percent capacity factor for plants, 40-year lifetime for coal and nuclear plants, for other plants lifetime come from country level responses, fuel price projection based on each country's models.

**Investment cost for nuclear power includes decommission cost.

***calculated as average gas producing country's gas price as a percent of the lowest and highest price in the OECD gas price data available; "Projected Cost of Generating Electricity 2005 Update"—NEA/IEA.

**Figure 9-12. Natural Gas Costs and Resulting Cost Comparison of Natural Gas and Nuclear Generation in Egypt.**

Egypt also faces the potential opportunity cost of exporting liquified natural gas (LNG) versus burning the gas for domestic consumption. Currently, Egypt produces enough gas to both export as LNG and satisfy domestic consumption. There may be a time in the future where supply of gas is not enough to satisfy both, and a choice has to be made between export and domestic use. However, the analysis of opportunity cost for Saudi Arabia shows that current conditions would still favor the use of gas for domestic energy generation until gas prices rise by a significant amount.

**Energy Security and Location of Energy Source.**

Egypt has significant natural gas reserves, and has also redoubled its efforts in exploration, with the goal of adding 30 trillion cubic feet of additional reserves by 2010.[40] Given its large indigenous supply, energy security is relatively high. Most of Egypt's gas fields are near the Nile River delta and the Sinai Peninsula, near its densest population centers.[41] There are also several gas fields, in addition to significant potential, in the Western Deserts which is connected to Cairo via an existing pipeline.[42] (See Figure 9-13.)

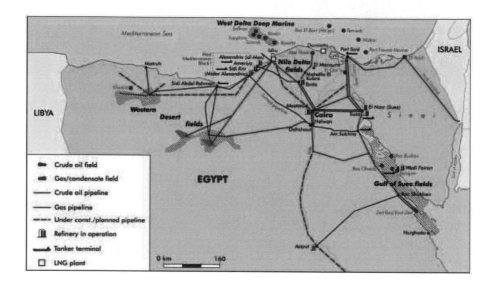

Source: *World Energy Outlook 2005* — IEA.

**Figure 9-13. Map of Reserves and Pipeline System
in Egypt.**

**Conclusion.**

In the foreseeable future, Egypt is likely to be able
to rely on its abundant and growing natural gas resources to power its economy. Natural gas supply disruptions are extremely unlikely in the medium term
due to its indigenous supply. The cost of natural gas
in Egypt is significantly lower than in other nations, so
that gas-fired generation is extremely cost competitive
versus other power sources, including nuclear. Thus,
there is no rush for Egypt to develop nuclear energy
from a resource and economic perspective.

# TURKEY

## Total Potential Capacity.

In 2004, Turkey had 36.8 gigawatts of electricity generation capacity.[43] The largest user of electricity is the industrial sector, which in 2004 accounted for 56.9 percent of electricity consumption.[44] Residential and commercial usage contributed to 22.8 percent and 12.9 percent of demand, respectively, with the remaining 7.4 percent of electricity consumption in government and public illumination.[45] Growth projections vary widely, with a low estimate showing total electricity generating capacity required by 2020 growing at 6.4 percent per annum and the high estimate showing growth to be 8.5 percent per annum.[46] This would imply that that 63 gigawatts to 99 gigawatts of generating capacity needs to be added between 2004 and 2020. This variation in electricity demand projections made by the Turkish Planning Commission is driven by the demand growth scenarios, with lower estimates assuming a higher mix of low energy intensive industries as part of Turkey's future GDP.

A variety of energy sources can be drawn on to meet the electricity generation demand. In 2004, the balance of sources used to generate electricity was pretty evenly spread between natural gas (41.3 percent), hydro (30.6 percent), and coal/lignite (22.9 percent), with the remaining 5.2 percent generated by oil and other resources.[47] Within these sources, natural gas has seen the most rapid growth in the past decade. In the coming decades, Turkey will likely diversify its energy supply so as not to rely on any one resource to meet its future electricity demand. See Figure 9-14.

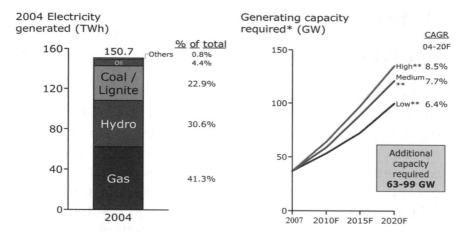

Notes:

*Generating capacity required projection based on growth rate in electricity generated projection.

**MAED-Model for Analysis of Energy Demand by the International Atomic Energy Agency—Scenario 1 is based on GNP growth by the Turkish Planning Organization (DPT) in May 2002, Scenario 2 based on GNP growth planned in Apr 2004, Scenario 3 assumes different production industry prediction than Scenario 2 (TBD), from Report of Turkey Long Term Electric Energy Demand), 2004. Ministry of Energy and Natural Resources.

Source: *Turkey Statistical Yearbook*, 2006.

**Figure 9-14. Turkey's Current and Projected Generating Capacity.**

**Natural Gas.**

Natural gas will most likely be the largest contributor to new electricity generation in the medium term. Turkey faced a financial crisis in 2001 which led to a contraction of the economy and subsequently a downward revision of its natural gas demand. This meant, however, that Turkey now has a significant over-supply of natural gas, stemming from a large number of natural gas import contracts signed previous to the economic contraction.[48] Thus, until the 2015-20 time period, tightness in supply of natural gas should not be a problem for Turkey. Given the numerous advantages of natural gas, from low emissions to ease of use as peak-load resource, the Turkish government is supportive of continuing to build more gas-fired power plants.

**Hydro.**

Hydropower will also likely be a main contributor to new generation capacity. Turkey is extremely well-endowed with hydro power resources, which currently contributes almost a third of its electricity generation. The government has ambitious plans to continue to develop the sector. It is currently undertaking a number of large projects, particularly the $32 billion Southeastern Anatolia Project (GAP) in the Tigris and Euphrates basin. The project would add a total of 22 dams and 19 hydro stations, representing 7.5 gigawatts of generation power, as well as tremendous irrigation capacity.[49] Phase I was completed in 2005 and the entire project should be completed by 2010. Some experts forecast even greater potential for hydro power, claiming that total generating capacity from hydro can reach 45 gigawatts by 2020, which would represent between 33-46 percent of the generation capacity required by 2020, depending on the growth scenario.[50]

**Coal.**

Coal and lignite will continue to be a part of the electricity generation scenario but more likely to be used as backup resources, perhaps to hedge against higher future gas prices. Two main reasons underlie why the government and the Turkish power sector has started to deemphasize coal. First, the indigenous coal in Turkey is of a very poor quality. Less than 7 percent of its total reserve of coal is "hard coal,"[51] whereas most comes in the form of lignite. Burning lignite and low quality coal is an extremely inefficient method of extracting energy. Second, Turkey's accession to the European Union (EU) has significantly influenced how it thinks about its carbon emissions. Understanding that as a full fledged member of the EU it will have to participate in a regime of capping emissions, the government has chosen to divert investment away from developing carbon-intensive coal power plants.

**Renewables.**

Renewable resources in Turkey have tremendous potential, but the pace of developing these resources is uncertain. Turkey's long coastline provides some of the best geographic conditions for exploiting wind energy. It is estimated that there are ~90 gigawatts worth of wind power in Turkey, with at least ~10 gigawatts able to be commercially and economically viable by 2020.[52] Turkey also enjoys one-eighth of the entire world's geothermal energy potential, estimated at 4.5 gigawatts of electricity and 31.1 gigawatts of thermal capacity.[53] While Turkey only has one geothermal power plant, exploratory projects are planned as are plans on how to use the thermal power of these sites to lower electricity demand. Turkey also enjoys high solarization levels and thus quite substantial potential for solar power development.[54] This potential has largely gone untapped, with only one PV based-grid connected solar electricity project. These renewable resources will likely be more fully developed in the future, since Turkey recently passed a Renewable Energy Law.

**Nuclear.**

Nuclear power has been proposed as one development path to decrease Turkey's reliance on imported hydrocarbon power. The government has a goal of building three nuclear power plants by 2012 with a total generating capacity of 4.5-5 gigawatts.[55] The first of these planned plants, to be built in the city of Mersin, went to bid by private contractors in March of 2008.[56] It remains to be seen whether these plants can be completed in time given the high likelihood of delays in nuclear power plant projects. If successful, nuclear power could provide a small portion of Turkey's generating capacity by 2020.

**Energy Efficiency.**

Energy efficiency efforts also hold considerable promise in Turkey. Tanay Sýdký Uyar, Vice President of the World Wind Energy Association and Associate Professor of Renewable Energy at Marmara University, estimates that "Turkey can cut its electricity needs by 50 percent if it uses more up-to-date energy efficient technology."[57] If Turkey can galvanize around policies that promote the use of more efficient technology in its industrial and building sector, it can be a large source of "negawatts."

**Relative Cost.**

Not being a significant producer of natural gas, Turkey does not enjoy the significantly lower price of gas generation as Saudi Arabia and Egypt, but renewable energy resources are likely more cost competitive. According to the IEA, the estimated cost of nuclear power in Turkey could be on par with that of gas generation.[58] However, given Turkey's geographic advantages in wind and geothermal, its renewable energy sources can be potentially more cost effective than in other countries. Country specific estimates place hydro and geothermal power lower than both gas and nuclear, with wind energy on par with gas and nuclear.[59] As such, nuclear power does not have any particular cost advantage in Turkey.

**Energy Security and Location of Energy Source.**

Other than low quality coal and lignite, Turkey does not have significant deposits of other fossil fuel resources, requiring it to import all of its natural gas and oil. Turkey's largest suppliers of natural gas are Russia and Iran, but it has actively looked to diversify its sources and began importing gas from Azerbaijan and Egypt, as well as LNG from Algeria and Nigeria in the late 1990s.[60] This move to diversify should give Turkey relatively more energy security regarding to importation of natural gas. The second factor that bolsters Turkey's energy security is that it is conveniently located so as to serve as the hub of energy transportation—both for shipping between the Black Sea and Mediterranean Sea and for natural gas pipelines from Russia, Central Asia, and the Middle East which supply continental Europe. Its position should provide Turkey relatively more bargaining power in securing oil and gas supplies. Domestically, the extensive system of pipelines will also ensure that energy supply can be easily transported to where they are needed. See Figure 9-15.

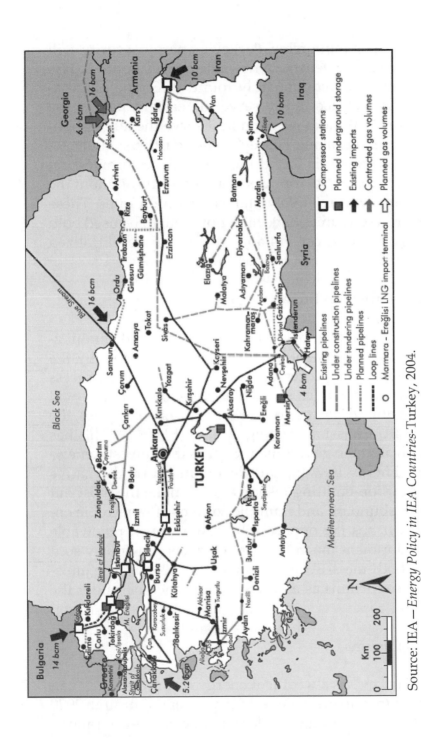

Source: IEA – *Energy Policy in IEA Countries*-Turkey, 2004.

Figure 9-15. Map of Pipeline System in Turkey.

Further reinforcing Turkey's energy security will be Turkey's extensive renewable energy sources. Hydro power, predominantly found in Southeast Anatolia, is well located to serve the eastern part of the country.[61] Wind power can best serve the large population centers along Turkey's long coastline,[62] and geothermal power is found mostly in the Southwest region.[63] Finally, solar power is available throughout the country with relatively high solarization rates. These renewable sources have high potential to be used both for population centers and to provide decentralized generation for dispersed population.

**Environmental Impacts.**

Turkey has moved increasingly toward minimizing its environmental impacts, largely driven by the EU accession process. As the EU accession process draws nearer to close, Turkey anticipates that it will be required to reduce carbon emissions under the next EU Emission Trading System (EU ETS). This has motivated the government to pass the newest Renewable Energy Law to spur development of alternative noncarbon-emitting resources and make the most out of its abundant and multiple sources of renewable energy. It has the opportunity now to push renewable development much faster than it has previously so it can both meet EU emission standards and use emission reductions as a new source of income under the ETS.

**Conclusions.**

Turkey has a wealth of energy resources that it can draw upon to meet its electricity demand through 2020 without requiring considerable nuclear development.

From a supply perspective, Turkey has relatively secure natural gas import security in the medium term, and can leverage its unique status as an energy transport hub to ensure security of the supply in the long term. Turkey also has much more abundant renewable energy resources that can be explored to meet demand. From a cost perspective, most of the generating sources in Turkey are on par with price, with certain renewable sources actually on the lower end of the cost comparison. Nuclear power is neither necessary to meet Turkey's future energy demand and neither does it have an overwhelming cost advantage.

## CONCLUDING REMARKS

The analysis of the economic and resource arguments for nuclear power energy in Saudi Arabia, Egypt, and Turkey shows that they are not as strong as politicians have articulated. In all three countries, there are alternative resources, either indigenous or comparatively secure, that, fully developed, can meet growing energy demands without additional investment in nuclear power. There is little confidence in the supposed cost advantage of nuclear power generation, given that cost calculation of nuclear plants, even in the United States and France, is obscured by a tradition of heavy government subsidization and existence of long lists of ancillary costs. This cost advantage becomes negative when evaluated against gas generation in many of the Middle Eastern nations that have significant gas production. Furthermore, nuclear development requires a considerable degree of public and private sector cooperation, which is best served by a high degree of government effectiveness, considerable regulatory strength, and a tight control on corruption. When even countries with such advantages struggle with nuclear development, it suggests that countries with fewer advantages may find it more challenging to develop nuclear power safely and effectively.

One key consideration in the potential development of nuclear power is when countries need to make a decision. Given that the countries under evaluation have secure and cost effective options in the medium term, a decision to develop nuclear power is not required at this time. The nuclear industry is seeking to initiate a revival in the West, with potentially safer and more cost effective designs which would make nuclear power more attractive on both a cost and an environmental basis. Developments are also underway on a variety of renewable energy sources as alternatives which could significantly lower their price and make them more feasible. With other resources to exploit; Saudi Arabia, Egypt, and Turkey can postpone decisions on nuclear power development and focus instead on developing the regulatory quality and governance needed to successfully execute such development. Furthermore, given the current economic and financial challenges facing the world, projects that require less upfront capital and allow the incremental building of generation capacity may be preferred by investors and governments.

## ENDNOTES - CHAPTER 9

1. Dan Murphy, "Middle East racing to nuclear power," *Christian Science Monitor*, November 1, 2007.

2. *Ibid.* "To have 13 states in the region say they're interested in nuclear power over the course of a year certainly catches the eye," says Mark Fitzpatrick, a former senior nonproliferation of-

ficial in the U.S. State Department who is now a fellow at the International Institute for Strategic Studies in London. "The Iranian angle is the reason." *Christian Science Monitor*, November 1, 2007.

3. "They feel politically threatened by Iran's nuclear program, they've pointed out rightly that Israel [hasn't been] a member of [nonproliferation] treaties for many years," says Jon Wolfsthal, a nonproliferation expert at the Center for Strategic and International Studies in Washington." *Christian Science Monitor*, November 1, 2007.

4. Rhoula Khalaf, "UAE set to launch nuclear programme," *Financial Times*, January 21, 2008, p. 4.

5. "Chain Reaction: Avoiding a Nuclear Arms Race in the Middle East," *Report to the Committee on Foreign Relations*, Washington, DC: U.S. Senate, February 2008, p. 15, available from *www.cfr.org/content/publications/attachments/BradleyBowman.pdf*.

6. Based on expert comments during the Nonproliferation Policy Education Center Conference, Prague, Czech Republic, March 18, 2008.

7. Jim Harding, "Reactor Economics in a Carbon Constrained World," November 2007, available from *www.npec-web.org/Presentations/DRAFT-20071105-Harding-ReactorEconomics.pdf*.

8. Communications: U.S. National Renewable Energy Laboratory (NREL) PowerPoint Presentation Template with Light Background, available from *www.nrel.gov/analysis/docs/cost_curves_2005.ppt*.

9. *Saudi Arabia Electricity and Cogeneration Regulatory Authority Annual Report 2006*, International Energy Agency Statistics; *World Energy Outlook 2005: Middle East and North Africa Insights*, Paris, France, 2005.

10. *Ibid.*

11. *Ibid.*

12. Projection based on base case scenario in *World Energy Outlook 2005*, from the *Saudi Arabia Electricity and Cogeneration Regulatory Authority Annual Report 2006*.

13. Telephone discussion with industry expert, February 2008.

14. This figure was derived from independent analysis conducted by Dalberg and Associates.

15. *Ibid.*

16. "Country Analysis Brief — Saudi Arabia," Energy Information Administration, 2006.

17. *World Energy Outlook 2005*.

18. *Ibid.*

19. *Ibid.*

20. *Egyptian Electric Holding Company Annual Report 2005/06*, available from *www.egelec.com/*.

21. *Ibid.*

22. *Ibid.*

23. *Ibid.*

24. *World Energy Outlook 2005*.

25. *Ibid.*

26. *Ibid.*

27. *Egyptian Electric Holding Company Annual Report 2005/06*.

28. *Ibid.*

29. *World Energy Outlook 2005*.

30. "Country Analysis Brief — Egypt," Energy Information Administration, 2006.

31. *World Factbook*, "Egypt," Central Intelligence Agency (CIA), 2007.

32. *Ibid.*

33. *Egyptian Electric Holding Company Annual Report 2005/06.*

34. *World Energy Outlook 2005.*

35. *Ibid.*

36. *Egyptian Electric Holding Company Annual Report 2005/06.*

37. *Country Profile: Egypt*, Economist Intelligence Unit N.A. Inc., 2008.

38. "Egypt approves plans to build three new nuclear power stations," BBC, September 26, 2006.

39. *Ibid.*

40. "Country Analysis Brief—Egypt."

41. *Ibid.*

42. *World Energy Outlook 2005.*

43. "Power Installed of Power Plants, Gross Generation and Net Consumption of Electricity," *Turkey Statistical Yearbook*, 2006, available from *www.turkstat.gov.tr/VeriBilgi.do?tb_id=11&ust_id=3.*

44. *Ibid.*

45. *Ibid.*

46. O. Yuksek, M. I. Komurcu, I. Yuksel, K. Kaygusuz, "The role of Hydropower in meeting Turkey's electric energy demand," *Energy Policy*, Vol. 34, 2006, pp. 3093-3103.

47. "Power Installed of Power Plants, Gross Generation and Net Consumption of Electricity."

48. "Country Analysis Brief—Turkey," Energy Information Administration, 2006.

49. "Energy Policy in IEA Countries—Turkey," IEA, 2005.

50. Yuksek, Komurcu, Yuksel, and Kaygusuz, pp. 3093-3103.

51. "Country Analysis Brief—Turkey."

52. "Energy Policy in IEA Countries—Turkey," IEA, 2005, p. 123.

53. *Ibid.*, p. 121.

54. *Ibid.*

55. "Country Analysis Brief—Turkey."

56. "Turkey inaugurates first nuclear plant tender," *Associated Press,* March 24, 2008.

57. Yuksek, Komurcu, Yuksel, and Kaygusuz, pp. 3093-3103.

58. Erkan Erdogdu, "Nuclear power in open energy markets: A Case Study of Turkey," *Energy Policy*, Vol. 35, 2007, pp. 3061-3073.

59. *Ibid.*

60. "Energy Policy in IEA Countries—Turkey."

61. "Country Analysis Brief—Turkey."

62. "Energy Policy in IEA Countries—Turkey."

63. *Ibid.*

## CHAPTER 10

## CIVILIAN NUCLEAR POWER IN THE MIDDLE EAST: THE TECHNICAL REQUIREMENTS

James M. Acton
Wyn Q. Bowen

For a developing country contemplating the construction of its first nuclear power plant (NPP), the technical requirements alone can appear daunting. This is before the legal, regulatory, economic, and political dimensions are brought into the mix. As the International Atomic Energy Agency (IAEA) noted in a recent report, launching an NPP "is a major undertaking requiring careful planning, preparation and investment in a sustainable infrastructure that provides legal, regulatory, technological, human and industrial support to ensure that the nuclear material is used exclusively for peaceful purposes and in a safe and secure manner."[1]

Against this background, this chapter seeks to examine the feasibility of three proposed new nuclear power programs in the Middle East: in Egypt, Saudi Arabia, and Turkey. Its aim is to explore the extent to which each of these countries currently has "what it takes," i.e., meets the technical and regulatory requirements, to build and operate an NPP. These three states have been chosen because, of all the states that have recently shown an interest in nuclear power, there has been the most speculation about their intentions. Moreover, these three states are useful case studies. Turkey, for instance, which has both a relatively strong economy and a relatively well-developed nuclear program, is representative of Libya. Saudi Arabia, like the United Arab Emirates (UAE) and Qatar for instance, is very rich but has comparatively little extant nuclear expertise. Egypt, a relatively poor state which already has extensive nuclear expertise, lies at the other end of the spectrum and, in this respect, appears to be unique among Middle Eastern states.

All three states have shown interest in nuclear energy at various times over the past half century. As discussed below, Egypt and Turkey have made repeated attempts to acquire nuclear reactors but without success. This chapter aims to shed some light on the question of whether it will be different this time and hence to contribute to the broader debate about the management of the "nuclear rennaissance." In spite of the growing literature about the intentions of states seeking to develop nuclear power, remarkably little has been written on this question of capabilities.[2] The answers that this chapter provides should be regarded as preliminary. They are based on an analysis of available open-source literature, and consequently there are a number of issues that we fully acknowledge are not authoritatively addressed. Issues requiring further study are indicated in the text.

The chapter begins with a framework for analysis which sets out in generic terms the technical, legal, and regulatory requirements to build and operate an NPP with a capacity of approximately 1 gigawatt (GWe). The framework is then applied to Egypt, Saudi Arabia, and Turkey. The chapter identifies the areas in which the three countries are currently deficient and, in doing so, generates greater understanding of the feasibility of proposed new nuclear power programs in the Middle East.

## A FRAMEWORK FOR ANALYSIS

The difficulty of developing a first NPP clearly depends on a number of factors including, most importantly, the degree to which the host state uses external assistance. In this chapter, we make the following assumptions about the nuclear power project:

1. The NPP will be of approximately 1 GWe capacity;

2. The NPP will be supplied by a major supplier state (such as Russia, the United States, France, Germany, South Korea, Canada, etc.)

3. The NPP will be either a pressurized water reactor (PWR) or a boiling water reactor (BWR);

4. The NPP will be procured from an external supplier under a "turn-key" contract, which includes the provision of fuel and the repatriation of spent fuel. Whether, in practice, any supplier state is willing to take back spent fuel (repatriation) remains to be seen. Nevertheless, this assumption is still included because it leads to conservative conclusions;

5. The NPP contract will include a technology transfer clause, including training, to help the host state establish a domestic skills base, as well as local suppliers capable of supporting the nuclear power sector; and,

6. The NPP will be run by the host nation through an operating organization with the requisite "rigor, culture, ethics and discipline needed to effectively manage nuclear power technology with due regard to the associated safety, security, and nonproliferation considerations."[3]

This set of assumptions is *not* a prediction about how any given state will choose to develop nuclear power; it is a model that might be adopted and that we utilize here to make the ensuing discussion more concrete. There are, of course, other plausible models. For instance, a state could further reduce the challenge of developing an NPP by contracting out its operation, as well as construction, to an external supplier.

Based on this model, the analytical framework is broken down into the following components: (1) the staffing requirements for the operation and maintenance of the NPP; (2) the legal and regulatory framework for the siting, construction, commissioning, operation, and decommissioning of the NPP; (3) the suitability and reliability of the electricity grid for the NPP and its proper and safe operation; and (4) the waste management and decommissioning requirements. The extra requirements imposed by desalination, which has been cited as one possible use for an NPP, have not been included since they are very modest. Typically, for instance, desalination represents "less than 5% of the total plant cost."[4]

The framework is largely based on IAEA guidance. To test whether this guidance is actually reflective of state practice, the framework is illustrated with the case of Slovenia. This country was selected because it has relatively modest national resources and operates just one NPP (a PWR) at Krško (jointly with its neighbor, Croatia). The Westinghouse-supplied reactor is rated at 730 megawatt (MWe) gross[5] and the fuel is provided by the supplier.[6] The plant commenced commercial operations in 1983,[7] and as a result of the dissolution of Yugoslavia, the Krško NPP is jointly owned by Slovenia and Croatia with half of the electricity generated going to each country.[8]

## STAFFING REQUIREMENT FOR THE CONSTRUCTION AND OPERATION OF AN NPP

When first contemplating the development of nuclear power, several staffing requirements need to be taken into account. Most obviously, there is the need for "a fully staffed nuclear power plant operation, maintenance, and technical support organization."[9] The IAEA estimates that this requires between 200 and 1,000 staff.[10] Indeed, in 2007 the Krško plant employed 573 staff.[11] However, a report published by the Office of Technology Assessment in 1993 noted that staffing at "single unit nuclear plants" in the United States increased from an average of about 150 employees to over 1,000" from 1977 to 1990. This expansion in the number of operating personnel occurred partially as a result of larger plants going online but also because of growing regulatory requirements, among other things.[12] Moreover, it is unlikely that the figure of 1,000 includes the personnel required to refuel and refit nuclear power plants every 12 to 24 months. For example, a July 1975 study notes that the annual refueling operation performed by General Electric at Boston Edison's 690 MWe Pilgrim nuclear power plant in Plymouth, MA, took "about 6 weeks" with General Electric (GE) bringing in about 40 specialized personnel to oversee the work along with a further 80 staff employed by subcontractors to provide assistance.[13] It is quite probable that these numbers have grown since the mid-1970s in part because of expanding regulatory requirements.

In order to operate an NPP in an effective and safe manner, the workforce will require technical skills in a range of disciplines including nuclear engineering, instrumentation and control, electrical engineering, mechanical engineering, radiation protection, chemistry, emergency preparedness, refueling and refitting operations, and safety analysis and assessment.[14] Creating this expertise requires "enhanced educational opportunities for nuclear science and technology."[15] At Krško, for instance, more than one-third of the 573 staff have what is described as, "higher, high, or university education."[16] On top of the relevant "scientific, engineering, and other technical education," NPP staff are usually expected to have "3 or more years of specialized training and experience prior to the initial fuel loading" of the plant.[17] In terms of staffing requirements for the operation of a first NPP then, a credible plan will require education and training programs to produce the human resource base to ensure that there is "a continuing flow of qualified people to all areas of the programme. . . ."[18]

Under our model, a great deal of the necessary training and experience is initially provided by the external supplier of the NPP as part of the contract (again, using this assumption leads to conservative conclusions; whether all suppliers are willing and able to provide this level of service in practice remains to be seen.)[19] For reasons of long-term sustainability, however, we also assume that the host state also wants to establish a domestic skills base and local suppliers who are capable of supporting the NPP in the future. Here, international assistance may also be useful.[20] In Slovenia, for example, the Krško plant is a member of Westinghouse's Pressurised Water Reactor Owners' Group.[21] It is also a member of the Nuclear Maintenance Experience Exchange (NUMEX) in order to further assist with relevant knowledge transfer.[22] Moreover, to address the challenge of an aging workforce, the IAEA recently participated in a joint mission with the World Association of Nuclear Operators at Krško to capture tacit knowledge from retiring workers.[23] Based on the available literature, it is impossible to assess the effectiveness of these programs.

## LEGAL AND REGULATORY REQUIREMENTS

A country embarking on a nuclear power program needs to establish a comprehensive legislative framework encompassing all issues related to the application of nuclear energy. The framework needs to cover site selection, licensing, commissioning, decommissioning, safety, security, safeguards, transport, and liability as well as "the commercial aspects related to the use of nuclear material."[24] While the legal frameworks of other countries can be used as a guide, there is a need to localize the framework by taking into account the existing constitutional and legislative base of the country, "cultural traditions, scientific, technical and industrial capacities, and financial and human resources."[25]

When planning a new nuclear power program, a process needs to be established for the regulatory organization to authorize the siting, commissioning, and operation of the NPP. This process can (but does not have to) incorporate an independent safety review of the reactor design—a particularly challenging task. The regulator will also require "the capabilities to plan and implement the review and safety assessment activities of the proposed facility throughout its life."[26] There is also some "shared functions" including, for example, "emergency preparedness and response, national and international cooperation, dissemination of technical and scientific information, environmental assessment, and communication with the public and other stakeholders."[27] The regulator should be able to fulfill its mandate in an independent manner with "clear authority and adequate human and financial resources."[28] However, the principal responsibility for the safety of facilities lies with the operating organization.[29]

Legislation covering liability in case of an accident is particularly important but also complex to implement. Liability is discussed in depth in another chapter in this volume, but one issue worth discussing here, because it has the potential to significantly affect the development of nuclear power, is the 1997 Convention on Supplementary Compensation for Nuclear Damage (CSC). The CSC links "countries with strong nuclear liability systems . . . [to] distribute the economic burden among several countries through a system of contributions by the member States in the unlikely event there were another catastrophic nuclear accident."[30] As of May 2008, only four states (Argentina, Morocco, Romania, and the United States) had the ratified the convention, and it had not yet entered into force.[31] However, U.S. nuclear firms have made it clear that they will not trade with states that have not adopted the CSC.[32] In contrast, French and Russian firms probably will. Preparing and implementing liability legislation is a significant challenge for a regulator but does affect the nuclear assistance that states can receive.

Based on an examination of existing national regulatory structures, the IAEA estimates that a nuclear regulatory organization comprising "30–50 staff members would be necessary for starting the implementation of a nuclear power plant programme."[33] It is important that "the technical training, knowledge, and capabilities" of these employees is "adequate for competent interaction with the owner/operator, supplier organizations and consultants."[34] External help is generally available to develop the required human resources.

The Slovenian regulator, the Slovenian Nuclear Safety Administration (SNSA), employed 34 people in 1999 comprising engineers, physicists, and other technical and administrative staff.[35] Croatia's State Office for Nuclear Safety commenced work on June 1, 2005.[36] As of March 29, 2007, 12 of the Office's 18 staff positions had been filled.[37] This makes a total of 52 staff devoted to the regulation of nuclear safety in both countries, although it does not appear that staff from Croatia's State Office for Nuclear Safety are connected to the Krško plant. Interestingly, whether the figure is 34 or 52 staff, this appears to be relatively staff heavy when only one nuclear power plant is involved. In the United Kingdom (UK), for example, some 250 staff work in the safety activities of the Nuclear Directorate in the Health and Safety Executive (HSE) which regulates the country's nuclear industry; three fifths of this figure are technical staff.[38] The HSE is responsible for regulating nuclear safety at 10 nuclear power plants (19 reactors in total) as well as conversion, fuel fabrication, enrichment, and reprocessing facilities at Springfields (conversion, fuel fabrication), Capenhurst (enrichment) and Sellafield (reprocessing, MOX fuel fabrication).

The SNSA's 1999 budget was €1·4 million.[39] As a result of purchasing an American NPP, Slovenia has acquired a good knowledge of American regulations and benefits from the U.S. training regime; SNSA inspectors are trained at the U.S. Nuclear Regulatory Commission, and the SNSA receives information on modifications to NRC regulations when they are implemented.[40] The Slovenian experience would suggest, therefore, that the choice of country from which to buy an NPP should not be seen as related solely to initial design and construction characteristics, but rather as a long-term partnership including the emulation of that country's regulatory standards. However, the extent to which different suppliers are willing and able to offer this degree of cooperation is an open question.

## ELECTRICAL GRID REQUIREMENTS

A country's existing and planned electrical grid must also be taken into account when contemplating the initiation of an NPP. The main issues in this respect include whether the grid is stable and large enough to absorb the planned output from the NPP, and whether it is sufficiently reliable to ensure the steady and safe operation of the plant. It is generally accepted that an individual power plant should not constitute over 5-10 percent of a grid's total installed capacity.[41] Moreover, it is also necessary to have two independent and reliable sources of electricity for the plant to assure the continued operation of the reactor control systems. This could turn out to be a particular challenge for developing countries, even those with large electricity grids.

A major problem with electrical grids in many developing countries is that they are both small and unreliable. In such circumstances, any serious effort to launch a NPP requires a plan to improve the grid's reliability. It also entails a plan to increase the grid's size or to opt for a smaller NPP (although for the purposes of this chapter, we assume the former).[42] The grid size could be increased by domestic expansion or integration with grids in neighboring countries or on a wider regional basis. By integrating at the regional level, the resultant grid is likely to be significantly larger than a national system and probably more reliable as a result.[43]

Slovenia offers an interesting example. The Krško plant generates 40 percent of the total electricity produced in Slovenia.[44] On face value, this figure would suggest that Krško's contribution to the national grid is well above the level recommended by the IAEA. However, Slovenia is connected to the Union for the Coordination of Transmission of Electricity (UCTE) grid. Indeed, Krško reportedly plays a significant role in stabilizing voltages for UCTE as a whole.[45]

## WASTE MANAGEMENT AND DECOMMISSIONING REQUIREMENTS

Any politically-acceptable plan to initiate an NPP needs to account for waste management and decommissioning. In terms of high-level radioactive waste, one of the key assumptions highlighted earlier is that the external supplier is responsible for repatriating spent fuel (after a period of initial storage in proximity to the NPP). This requires only the construction of interim storage for spent fuel, not a long-term repository. An agreement with the external supplier to repatriate spent fuel obviously lessens the disposal demands on the recipient and generates international confidence in nonproliferation. The alternative to the repatriation of high-level waste is the construction of a national disposal site which "for a small nuclear programme could be prohibitively expensive."[46]

The quantity of low-level waste (LLW) produced by a nuclear reactor depends highly on the type of reactor and the way that waste is treated. For instance, in 1979, typical annual disposal volumes for LLW from commercial light water reactors were in the range of 500-1,500 m³.[47] Since then, Western reactors have typically reduced disposal volumes by "at least an order of magnitude," whereas disposal volumes from Russian VVER reactors have remained essentially unchanged.[48] In the United States in 1997, for instance, the average disposal volume per reactor was about 55 m³.[49] A state has to develop a repository for the final disposal of this LLW, which it would presumably handle "in accordance with the procedures that have been established for the management of existing radioactive materials, such as radioactive sources and radioactive waste generated by medical use of radioactive substances."[50]

Planning for the decommissioning of an NPP should be integral to the process of developing it. This involves planning for the clean up of all radioactivity associated with the NPP as well as its dismantling. While some 99 percent of the radioactivity is associated with spent fuel, the remainder involves "surface contamination of plant" and radioactivity from "'activation products' such as steel components that have long been exposed to neutron irradiation."[51] About 6,200 tons of radioactive material can be expected to result from the decommissioning of a 1 GWe PWR.[52] This requires a suitable site for final disposition.

One important reason for planning for decommissioning in advance is so that a country is able to opt for pre-payment by, for example, depositing money "in a separate account to cover decommissioning costs even before the plant begins operation."[53] Alternatively, as in the United States, an "external sinking fund (Nuclear Power Levy)" is the preferred method by which such a fund "is built up over the years from a percentage of the electricity rates charged to consumers."[54]

In the Slovenian context, responsibility for waste disposal and decommissioning is shared between Slovenia and Croatia.[55] Like many other countries, there is a debate in Slovenia over what to do with Krško's high level waste as the deal to supply the reactor did not include provisions for the repatriation of spent fuel.[56] Krško's waste and spent fuel is stored at the plant itself and existing capacity is reported to be sufficient until 2023.[57] On the decommissioning front, the Croatian government drafted a plan in September 2007 to set up a fund of €350 million for the dismantling of the Krško NPP to be funded equally by Croatian and the Slovenian governments.[58] Depending on the method of decommissioning Krško, roughly half the costs are attributable to spent fuel management.[59] While the Slovenian and Croation governments are obviously planning for the decommissioning of the plant, it would appear that the issue of long-term waste disposal has yet to be resolved, which illustrates the importance of factoring this in at the very outset of the planning stages.

## PLANNED NUCLEAR POWER PROGRAMS IN EGYPT, SAUDI ARABIA, AND TURKEY

Having set out a framework for examining the technical and regulatory requirements for the development of an NPP, we proceed to examine what that framework indicates about the challenges that Egypt, Turkey, and Saudi Arabia need to confront. It is informative to begin by providing some background on each of the country's nuclear programs and ambitions.

**Egypt.**

Egypt has had an interest in developing nuclear power since the 1960s. It entered into numerous sets of negotiations and even signed contracts for the provision of nuclear reactors with, among others, Siemens and Westinghouse, but without any significant results.[60] In September 2006, however, the Egyptian government announced that it was reinvigorating its civil nuclear power program. Currently the country's power requirements are fulfilled largely by oil and gas, but Egypt has been experiencing supply shortages at a time of rapidly increasing demand. The Energy and Electricity Minister Hassan Younis stated in March 2007 that under current projections Egypt will build "10 nuclear-powered electricity-generating stations across the country."[61] El Dabba is reported to be the location for the first NPP.[62] Several countries have recently said they would work with Egypt in the context of providing the technology and materials to launch a nuclear power program. These include Canada, China, France, Germany, Russia, South Korea, and the United States.

**Saudi Arabia.**

Saudi Arabia is perceived to be the prime motivator of the announcement by the Gulf Cooperation Council (GCC) in December 2006 which states that the organization is launching a "joint programme in nuclear technology for peaceful purposes, according to international standards and arrangements."[63] Their reported plan is to start developing a first joint NPP by 2009—a target that is certain not to to be met. The UAE appears to have made the most progress so far of any of the GCC states, and recently published a white paper on nuclear energy which "renounc[ed] any intention to develop a domestic enrichment and reprocessing capability."[64] For its part, Saudi Arabia has demonstrated an interest in developing a nuclear power capability since the 1970s, which has been motivated, in part at least, by its potential application in the field of desalination.[65] Several NPP supplier countries have offered their services to the GCC as a whole as well as to individual members. In addition to a commitment from the IAEA to provide technical expertise, Saudi Arabia has received offers of assistance from Russia, France, and the United States.

A "US-Saudi Memorandum of Understanding (MOU) on Civil Nuclear Energy Cooperation" signed in May 2008 commits the United States to:

> assist the Kingdom of Saudi Arabia to develop civilian nuclear energy for use in medicine, industry, and power generation and will help in development of both the human and infrastructure resources in accordance with evolving International Atomic Energy Agency guidance and standards.[66]

Under the MOU, Saudi Arabia also "stated its intent to rely on international markets for nuclear fuel and to not pursue sensitive nuclear technologies, which stands in direct contrast to the actions of Iran."[67] In doing so, Riyadh has committed itself not to develop uranium enrichment and plutonium reprocessing capabilities if it accepts U.S. assistance. This commitment is not binding if Saudi Arabia opts not to receive American assistance and deals with other suppliers instead (indeed, some have questioned the likelihood of the United States insisting upon this condition being written into a reactor procurement contract).

**Turkey.**

Like Egypt, Turkey has tried on multiple previous occasions to develop nuclear power. These attempts have failed because of the economic costs involved as well as environmental, safety and proliferation concerns.[68] Today, Turkey is a net energy importer and nuclear power again appears attractive, given that its electricity consumption is increasing at a time of rising energy prices. In March 2008, Turkey issued a tender, calling for bids to construct the country's first NPP at Akkuyu on the Mediterranean coast. Potential technology suppliers for Turkey's renewed program include South Korea, Canada, Germany, and the United States.

## STAFFING REQUIREMENTS FOR THE CONSTRUCTION OF AN NPP

**Research Reactors.**

Research reactors are very useful for training a workforce in most, if not all, skills that are needed for an NPP. Although research reactor staff require additional training before being able to operate a power reactor, they are nonetheless among the most usefully-skilled personnel in a state constructing its first NPP.

From this perspective, the best prepared state is Egypt, which has two research reactors: the ETRR-1 (a 2 MWt Russian-supplied tank-type reactor) and the ETRR-2 (an Argentine-supplied 22 MWt pool-type reactor).[69] In particular, as a relatively high-powered research reactor, the ETRR-2 is especially relevant for training power reactor operators.[70] Nevertheless, given that between them, Egypt's two research reactors only employ a total staff of 60 (of which 22 are operators), it would be a challenge for Egypt to train the 200-1,000 personnel required for the operation of an NPP.[71]

Turkey definitely has one operational research reactor: the ITU-TRR (a U.S.-supplied 250 kW, TRIGA Mark II reactor).[72] This reactor, located at the Institute for Nuclear Energy at Istanbul Technical University, is less suited than Egypt's to training power reactor operators because it has a much smaller staffing complement (two operators and a further four staff). In addition, Turkey does have a second reactor, the TR-2 (a 1 MWt U.S.-supplied pool-type reactor that was subsequently upgraded to 5 MWt by Belgium), but reports on whether it is currently operational are contradictory.[73] Located at the Çekmece Nuclear Research and Training Center, it has been used for training, research, and isotope production.

Saudi Arabia does not have a research reactor. Saudi scientists have conducted theoretical studies into research reactor design and some very specific aspects of power reactor technology (such as the best type of concrete to use as shielding).[74] In addition, King Abdul Aziz University has reactor simulator software for use in training students.[75] Although of some relevance, such theoretical training cannot compensate for hands-on experience. If Saudi Arabia is to develop a nuclear power program, the purchase of a research reactor would likely be a useful investment.

**Nuclear Activities.**

A second key group of skilled personnel are those with experience of other relevant areas of industry or academia. Some of these could be of direct use in an NPP. For instance, although the model of nuclear power development used in this chapter assumes both fuel provision and take back, a state with experience with other parts of the fuel cycle (such as waste disposal or fuel fabrication for research reactors) would have personnel with training in disciplines such as radiation protection or chemistry who could be retrained to work in an NPP. More generally, the following material is intended to provide an indication of the overall level of nuclear expertise in a state. For instance, the fact that there are a relatively small number of Ph.D.s working in Saudi Arabia's premier nuclear research institution, the Atomic Energy Research Institute (AERI), is significant because it is presumed to be indicative of a general lack of nuclear expertise, not because a large number of doctoral-level scientists is necessarily required to operate a nuclear reactor. We have not attempted to develop a quantitative metric for the level of expertise required to operate a nuclear reactor, although it would be very valuable to do so.

Based on a survey of open-source literature, the fuel cycle activities conducted by Egypt, Saudi Arabia and Turkey are summarized in Table 10-1. Waste management has not been included but is discussed here.

| | Egypt | Turkey | Saudi Arabia |
|---|---|---|---|
| Mining | Exploratory mining[76] | Some research, survey work and feasibility studies into uranium; interest in thorium[77] | Some research, survey work and feasibility studies[78] |
| Milling | Significant research[79]<br><br>Facilities: Inshas Pilot Plant (used 1990—1996; little current information)[80]; Phosphoric Acid Purification Plant (Inshas, designed to extract uranium from phosphate ore; used for non-nuclear purposes)[81] | Significant research[82]<br><br>Facility: Koprubasi Uranium Pilot Plant (little information available; status unclear)[83] | Occasional research of slight relevance[84] |
| Conversion | Bench-scale experiments before 1982 (when Egypt's safeguards agreement with the IAEA entered into force);[85] one more recent publication of potential relevance[86]<br><br>Facilities: Nuclear Chemistry Building (Inshas) | Significant bench-scale research[87]<br><br>Facility: Nuclear Fuel Pilot Plant (Instanbul) | No activities identified |
| Enrichment | One bench-scale project of potential relevance[88] | One bench-scale project of potential relevance[89] | No activities identified |

**Table 10-1. Fuel-Cycle Activities in Egypt, Saudi Arabia, and Turkey.**

| | Egypt | Turkey | Saudi Arabia |
|---|---|---|---|
| Fuel Fabrication | Capability to fabricate fuel for the ETRR-2[90]<br><br>Facilities: Fuel Manufacturing Pilot Plant (Inshas); Nuclear Fuel Research Laboratory (Inshas)[91] | Significant bench-scale research<br><br>Facility: Nuclear Fuel Pilot Plant; Nuclear Applications Laboratory (METU)[92] | No activities identified |
| Reprocessing (including irradiation experiments and isotope separation facilities) | Continuous but low intensity bench-scale research from before 1982 to 2003;[93] some research conducted by the Atomic Energy Authority is potentially relevant[94]<br><br>Facilities: Hot cells at the ETRR-1 and ETRR-2 (Inshas) and the Hot Laboratory and Waste Management Center (HLWMC; Inshas);[95] Hydrometallurgy Pilot Plant;[96] Nuclear Chemistry Building;[97] Radioisotope Production Facility (under construction)[98] | Sporadic largely-theoretical research since 1980s[99] | No activities identified<br><br>Facilities: Hot cells at the King Faisal Specialist Hospital and research Center (Riyadh); separation laboratories at the Atomic Energy Research Institute[100] |

Based on an article by James M. Acton and Wyn Q. Bowen, "Nurturing Nuclear Neophytes," *Bulletin of the Atomic Scientists*, September-October 2008.

## Table 10-1. Fuel-Cycle Activities in Egypt, Saudi Arabia, and Turkey. (cont.)

Egypt has the most impressive track-record record having conducted significant research and development (R&D) activities across the whole fuel cycle (except enrichment). Probably the most sophisticated fuel cycle facility in Egypt is the Fuel Manufacturing Pilot Plant at Inshas. This is a semi-pilot facility provided by Argentina to produce the fuel elements for the ETRR-2 reactor.[101] Although some of Egypt's activities were undeclared and subject to an IAEA investigation from 2004-05, its past activities demonstrate that it possesses a range of skilled nuclear workers and the means to train them.

Like Egypt, Turkey has conducted research into many stages of the fuel cycle. From open-source literature, the current status of many of these activities is hard to determine but, on balance, it appears that Turkish research and development efforts generally lag slightly behind those of Egypt. Nonetheless, the range of nuclear activities conducted by Turkey and the in-depth nature of some of this research clearly indicates that Turkey starts from the position of a relatively strong nuclear sciences base.

In contrast, Saudi Arabia has only limited experience of nuclear activities. In fact, the survey revealed that the only part of the fuel cycle of which Saudi Arabia has significant experience is mining and milling—the least relevant part from the perspective of developing nuclear power.

## NATIONAL TRAINING AND RESEARCH INFRASTRUCTURE

Based on the analysis provided above, it is evident that the three states considered by this chapter need to educate and train substantial numbers of scientists, engineers, and technicians. To accomplish this, they would require a strong university sector capable of producing suitable graduates. In addition, a strong national infrastructure capable of coordinating and implementing a national strategy would be a significant asset.

Turkey has a strong nuclear infrastructure based around several national organizations and universities.[102] Turkey's relative strengths in this respect are partly the result of the country's previous failed attempts to set up nuclear power plants. Turkey's biggest strength is perhaps its university sector. Our survey identified 11 universities that have significant teaching and/or research experience relevant to the development of nuclear power (in addition, a number of other universities also appear to have some kind of relevant expertise). Moreover, there are very strong interconnections between the universities and the Turkish Atomic Energy Commission (TAEK). These should enable the country to implement a coherent national strategy. Indeed, according to the OECD, in 2002 the Turkish government's energy R&D budget (excluding TAEK activities) was $3.33 million (and projected to rise to $5.51 by the following year).[103] By contrast the budget of TAEK was about $50 million. Although TAEK is responsible for conducting many activities other than R&D, these figures clearly demonstrate the importance accorded by the Turkish government to nuclear power as part of the state's potential future energy mix.

In addition to its universities, Turkey has eight national research institutions or facilities of potential relevance to a nuclear power program (although some, such as those that focus on the use of radioisotopes in agriculture, are only of tangential relevance).[104] Some of these institutions are listed in Table 10-1, or discussed elsewhere in this chapter. However, of particular relevance for developing a skills base is the Çekmece Nuclear Research & Training Centre's (CNAEM-CNRTC) which has a "programme of work" that is "coordinated with TAEK's nuclear programme in support of the national economy, and focuses on nuclear technology, applications and training."[105]

Egypt's nuclear infrastructure is broadly similar to Turkey's, if not quite as extensive. Seven universities in Egypt were identified as having significant teaching and/or research experience in fields relevant to the development of a nuclear energy.[106] Academics from a further two universities have published nuclear-related papers.[107] Of Egypt's universities, Cairo University, which offers postgraduate courses in nuclear reactors and radiation physics, appears to be the most important from a teaching perspective. As in Turkey, there are strong interconnections between Egypt's Atomic Energy Authority (AEA) and Egyptian Universities. There are six state-sponsored research centers in Egypt that are potentially relevant to the development of an NPP.[108] Of these, by far the most significant is the Nuclear Research Center at Inshas, where most of Egypt's key nuclear facilities, including its two research reactors, are located.

Saudi Arabia's nuclear infrastructure is considerably weaker than either Turkey's or Egypt's. Only King Abdul Aziz University was positively identified

445

as offering relevant courses. In particular, the courses offered by the Nuclear Engineering Department include nuclear instrumentation, nuclear reactor safety, and nuclear desalination. It seems probable, however, that the other two universities in Saudi Arabia which have conducted significant nuclear-related research (King Fahd University of Petroleum and Minerals and King Saud University) also offer relevant courses (although no specific information on them was obtained). Faculty from at least four other universities in Saudi Arabia have been co-authored on at least one nuclear-related publication, suggesting that these universities may perhaps offer nuclear-related courses as part of their scientific curricula.[109]

The Atomic Energy Research Institute (AERI) is Saudi Arabia's premier state-sponsored institution for research into nuclear energy. It has four programs of potential relevance to a nuclear power program: industrial applications of radiation and radioactive isotopes, nuclear power and reactors, nuclear materials, and radiation protection. Based on its publications, however, the focus of its work appears to be on radioactive waste storage and environmental monitoring. Moreover, according to an interview in 2001 with the General Inspector of AERI, there are only 15 Saudi nationals working there who hold Ph.D.s in relevant subjects (in addition to a number of highly trained foreign workers).[110] Moreover, there are two Saudi state-sponsored institutes that focus on the civilian use of radioisotopes: King Faisal Specialist Hospital and Research Center (which has a cyclotron and hot cells for radioisotope production) and the National Research for Agriculture and Animals Resources Center (which has a Radiation Measurement Division).[111]

446

Egypt, Saudi Arabia, and Turkey have all initiated various technical cooperation projects with the IAEA to develop their skills base. It appears that all the projects that are of direct relevance to nuclear power are related to regulation, waste management, or desalination and are discussed below. One partial exception is a very general ongoing project with Egypt "[t]o enhance the National Information and Documentation Centre (IDC) to become the main national nuclear information services centre in Egypt."[112] It is interesting to note that Egypt has been by far the most active in its use of technical cooperation, having initiated 33 projects since 2000 (by contrast, Turkey has initiated 19 and Saudi Arabia, 13). This is suggestive of a very deliberate strategy by Egypt to improve its skills base. Moreover, from the information that is available, it appears that the technical cooperation projects with Egypt and Turkey are more narrowly-focused and more technically demanding than those conducted with Saudi Arabia, again suggesting the more developed state of the Egyptian and Turkish nuclear skills base.

Finally, we were unable to find out whether any of the three states were also attempting to develop their nuclear expertise by sending personnel to work on foreign nuclear power programs. This is significant because foreign programs could be a useful way of building a relevant skills base.

## LEGAL AND REGULATORY REQUIREMENTS

Turkey already has a very well-developed legal and regulatory framework for the development of nuclear power which reflects its long history of contemplating the nuclear power option. The IAEA notes that the only significant omission is legislation relating to decommissioning.[113] However, such legislation is expected to result from a project initiated in 2000 to revise and update legislation in line with IAEA standards. Within Turkey's existing legislative framework, the "Decree Pertaining to Issue License for Nuclear Installations [RG No. 18256 of December 19, 1983]" is of particular importance. The licensing process for NPPs it sets out is the responsibility of the Nuclear Safety Department of TAEK and is a three stage process covering site selection, construction, and commissioning.[114] One site — Akkuyu, the site of Turkey's proposed NNP — is already licensed.

TAEK, which was established in 1982, is also responsible for all other aspects of licensing and regulation in Turkey. Its specific responsibilities are described as:

- "defining safety measures for all nuclear activities and for drawing up regulations concerning radiation protection and the licensing and safety of nuclear installations;"
- "issuing licences to both private and state enterprises conducting various activities involving radioactive materials, supervising the radiological safety of such enterprises, and ensuring compliance with licence conditions;"
- "issuing authorisations, permits and licenses related to the siting, construction, operation and environmental safety of nuclear installations;"
- "performing the necessary reviews, assessments, and inspections of these installations;"
- "limiting the operating authorisation in the event of non-compliance with the permit or licence;"

- "revoking licences and/or permits issued previously either temporarily or permanently, and submitting recommendations to the Prime Minister on the closure of installations covered by such authorizations;" and,
- "preparing the necessary rules and regulations governing the above operations."[115]

In terms of liability arrangements, Turkey is a party to the 1960 Paris Convention on Nuclear Third Party Liability, one of the two main international liability agreements.[116] It is also party to the 1998 Joint Protocol Relating to the Application of the Vienna Convention and the Paris Convention (the Joint Protocol), designed to link the Paris convention to the 1963 Vienna Convention on Civil Liability for Nuclear Damage, the other main agreement.[117] However, it has neither signed nor ratified the CSC.

Not only does Turkey already appear to have a comprehensive legal and regulatory framework in place, but it has been cooperating with the IAEA to improve it further. Of particular relevance is one ongoing technical cooperation project which aims "to increase the effectiveness of the regulatory activities in nuclear safety through enhancing the expertise and reviewing draft regulations on nuclear installation safety with respect to international standards."[118] However, as important as the regulations themselves is the ability of the regulator to assess compliance and enforce its decisions. The effectiveness of TAEK in this regard is hard to assess. TAEK has been working with the IAEA to improve its inspection and enforcement capabilities, but not so actively as in the field of legislation. For instance, a second aim of the project discussed above was "to increase the hands-on experience of some staff of the Department of Nuclear Safety (DNS) through on-the-job training."[119]

In practice, the success of an NPP depends as much on the project manager as it does the regulator. Here the situation is complicated by ongoing reform in the Turkish electricity market.[120] In 1993, the Turkish Electricity Authority which was split in two to form the Turkish Electricity Generation and Transmission Company (TEAŞ) and the Turkish Electrical Distribution Company (TEDAŞ). In 2001, TEAŞ was further split into the Electricity Generation Company (EÜAŞ), the Turkish Electricity Transmission Company (TEIAŞ) and the Turkish Electricity Trading and Contracting Company (TETAŞ). All these companies are state owned. Further unbundling followed by privatization is planned but it is proceeding slowly.[121] Under existing legislation EÜAŞ's role is limited to operating certain existing plants. It can only build new plants if the market is unable to meet electricity demand. Indeed, the tender for the prospective NPP at Akkuyu was issued by TETAŞ and calls for a private company to build and operate the facility.[122] Thus, in practice, it appears that Turkey plans to follow a slightly different model from the one considered in this chapter, in which EÜAŞ would oversee construction and then operate the plant. However, analyzing the ability of EÜAŞ to manage an NPP project is of more than just academic interest because EÜAŞ might be required to do so if the market cannot.[123]

The extent to which EÜAŞ could competently manage an NPP project probably depends on how much expertise it has inherited from TEAŞ. According to a study by staff of the Nuclear Power Plants Department of TEAŞ, one result of the three previous unsuccessful attempts to construct an NPP at Akkuyu is that TEAŞ "has gained the following capabilities and experience: site selection capability . . .; technical, administrative, commercial, and economical evaluation experience; capability of carrying out contract negotiations, and preparation of contract documents; experienced staff to prepare the bid specifications; trained and experienced technical staff to start and carry the project."[124] No information about whether this experience and knowledge has been transferred to EÜAŞ was available. Nevertheless, on balance, Turkey already appears close to meeting the regulatory and legal criteria laid out in the analytical framework.

Like Turkey, the regulatory structure of Egypt, shown in Figure 10-1, is primarily split between two bodies: The Nuclear Power Plants Authority for Electricity Generation (NPPA) and the Center for Nuclear Safety and Radiation Control (CNSRC), although both are part of the Ministry of Electricity and Energy.[125] The former body (set up by Law No. 13 of 1976) is responsible for proposing NPP projects, determining the bid specifications and overseeing the implementation of the project. It is hard to assess its effectiveness but, like the now-defunct TEAŞ, it has probably gained useful hands-on experience through its involvement in previous unsuccessful projects. Moreover, it has been working very actively with the IAEA to improve its capabilities. It has initiated four relevant technical cooperation projects since 2000 (two of which have now been completed), which, between them, have covered most of the process for initiating an NPP from feasibility studies to site assessment through preparing a bid invitation specification.[126] Egypt is a party to the both the Vienna Convention and the Joint Protocol but not to the CSC.[127]

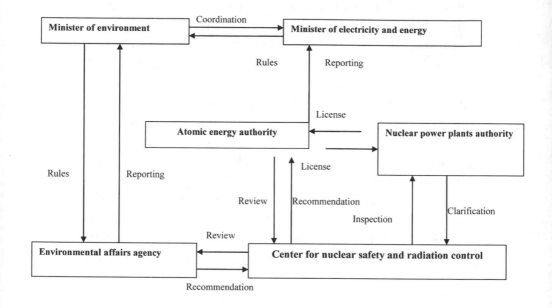

From "Egypt," *Country Nuclear Power Profiles*, IAEA, August 2005, available from *www-pub.iaea.org/MTCD/publications/PDF/ cnpp2004/CNPP_Webpage/countryprofiles/Egypt/Egypt2005.htm.*

**Figure 10-1. The Egyptian Regulatory System.**

CNSRC, which is a department of AEA, is a well-established body and already has responsibility for regulating both of Egypt's research reactors as well as all use of radioisotopes in Egypt.[128] Licensing in Egypt is a five-stage process divided into site approval, construction, fuel loading and commissioning, operating, and decommissioning. According to a recent IAEA survey "IAEA Nuclear Safety Standards (NUSS) shall be a main source of the Egyptian nuclear regulations" but that, in addition, Egypt "may also accept the safety criteria, codes, rules, and standards used in the vendor country."[129]

CNSRC is divided into the Regulatory Inspection and Enforcement Unit (RIEU) and the Review, Assessment and Licensing Unit (RALU).[130] A 2003 paper from CNSRC, which focuses on the work of RALU, outlined the Egyptian regulatory system, arguing that it is generally good and that some weaknesses previously identified with emergency planning have been rectified.[131] Since this chapter was published, Egypt has completed a technical cooperation project with the IAEA specifically focused on the work of RIEU.[132] Two additional projects which appear to focus on RIEU are currently underway, but little information is available on them.[133] All this suggests that CNSRC has identified inspection and enforcement capabilities as weak points in the Egyptian regulatory system, but that it is taking steps to rectify those deficiencies.

The national nuclear authority in Saudi Arabia is the King Abdul Aziz City for Science and Technology (KACST). Given the highly limited nature of nuclear activity in Saudi Arabia to date, it has only required a basic regulatory system, largely focused on the use of radioisotopes for medicine, industry, and agriculture. Moreover, there is evidence that existing regulation is generally weak. For instance, a 2001 study by King Faisal Specialist Hospital and Research Center and the University of North Texas that examined the use of gamma-ray cameras in hospitals found that ". . . only few centres performed acceptance testing on their cameras and few of these centres perform the minimum periodic quality control procedures for their gamma cameras."[134] Similarly, an undated study undertaken by King Abdul Aziz University discovered

that in many, if not most, establishments the level of radiation protection needs to be increased to meet safety requirements. The absence of emergency plans and the lack of proper training on the use of measuring instruments became apparent. It was also observed that interest in radiation protection improvement was low.[135]

Not only does Saudi Arabia therefore need to develop a comprehensive legal framework but it must also inculcate an appropriate safety culture.

Saudi Arabia's expertise to oversee the bidding, construction, and commissioning process appears to be limited to a few (fairly dated) academic projects, which examine issues such as the siting of an NPP.[136] In addition, Saudi Arabia has no experience of IAEA inspections or of the accounting and control of nuclear materials. Indeed, although Saudi Arabia is a signatory to the Non-Proliferation Treaty, it currently does not have a safeguards agreement with the IAEA in force.[137] Neither it is a signatory to any of the main nuclear liability agreements.

## ELECTRICAL GRID REQUIREMENTS

Table 10-2 summarizes how installed capacity in Egypt, Saudi Arabia, and Turkey has changed in the most recent years for which information is available (2002-05 for Egypt, 2000-04 for Saudi Arabia, and 2004-07 for Turkey).[138] It shows: (1) total installed capacity for the start and end dates; (2) the average annual increase in installed capacity over that period; (3) the actual amount of electricity produced as a percentage of installed capacity (the capacity factor) in 2005; and (4) the percentage of the 2005 installed capacity that a 1 GWe NPP would have represented.

| | Start of period | | End of Period | | Average annual increase (%) | Capacity factor (2005, %) | % of IC (2005) represented by 1 NPP |
|---|---|---|---|---|---|---|---|
| | IC (GW) | Year | IC (GW) | Year | | | |
| Egypt | 16.7 | 2002 | 18.4 | 2005 | 3.3 | 67 | 5.4 |
| SA | 22.9 | 2000 | 29.1 | 2004 | 6.2 | 65 | 3.2 |
| Turkey | 36.8 | 2004 | 40.8 | 2007 | 3.5 | 48 | 2.6 |

**Table 10-2. Electricity statistics for Egypt (2002-05), Saudi Arabia (2000-04), and Turkey (2004-07).**

Clearly, demand for electricity is increasing rapidly in each of the three states considered by this chapter. All three states expect this trend to continue.[139] Although demand might conceivably decline if fuel subsidies in these states were abolished; realistically, all three states will continue to meet the IAEA's criterion that no one NPP should represent more than 5-10 percent of installed capacity. No information is available on the reliability or stability of the electricity grids in any of the three states under consideration. There have recently been electricity shortages in both Egypt and Saudi Arabia but these have been put down to a lack of capacity rather than transmission problems.[140] Likewise, no information was obtained on the availability of two independent power sources for an NPP.

One trend of interest is the emergence of electricity links between the grids of regional states. Turkey's grid, for instance, is currently connected to those of Azerbaijan, Armenia, Bulgaria, Georgia, Iran, Iraq, and Syria.[141] Saudi Arabia and Egypt are also part of emerging regional networks and are considering installing a direct connection between each other.[142] Although this trend does not affect the three states considered in this chapter with regard to their development of nuclear power, it would help the other states in the region (such as the other GCC states) that individually lack sufficiently large electricity grids to develop NPPs on a collective basis.

## WASTE MANAGEMENT AND DECOMMISSIONING REQUIREMENTS

Turkey possesses a relatively sophisticated waste management infrastructure for dealing with waste from medicine, industry, academia, and research reactor operation. All such waste is collected, treated, and stored at the Radioactive Waste Processing and Storage Facility (CWPSF) of the Çekmece Nuclear Research and Training Center. CWPSF was established with technical support from the IAEA and subsequently upgraded by Turkey. The activities carried out by CWPSF include the treatment of liquid waste in a chemical processing unit and the compression of compactable solids in a compaction cell.[143] This experience is relevant for handling the LLW from an NPP. Although the volume currently available for LLW disposal at CWPSF was not identified, it seems probable that Turkey would have to increase it to handle the quantity of LLW that might be expected from an NPP, but this is unlikely to pose a significant challenge.

Egypt's radioactive waste facility is the Hot Laboratory and Waste Management Center (HLWMC) at Inshas.[144] Facilities at the center include a low and intermediate level liquid waste station and a radioactive waste disposal site. Like Turkey, Egypt would probably have to increase the volume available for disposal.

AERI is responsible for all radioactive waste disposal in Saudi Arabia and is reported to be preparing national standards for radioactive waste disposal.[145] Saudi Arabia currently has one waste storage facility, the Temporary Radioactive Waste Storage Facility. This facility consists of one room with a volume of 40 m³.[146] This is smaller than the volume of LLW produced annually by a well-run nuclear reactor. Building a larger waste depository is, however, relatively straightforward. Harder might be developing the appropriate culture and skills for handling radioactive waste. A 1997 study, for instance, found that considerable amounts of radioactive iodine were disposed of in the domestic sewage system.[147] This suggests that Saudi Arabia currently lacks the appropriate safety culture for operating an NPP.

There is little evidence of any serious planning for decommissioning a nuclear reactor in Egypt, Turkey, or Saudi Arabia. This is a very significant omission. As argued above, it is important that plans for decommissioning form an integral part of nuclear power development plans.

## CONCLUSION

For any state, developing its first NPP is a daunting challenge. The only prerequisite which is currently met by any of three states is the size of their electricity grids (and this criterion is met by all three states). Of the remaining requirements, low level waste disposal is likely to be the easiest challenge to meet. Developing plans for decommissioning may prove more difficult. However, the most significant challenges are staffing and regulation. Here Egypt, Saudi Arabia, and Turkey do not start from the same position.

All three states require significant additional numbers of personnel before their first NPPs become operational. Egypt and Turkey are, however, in a much stronger position than Saudi Arabia to accomplish this. Both Egypt and Turkey possess a sizeable and sufficiently specialized national nuclear infrastructure, including a well-established higher education sector. Both states have experience working with nuclear materials in the context of research reactors and other parts of the fuel cycle. The principal difference between these states is that Egypt's research reactor program could probably provide more training opportunities than Turkey's. Saudi Arabia starts from a much weaker position. Not only does it have a significant shortage of skills in most, if not all, relevant areas but it also appears to have comparatively limited capacity to train skilled personnel. In addition, nuclear-related research in Saudi Arabia is relatively fragmented and, in contrast to Egypt and Turkey, there does not appear to be much of a coherent national strategy.

In terms of regulation, Turkey has the best developed framework. With the important exception of decommissioning legislation, it has no significant omissions in its legislative framework. In no small part, this is a result of the country having seriously contemplated the nuclear power option on several occasions over the past 30 years. Egypt does not appear to lag far behind. Its regulatory structure does not appear to be quite as comprehensive as Turkey's, but it has been cooperating very actively with the IAEA to rectify weaknesses. In both cases, it is hard to ascertain the effectiveness of the inspection and enforcement arms of the relevant regulatory bodies. Saudi Arabia has a significant distance to go before it has the legal and regulatory structure required even to start developing an NPP. Beyond the legislation itself, it also needs to develop, almost from scratch, a safety culture appropriate for the handling of nuclear materials and the human resources to manage an NPP contract and regulate all aspects of the plant's construction and operation.

## ENDNOTES - CHAPTER 10

1. *Considerations to Launch a Nuclear Power Programme*, Vienna, Austria, International Atomic Energy Agency (IAEA), 2007, GOV/INF/2007/2/Colour, pp. 1-3, available from *www.iaea.org/ NuclearPower/Downloads/Launch_NPP/07-11471_Launch_NPP.pdf*.

2. To our knowledge, the only recent study that addresses the capabilities question in detail is *Nuclear Programmes in the Middle East: In the Shadow of Iran*, London, UK: International Institute of Strategic Studies (IISS), 2008.

3. *Considerations to Launch a Nuclear Power Programme*, pp. 1-3.

4. N. Bouzguenda, S. Nisan, and M. Albouy, "Financing of an integrated nuclear desalination system in developing countries," *Desalination*, Vol. 205, Nos. 1-3, February 2007, p. 320.

5. "Power Reactor Information System," IAEA, available from *www.iaea.org/programmes/a2/*.

6. Andrew Koch, "Yugoslavia's Nuclear Legacy: Should We Worry?" *The Nonproliferation Review*, Vol. 4, No. 3, Spring-Summer 1997, p. 125, available from *cns.miis.edu/pubs/npr/vol04/43/ koch43.pdf*.

7. "Power Reactor Information System."

8. Krško Nuclear Power Plant, *Annual Report 2007*, p. 58, available from *www.nek.si/uploads/documents/lp_ang_2007.pdf*; Irena Mele, "What Can a Country with Small Nuclear Program Learn From Other European Countries?" *Nuclear Engineering and Design*, Vol. 176, Nos. 1-2, November 1997, p. 145.

9. *Milestones in the Development of a National Infrastructure for Nuclear Power*, IAEA Nuclear Energy Series, NG-G-3.1, IAEA, Vi-

enna, Austria, 2007, pp. 41-44, available from *www-pub.iaea.org/ MTCD/publications/PDF/Pub1305_web.pdf.*

10. *Considerations to Launch a Nuclear Power Programme.*

11. Krško Nuclear Power Plant, *Annual Report 2007*, p. 58.

12. *Aging Nuclear Power Plants: Managing Plant Life and Decommissioning*, OTA-E-575, Washington, DC: Office of Technology Assessment, September 1993, pp.91-92.

13. Alice W. Shurcliff, *Local Economic Impact of Nuclear Power Plants* (No. MIT-EL 75-005WP), DSpace, MIT Libraries, July 1975, p. 2.

14. *Considerations to Launch a Nuclear Power Programme*, pp. 7-8.

15. *Milestones in the Development of a National Infrastructure for Nuclear Power*, pp. 41-44.

16. Krško Nuclear Power Plant, *Annual Report 2007*, p. 58.

17. *Considerations to Launch a Nuclear Power Programme*, pp. 7-8.

18. *Milestones in the Development of a National Infrastructure for Nuclear Power*, pp. 41-44.

19. *Ibid.*

20. *Considerations to Launch a Nuclear Power Programme*, pp. 7-8.

21. Krško Nuclear Power Plant, *Annual Report 2007*, p. 41.

22. *Ibid.*, p. 40.

23. "Nuclear Energy: 21st Century Promise," statement of IAEA Director General Dr. Mohamed El Baradei, October 10, 2005, available from *www.iaea.org/NewsCenter/Statements/2005/ ebsp2005n014.html.*

24. *Considerations to Launch a Nuclear Power Programme*, p. 3.

25. "Nuclear Power in Developing Countries," 50th IAEA General Conference, Vienna, Austria, September 18-22, 2006, p. 7, available from *www.iaea.org/About/Policy/GC/GC50/GC50Inf Documents/English/gc50inf-3-att3_en.pdf*.

26. *Considerations to Launch a Nuclear Power Programme*, pp. 8-9.

27. *Ibid.*

28. *Ibid.*

29. "Nuclear Power in Developing Countries," p. 7.

30. Omer F. Brown II and Nathalie L. J. T. Horbach, "Nuclear Liability: A Continuing Impediment To International Trade," World Nuclear Fuel Cycle 2004 Conference, Madrid, Spain, April 1, 2004, p. 3.

31. "United States Begins Push for Wider CSC Ratification," United States Mission to International Organizations in Vienna, Austria, May 28, 2007, available from *vienna.usmission.gov/08-05-21_csc/*.

32. Henry Sokolski, "The U.S.-Russia Nuclear Cooperation Agreement: The Case for Conditioning," testimony to the U.S. House of Representatives Committee on Foreign Affairs, June 12, 2008, available from *foreignaffairs.house.gov/110/sok061208.pdf*.

33. *Considerations to Launch a Nuclear Power Programme*, pp. 8-9.

34. *Milestones in the Development of a National Infrastructure for Nuclear Power*, pp. 34-47.

35. "Number of Staff of the SNSA," in IAEA, *Report of the International Regulatory Review Team (IRRT) to Slovenia, 1999*, Appendix III, available from *www.ursjv.gov.si/en/info/reports/porocila_eu_in_strokovnih_misij/irrt_report/*.

36. Statement by the Head of Croatian Delegation Mr. Matjaž Prah, Director of the State Office for Nuclear Safety, to the 49th General Conference of the IAEA, September 2005, available from *www.iaea.org/About/Policy/GC/GC49/Statements/croatia.pdf.*

37. Chap. 15, "Energy, European Commission," *Screening Report, Croatia,* March 29, 2007, available from *ec.europa.eu/enlargement/pdf/croatia/screening_reports/screening_report_15_hr_internet_en.pdf.*

38. "Nuclear Directorate-Who we are," Nuclear Directorate, UK Health and Safety Executive, available from *www.hse.gov.uk/nuclear/nsd1.htm#staff.*

39. V. Ranguelova and L. Matteocci, "Legislative and Governmental Responsibilities," in IAEA, *Report of the International Regulatory Review Team (IRRT) to Slovenia, 1999,* Section 1, available from *www.ursjv.gov.si/en/info/reports/porocila_eu_in_strokovnih_misij/irrt_report/.*

40. V. Ranguelova and L. Matteocci, "Organization of the Regulatory Body," in IAEA, *Report of the International Regulatory Review Team (IRRT) to Slovenia, 1999,* Section 3, available from *www.ursjv.gov.si/en/info/reports/porocila_eu_in_strokovnih_misij/irrt_report/;* Slovenian Nuclear Safety Administration.

41. This requirement is based on two factors. First, NPPs "are most efficiently run as base load generation and the dispatching of their full capacity should be possible." Moreover, nuclear safety necessitates a reliable and independent (off-site) electricity supply; *Milestones in the Development of a National Infrastructure for Nuclear Power,* pp. 39-41.

42. The IAEA notes that there are many designs for small and medium sized reactors (SMRs), some of which are likely to become available commercially within the next 10 years, that could be more readily incorporated in a small grid. SMRs do not benefit from economies of scale, but lend themselves to a number of alternative approaches to reducing unit costs: system simplification, component modularization, factory fabrication, direct site installation, the possibility of staggered construction of multiple modules, and standardization and construction in series. "Nuclear Power in Developing Countries," p. 9.

43. See *Milestones in the Development of a National Infrastructure for Nuclear Power*, pp. 39-41; "Nuclear Power in Developing Countries," p. 9.

44. Krško Nuclear Power Plant, available from *www.nek.si/en/about_nek/production/*; Westinghouse Electric Company, "Krsko Uprates to 2000MWt," *Westinghouse World View*, August 2002, p. 10, available from *www.westinghousenuclear.com/docs/news_room/worldview0802.pdf*.

45. Krško Nuclear Power Plant.

46. "Nuclear Power in Developing Countries," p. 10.

47. *Improvements of Radioactive Waste Management at WWER Nuclear Power Plants*, TECDOC-1492, IAEA, Vienna, Austria, 2006, p. 28, available from *www-pub.iaea.org/MTCD/publications/PDF/te_1492_web.pdf*.

48. *Ibid*.

49. Calculated from data in Ronald L. Fuchs, *1997 State-by-State Assessment of Low-Level Radioactive Wastes Received at Commercial Disposal Sites*, DOE/LLW-247, Washington, DC: U.S. Department of Energy, 1998, available from *www.osti.gov/bridge/servlets/purl/5310-trA182/webviewable/5310.PDF*.

50. "Nuclear Power in Developing Countries," pp. 11-12.

51. "Decommissioning Nuclear Facilities," World Nuclear Association, December 2007, available from *www.world-nuclear.org/info/inf19.html*.

52. Calculated from *Managing Low Radioactivity Material from the Decommissioning of Nuclear Facilities*, Technical Report Series, No. 462, IAEA, Vienna, Austria, 2008, p. 11, available from *www-pub.iaea.org/MTCD/publications/PDF/trs462_web.pdf*.

53. "Decommissioning Nuclear Facilities."

54. *Ibid*.

55. "Yugoslavia Nuclear Chronology 1947-2007," *Nuclear Threat Initiative*, November 2007, available from *www.nti.org/e_research/profiles/Yugoslavia/Nuclear/3994_3995.html*.

56. Irena Mele, "What Can a Country with a Small Nuclear Program Learn From Other European Countries?" *Nuclear Engineering and Design*, Vol. 176, Issue 1-2, 1997, p. 147.

57. Milena Cernilogar Radež, "Slovenia: The Positive Present and Future of Nuclear," *European Nuclear Society e-news*, Issue 12, Spring 2006, available from *www.euronuclear.org/e-news/e-news-12/nuclear-field-in-slovenia-print.htm*; Slovenian Nuclear Safety Administration, "Republic of Slovenia National Report on Fulfilment of the Obligations of the Convention on Nuclear Safety: The First Slovenian Report in Accordance with Article 5, August 1998," available from *www.ursjv.gov.si/index.php?id=5464&type=98*.

58. "Government Drafts Bill on Fund for Dismantling Krsko Nuke," Government of the Republic of Croatia, September 7, 2007, avalable from *www.vlada.hr/en/naslovnica/novosti_i_najave/2007/rujan/vlada_predlozila_zakon_o_fondu_za_ne_krsko*.

59. "Program of NPP Krško Decommissioning and SF & LILW Disposal: Executive Summary" in Republic of Croatia, *National Report on Implementation of the Obligations Under the Joint Convention on the Safety of Spent Fuel Management and on the Safety of Radioactive Waste Management* (2nd Report), Republic of Croatia, Zagreb, 2005, available from *www.dzns.hr/_download/repository/JointConReport2-final.pdf*.

60. *Nuclear Programmes in the Middle East: In the Shadow of Iran*, London, UK: International Institute of Strategic Studies (IISS), 2008, pp. 17-23.

61. "Minister: Egypt Peaceful Nuclear Programme Receives World-wide Acclaim," OSC document GMP20070315950037, MENA, March 15, 2007.

62. "Egypt unveils nuclear power plan," *BBC News Online*, September 25, 2006, available from *news.bbc.co.uk/1/hi/world/middle_east/5376860.stm*.

63. William J. Broad and David E. Sanger, "With an eye on Iran, rivals also want nuclear power," *New York Times*, April 15, 2007, available from *www.nytimes.com/2007/04/15/world/middleeast/15sunnis.html*; "Summit Ends with Call for N-Capability," *Times of Oman*, December 11, 2006.

64. UAE, *Policy of the United Arab Emirates on the Evaluation and Potential Development of Peaceful Nuclear Energy*, p. 9, available from *mofa.gov.ae/pdf/UAE_Policy_Nuclear_Energy_ENGLISH_E.pdf*.

65. Wyn Q. Bowen and Joanna Kidd, "The Nuclear Capabilities and Ambitions of Iran's Neighbors," in Patrick Clawson and Henry Sokolski, eds., *Getting Ready for a Nuclear Iran*, Carlisle, PA: Strategic Studies Institute, U.S. Army War College, pp. 56-57, available from *www.npec-web.org/Books/Book051109Getting ReadyIran.pdf*.

66. "U.S.-Saudi Arabia Memorandum of Understanding on Nuclear Energy Cooperation," U.S. State Department, May 16, 2008, available from *www.state.gov/r/pa/prs/ps/2008/may/104961.htm*.

67. *Ibid.*

68. Erkan Erdogdu, "Nuclear power in open energy markets: A case study of Turkey," *Energy Policy*, Vol. 23, No. 5, 2007, pp. 3069-3070; Bowen and Kidd, p. 68; *Nuclear Programmes in the Middle East: In the Shadow of Iran*, p. 61.

69. "Research Reactor Details - ETRR-1," Research Reactor Database, IAEA, available from *www.iaea.org/worldatom/rrdb*; "Research Reactor Details - ETRR-2," Research Reactor Database, IAEA, available from *www.iaea.org/worldatom/rrdb*.

70. H.-J. Roegler, Chap. 10.1, "Training at Research Reactors: Requirements, Features, Constraints" in IAEA, *Utilization Related Design Features of Research Reactors: A Compendium*, Technical Report Series, No. 455, IAEA, Vienna, Austria, 2007, available from *www-pub.iaea.org/MTCD/publications/PDF/TRS455_web.pdf*.

71. "Research Reactor Details - ETRR-1"; "Research Reactor Details - ETRR-2."

72. "Research Reactor Details - ITU-TRR, TECH UNIV," Research Reactor Database, IAEA, available from *www.iaea.org/worldatom/rrdb*.

73. "Research Reactor Details-TR-2, Turkish Reactor 2," Research Reactor Database, IAEA, available from *www.iaea.org/worldatom/rrdb*; *Nuclear Programmes in the Middle East: In the Shadow of Iran*, pp. 63-64.

74. See, for instance, Youssef Shatilla, "A pressure-tube advanced burner test reactor concept," *Nuclear Engineering and Design*, Vol. 238, No. 1, 2008, pp. 102-108; Y. A. Shatilla and E. P. Loewen, "A fast spectrum test reactor concept," *Nuclear Technology*, Vol. 151, No. 3, 2005, pp. 239-249; W. H. Abulfaraj and Ɛ. M. Kamal, "Evaluation of Ilmenite Serpentine Concrete and Ordinary Concrete as Nuclear Reactor Shielding," *Radiation Physics and Chemistry*, Vol. 44, No. 1-2, July-August 1994, pp. 139-148.

75. Saudi Research Database.

76. "Egypt," *Uranium 1999: Resources, Production and Demand*, Paris, France: OECD Nuclear Energy Agency (NEA) and IAEA, 1999, pp. 126-127.

77. "Turkey," *Uranium 2001: Resources, Production and Demand*, Paris, France: OECD NEA and IAEA, 2002; S. Anaç, N. Birtek, N. Birsen, and M. S. Kafadar, "Exploitation Of Uranium Resources In Turkey: A Short Review in Retrospect of the Technical and Economical Aspects," I Eurasia Conference on Nuclear Science and its Applications, Izmir, Turkey, October 23-27, 2000, available from *kutuphane.taek.gov.tr/internet_tarama/dosyalar/cd/3881/Nuclear/Nuclear-13.PDF*.

78. For instance, "Saudi Geological Survey Authority Reveals the Presence of Raw Uranium," *Al-Sharq Al-Awsat*, in Arabic, March 2007, available from *www.asharqalawsat.com/details.asp?section=43&issue=10326&article=409388*.

79. For instance, E. A. Fouad, M. A. Mahdy, M. Y. Bakr, and A. A. Zatout, "Uranium Recovery from the Concentrated Phosphoric Acid Prepared by the Hemihydrate Process," Second Arab Conference on the Peaceful Uses of Atomic Energy, Cairo, Egypt, No-

vember 5-9, 1994, available from *www.etde.org/etdeweb/*; K. Shakir, M. Aziz, and Sh. G. Beheir, "Studies on Uranium Recovery from a Uranium-Bearing Phosphatic Sandstone by a Combined Heap Leaching-Liquid-Gel Extraction Process. 1--Heap Leaching," *Hydrometallurgy*, Vol. 31, Nos. 1-2, September 1992, pp. 29-40.

80. Abd El-Ghany, M. S. Mahdy *et al.*, "Pilot Plant Studies on the Treatment of El Atshan Uranium Ores, Eastern Desert Egypt," *Proceedings of the Second Arab Conference on the Peaceful Uses of Atomic Energy*, Cairo, Egypt, November 5-9, 1994, pp. 229-231.

81. IAEA, *Implementation of the NPT Safeguards Agreement in the Arab Republic of Egypt*, GOV/2005/9, February 14, 2005, para. 11, available from *www.globalsecurity.org/wmd/library/report/2005/egypt_iaea_gov-2005-9_14nov2005.pdf;* "Egypt," pp. 128.

82. For instance, S. Girgin, N. Acarkan, and A. Ali Sirkeci, "The Uranium (VI) Extraction Mechanism of D2EHPA-TOPO from a Wet Process Phosphoric Acid," *Journal of Radioanalytical and Nuclear Chemistry*, Vol. 251, No. 2, February 2002, pp. 263-271; Ö. Genç, Y. Yalçınkaya, E. Büyüktuncel, A. Denzili, M. Y. Arıca, and S. Bektaş, "Uranium Recovery by Immobilized and Dried Powdered Biomass: Characterization and Comparison," *International Journal of Mineral Processing*, Vol. 68, Nos. 1-4, January 2003, pp. 93-107.

83. S. Anaç, N. Birtek, N. Birsen and M. S. Kafadar, "Exploitation Of Uranium Resources In Turkey: A Short Review in Retrospect of the Technical and Economical Aspects," I Eurasia Conference on Nuclear Science and its Applications, Izmir, Turkey, October 23-27, 2000, available from *kutuphane.taek.gov.tr/internet_tarama/dosyalar/cd/3881/Nuclear/Nuclear-13.PDF*.

84. For instance, El Shabana and A. S. Al-Hobaib, "Activity concentrations of natural radium, thorium and uranium isotopes in ground water of two different regions," *Radiochimica Acta*, Vol. 87, Nos. 1-2, 1999, pp. 41-45; Project number SAU/3/003.

85. *Implementation of the NPT Safeguards Agreement in the Arab Republic of Egypt*, GOV/2005/9, paras. 9-10, IAEA, February 14, 2005, available from *www.globalsecurity.org/wmd/library/report/2005/egypt_iaea_gov-2005-9_14nov2005.pdf*.

86. B. S. Girgis and N. H. Rofail, "Reactivity of Various UO3 Modifications in the Fluorination to UH4 by Freon-12," *Journal of Nuclear Materials*, Vol. 195, Nos. 1-2, October 1992, pp. 126-133.

87. *Nuclear Programmes in the Middle East: In the Shadow of Iran*, p. 64; Mustafa Kibaroglu, "Turkey's Quest for Peaceful Nuclear Power," *The Nonproliferation Review*, Vol. 4, No. 3, Spring-Summer 1997, p. 41, available from *cns.miis.edu/pubs/npr/vol04/43/kibaro43.pdf*.

88. Sayed M. Badawy, "Uranium Isotope Enrichment by Complexation with Chelating Polymer Adsorbent," *Radiation Physics and Chemistry*, Vol. 66, No. 1, January 2003, pp. 67-71. The Egyptian National Scientific and Technical Information Network also lists the abstract from an MSc thesis conducted at the University of Alexandria into the theory of multi-component isotope separation in asymmetric cascades, available from *www.sti.sci.eg/*.

89. Z. E. Erkmen, "A Study on the Reaction of Yttria (Y2O3) in Flowing Uranium Hexafluoride (UF6) Gas at 900°C," *Journal of Nuclear Materials*, Vol. 257, No. 2, November 1998, pp. 152-161.

90. W. I. Zidan and I. M. Elseaidy, "General description and production lines of the Egyptian Fuel Manufacturing Pilot Plant," IGORR Conference 7, San Carlos de Bariloche, Argentina, October 26-29, 1999, available from *www.igorr.com/home/liblocal/docs/Proceeding/Meeting%208/ts5_elseaidy.pdf*.

91. *Implementation of the NPT Safeguards Agreement in the Arab Republic of Egypt*, para. 23.

92. "Nuclear Applications Laboratory, Middle East Technical University," available from *www.ia.metu.edu.tr/en/as/000l0009nuk leeruygulamalarlaboratuvari.htm*; *Nuclear Programmes in the Middle East: In the Shadow of Iran*, p. 64.

93. *Implementation of the NPT Safeguards Agreement in the Arab Republic of Egypt*, paras. 13-21.

94. "Division of Nuclear Fuel Research," *Egyptian Universities Network*, available from *www.frcu.eun.eg/www/homepage/aea/cent32.htm*.

95. H. R. Higgy and A. A. Abdel-Rassoul, "Mechanical Shielded Hot Cell," *Annals of Nuclear Energy*, Vol. 10, No. 7, 1983, pp. 383-387; "Reactor ETRR-2 (Egypt): Data Sheet," INVAP, available from *www.invap.net/nuclear/etrr-2/data-e.html*; "Atomic Energy Authority: HLWMC: Division of Nuclear Fuel Research," *Egyptian Universities Network*, available from *www.frcu.eun.eg/www/homepage/aea/cent32.htm*.

96. *Implementation of the NPT Safeguards Agreement in the Arab Republic of Egypt*, paras. 17-20.

97. *Ibid.*, paras. 10-16.

98. *Ibid.*, para. 21.

99. For instance, O. H. Zabunoğlu and T. Akbaş, "Flow Sheet Calculations in Thorex Method for Reprocessing Th-Based Spent Fuels," *Nuclear Engineering and Design*, Vol. 219, No. 1, January 2003, pp. 77-86.

100. Atomic Energy Research Institute, King Abdulaziz City for Science and Technology, Saudi Arabia, available from *www.kacst.edu.sa/en/institutes/ueri/index.usp*.

101. W. I. Zidan and I. M. Elseaidy, "General description and production lines of the Egyptian Fuel Manufacturing Pilot Plant," IGORR Conference 7, San Carlos de Bariloche, Argentina, October 26-29, 1999.

102. No information could be found to provide even a general idea of the number of academics qualified in nuclear science and engineering. The following universities were identified as having significant research and/or teaching experience in relevant fields: Aegean University (Nuclear Sciences Institute); Bilkent University (Physics Department); Bogazici University (Engineering Department, Graduate Studies in Nuclear Engineering Department, Physics Department); Canakkale Onsekiz Mart University (Physics Department); Cumhuriyet University (Chemical Engineering Department, Chemistry Department); Dokuz Eylul University (Graduate School of Natural and Applied Sciences); Ege University (Faculty of Engineering, Faculty of Science, Institute of Nuclear Sciences); Hacettepe University (Nuclear Engineering

Department); Istanbul Technical University (Institute for Nuclear Energy, Metallurgical Engineering Department, Mining Faculty, Physics Department); Middle East Technical University (METU) (Chemical Engineering Department, Chemistry Department, Mechanical Engineering Department, Metallurgical Engineering Department, Mining Department, Physics Department); Suleyman Demirel University (Renewable Energy Sources Research and Application Centre [YEKARUM]).

103. *Turkey: 2005 Review, Energy Policies of IAE Countries*, Paris, France: IAE and OECD, 2005, p. 163, available from *www.iea. org/textbase/nppdf/free/2005/turkey2005.pdf*. According to this report, the regulatory and R&D activities of TAEK will be separated in 2005 by creating an independent nuclear regulator. It is not known whether this has occurred.

104. Ankara Nuclear Agriculture and Animal Research Centre (ANTHAM); Ankara Nuclear Research & Training Centre (ANEAM); Çekmece Nuclear Research & Training Centre (CNAEM-CNRTC); Experimental Test Facility (based in METU); Nuclear Energy Institute (based at Instanbul Technical University); Scientific and Technical Research Centre of Turkey (TÜBİTAK); Synchrotron Light for Experimental Science and Applications for the Middle East (SESAME, located in Jordan); Turkish Speaking States Nuclear Cooperation, Research and Training Centre (TUDNAEM).

105. *Nuclear Legislation in OECD Countries: Regulatory and Institutional Framework for Nuclear Activities: Turkey*, Paris, France: OECD, 1999, p. 13.

106. No information could be found to provide even a general idea of the number of academics qualified in nuclear science and engineering. The following universities were identified as having significant research and/or teaching experience in relevant fields: Ain Shams University (Nuclear Physics Laboratory, Physics Department); Alexandria University (Faculty of Science, Nuclear Engineering Department); American University in Cairo (Physics Department); Assiut University (Faculty of Science); Cairo University (Physics Department, Faculty of Engineering); Mansoura University (Physics Department); Tanta University (Mathematics Department, Physics Department).

107. Al Azhar University and Zagazig University.

108. Atomic Energy Authority; Middle Eastern Regional Radioisotope Center for the Arab Countries; National Research Center; National Center for Radiation Research and Technology; Nuclear Materials Authority; Nuclear Research Center.

109. King Faisal University, Umm al-Qura University, Girls College of Education in Riyadh and Taif Teachers College.

110. "Interview with the General Inspector of the Atomic Energy Research Institute at the King Abdulaziz City for Science and Technology, Saudi Arabia," *Al-Jazirah Newspaper*, in Arabic, March 17, 2001, available from *www.al-jazirah.com/*.

111. In 2001, there were 29 nuclear medicine departments in the Kingdom and a total of 44 gamma cameras. R. Y. Al Mazrou, J. Prince, and A. Arafah, "Nuclear medicine services in the Kingdom of Saudi Arabia," *European Journal of Nuclear Medicine*, Vol. 28, No. 8, August 2001 (supplement), p. 1150. There is also a Co-60 irradiation facility at AERI.

112. Project number EGY/0/017. Details of all IAEA technical cooperation projects are available from *www-tc.iaea.org/tcweb/ default.asp*. Another possibility is a very general project on strengthening skills in Saudi Arabia (SAU/0/007). Because this project is still ongoing, very little information is available about it. However, a project with an identical description and title (SAU/0/005) has been completed and so is documented in more detail. This project was entirely about the peaceful uses of radioisotopes and so would have been of only tangential relevance to building skills for a nuclear power program.

113. "Turkey," Country Nuclear Power Profiles, IAEA, December 2004, available from *www-pub.iaea.org/MTCD/publications/ PDF/cnpp2004/CNPP_Webpage/countryprofiles/Turkey/Turkey2004. htm*.

114. *Ibid.*

115. *Nuclear Legislation in OECD Countries: Regulatory and Institutional Framework for Nuclear Activities: Turkey*, Paris, France: OECD, 1999, p. 11.

116. OECD Nuclear Energy Agency, "Paris Convention on Nuclear Third Party Liability: Latest Status of Ratifications or Accessions," November 20, 2007, available from *www.nea.fr/html/law/paris-convention-ratification.html*.

117. "Joint Protocol Relating to the Application of the Vienna Convention and the Paris Convention," IAEA, January 9, 2008, available from *www.iaea.org/Publications/Documents/Conventions/jointprot_status.pdf*.

118. Project number TUR/9/015. See also project numbers TUR/9/016 and TUR/0/006.

119. Project number TUR/9/015. See also project number TUR/0/009.

120. Erkan Erdogdu, "Nuclear power in open energy markets: A case study of Turkey," *Energy Policy*, Vol. 23, No. 5, 2007, pp. 3068-3069.

121. "Country Analysis Briefs: Turkey," Energy Information Administration (EIA), October 2006, pp. 10-11, available from *www.eia.doe.gov/cabs/Turkey/pdf.pdf*.

122. The tender documents are available from *www.tetas.gov.tr/nukleer/eng/nklr.htm*.

123. Indeed, EÜAŞ describes one of its responsibilities as "[i]n accordance with the relevant legislation about nuclear energy generation facility establishment, to accomplish the relevant procedures of acquiring placement, construction, operation and alike licences and permissions from the relevant authorities." EÜAŞ, *Yillik Rapor (Annual Report) 2007*, EÜAŞ, Ankara, Turkey, 2007, p. 16, available from *www.euas.gov.tr/_EUAS/Images/Birimler/apk/2007_yillikrapor.pdf*.

124. E. Lutfi Sarici, Serkan Yilmaz, Bora Sekip Guray, "Nuclear energy in medium and long term energy generation of Turkey," *TAEK*, available from *kutuphane.taek.gov.tr/internet_tarama/dosyalar/cd/3881/Nuclear/Nuclear-21.PDF*.

125. "Egypt," *Country Nuclear Power Profiles*, IAEA, August 2005, available from *www-pub.iaea.org/MTCD/publications/PDF/cnpp2004/CNPP_Webpage/countryprofiles/Egypt/Egypt2005.htm*.

126. Project numbers EGY/4/045, EGY/4/047, EGY/4/049, and EGY/4/053.

127. "Vienna Convention on Civil Liability for Nuclear Damage," IAEA, April 20, 2007, available from *www.iaea.org/Publications/Documents/Conventions/liability_status.pdf*; "Joint Protocol Relating to the Application of the Vienna Convention and the Paris Convention," IAEA, January 9, 2008, available from *www.iaea.org/Publications/Documents/Conventions/jointprot_status.pdf*.

128. M. A. Salama, "Nuclear Activities in Egypt," *Applied Energy*, Vol. 75, Nos. 1-2, May-June 2003, pp. 73-80.

129. "Egypt," *Country Nuclear Power Profiles*.

130. See Project number EGY/9/035.

131. Salama, "Nuclear Activities in Egypt," pp. 73-80.

132. Project number EGY/9/035.

133. Project numbers EGY/9/036 and EGY/9/037.

134. R. Y. Al Mazrou, J. Prince, and A. Arafah, "Nuclear medicine services in the Kingdom of Saudi Arabia," *European Journal of Nuclear Medicine*, Vol. 28, No. 8, August 2001 (supplement), p. 1150.

135. I. I. Kutbi, W. H. M. Abul-Faraj, S. A. A. Al-Zaidi, K. M. A Al-Soliman, and A. M. Al-Arfaj, "Radwaste Management Plan Implementation In Saudi Arabia," Saudi Research Database, available from *srdb.org*.

136. Ibrahim Ismail Kutbi, "A Pragmatic Pairwise Group-Decision Method for Selection of Sites for Nuclear Power Plants," *Nuclear Engineering and Design*, Vol. 100, No. 1, February 1987, pp. 49-63; F. M. Husein, M. A. Obeid and K. S. El-Malahy, "Site Selec-

tion of a Dual Purpose Nuclear Power Plant in Saudi Arabia,"
*Nuclear Technology*, Vol. 79, No. 3, December 1987, pp. 311-321.

137. "Current Safeguards Status," IAEA, March 5, 2008, available from *www.iaea.org/OurWork/SV/Safeguards/sir_table.pdf.*

138. "2005 Energy Balances for Egypt," IEA, available from *www.iea.org/Textbase/stats/balancetable.asp?COUNTRY_CODE=EG*; "2005 Energy Balances for Saudi Arabia," IEA, available from *www.iea.org/Textbase/stats/balancetable.asp?COUNTRY_CODE=SA*; "Country Analysis Briefs: Egypt," EIA, August 2006, pp. 6-7, available from *www.eia.doe.gov/cabs/Egypt/pdf.pdf*; "Country Analysis Briefs: Saudi Arabia," EIA, February 2007, pp. 15-18, available from *www.eia.doe.gov/cabs/Saudi_Arabia/pdf.pdf*; "Development in Installed Capacity (1980/1981-2004/2005)," Egyptian Electricity Holding Company, available from *www.egelec.com/english/dr-aw/develop-en.htm*; EÜAŞ, Yillik Rapor (Annual Report), 2007, p. 23, available from *www.euas.gov.tr/_EUAS/Images/Birimler/apk/2007_yillikrapor.pdf.*

139. "Country Analysis Briefs: Egypt"; "Country Analysis Briefs: Saudi Arabia"; Arif Hepbalsi, "Development and Restructuring of Turkey's Electricity Sector: A Review," *Renewable and Sustainable Energy Reviews*, Vol. 9, No. 4, August 2005, p. 319.

140. "Country Analysis Briefs: Saudi Arabia," p. 15 .

141. "National Energy Grid: Turkey," Global Energy Network Institute, available from *www.geni.org/globalenergy/library/national_energy_grid/turkey/index.shtml.*

142. "Country Analysis Briefs: Egypt," pp. 6-7; "Country Analysis Briefs: Saudi Arabia," pp. 17-18; Farqad AlKhala, Riad Chedidb, Zeina Itania, and Tony Karam, "An assessment of the potential benefits from integrated electricity capacity planning in the northern Middle East region," *Energy*, Vol. 31, No. 13, 2006, pp. 2317-2318.

143. Mark Hibbs, "Turkey Considers Spent Fuel Deal with Bulgaria, Hungary for Akkuyu," *Nuclear Fuel*, Vol. 22, No. 17, August 25, 1997; Aktuerk, Gerceker, and Dara, "Spent Nuclear Fuel Management: Trends and Back End Scenariors," Eurasia Confer-

ence on Nuclear Science and Its Application, October 23-27, 2000; A. I. Izmir and I. Uslu, "Non-fuel Cycle Radioactive Waste Policy in Turkey," *TAEK*, May 29, 2003, available from *www.taek.gov.tr/ taek/rsgd/yayinlarimiz/teknik/rad_waste.htm*; "Measures to Strengthen International Co-operation in Nuclear, Radiation and Waste Safety," GC(41)/14, IAEA, August 5, 1997, available from *www. iaea.org/About/Policy/GC/GC41/Documents/gc41-14.html*.

144. "Atomic Energy Authority: HLWMC: Division of Nuclear Fuel Research," *Egyptian Universities Network*, available from *www.frcu.eun.eg/www/homepage/aea/cent3.htm*.

145. "KACST—Saudi Arabia"s Foremost Research Centre," Embassy of Saudi Arabia in Washington, DC, available from *www.saudiembassy.net/Publications/MagSummer00/KACST.htm*.

146. Waled H. Abulfaraj, Tamim A. Samman, and Salah El-Din M. Kamal, "Design of a Temporary Radioactive Waste Storage Facility," *Radiation Physics and Chemistry*, Vol. 44, Nos. 1-2, 1994, pp. 149-156.

147. Rasheed Al-Owain and Yousef Aldorwish, "Medical radioactive waste management in Saudi Arabia," Waste Management Conference 2001, February 25-March 1, 2001, available from *www.wmsym.org/Abstracts/2001/21C/21C-29.pdf*.

# PART III

# MAKING AND DISPOSING OF NUCLEAR FUEL

# CHAPTER 11

## NUCLEAR FUEL: MYTHS AND REALITIES

### Steve Kidd

The revival of interest in nuclear power, apparent over the past few years, can be explained by a combination of three factors. First, the improvement in the perceived economic viability of running nuclear reactors to generate electricity (indicated by the renewed interest of the financial sector); second, the contributions that more nuclear power may make towards curbing global carbon emissions; and third, by its possible role in enhancing energy security of supply. This return to the spotlight for nuclear has not been without some controversy, and one area that has come under scrutiny is the fuel necessary to run the power reactors. There are some important questions worthy of detailed discussion, such as will there be enough uranium to satisfy rising future requirements (especially if the number of reactors doubles or even quadruples), does an increased quantity of nuclear fuel constitute a proliferation risk, could rising uranium prices threaten the economic viability of nuclear, and are the procedures within the nuclear fuel cycle adequate to protect workers and the general public from any possible incremental health risks? These are just some more obvious examples, but unwelcome answers could serve to prevent the inchoate nuclear renaissance from coming to fruition.

## THE NUCLEAR FUEL CYCLE

The most obvious point to make about the supply of nuclear fuel is that the underlying fuel cycle is rather complex, especially by comparison with the supply of such fossil fuels as coal, oil, and gas for electricity generating stations. Oil goes to a sophisticated refinery where the crude is divided into separate distillates to service the needs for electricity, transportation, and chemicals. In common with coal and gas, it is just a matter of getting it out of the ground, then onto a ship or train or into a pipeline to reach the generating station where it is burned to create the heat which drives the turbines. Nuclear is also a "thermal" mode of generating power, relying on heat, with much of a plant very similar to the fossil fuel powered stations. It is the process used to create the heat — nuclear fission rather than combustion — and the required fuel with its attendant production cycle which are distinctive.

The key features of the nuclear fuel cycle (see Figure 11-1) are worthy of some initial discussion.[1] Uranium is mined (via processes which give rise to waste streams, mainly tailings) and then converted, usually enriched (for 90 percent of the reactors around the world, the process entails increasing the share of the U-235 isotope beyond the natural 0.7 percent and creating depleted uranium of lower assay) before being fabricated into fuel to be introduced to the reactor. This phase is termed the "front end" of the cycle, before the generation of electricity in the reactor, is the most important stage as it brings in the only revenue — the sale of billions of kilowatt hours of electricity necessarily supports all the other activities, in the absence of any government subsidies.

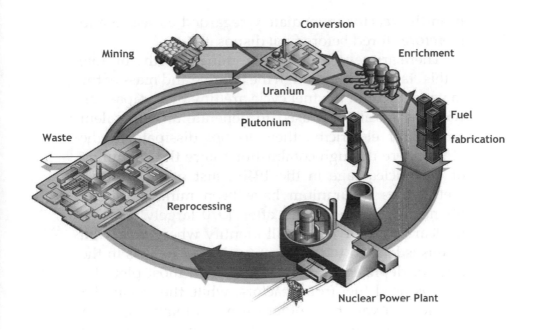

**Figure 11-1. The Nuclear Fuel Cycle.**

When the used fuel is unloaded from the reactor, it must initially be stored for cooling, but then there are effectively two choices regarding the "back end" of the cycle. Figure 11-1 shows a "closed" nuclear fuel cycle, with the used fuel going to a reprocessing plant. Here usable uranium and plutonium can be separated out and then recycled within the cycle to supplement supplies of fresh uranium, in the form of reprocessed uranium (RepU) and mixed oxide (MOX) fuel, respectively. What cannot be recycled becomes a waste stream from the reprocessing plant and can be vitrified (encased in plastic) before being stored "temporarily" in advance of disposal in a deep geological repository. The alternative "closed" cycle skips the reprocessing stage, with all the used fuel from the reactor immediately regarded as waste and therefore stored before final disposal.

There are several additional things worthy of note at this stage. First, although the volume and mass of the materials within the fuel cycle are tiny by comparison with the fossil fuels used to generate an equivalent amount of electricity, they do not dissipate in the atmosphere through combustion. Since the beginning of the nuclear age in the 1940s, just over 2 million metric tons of uranium have been mined, initially for nuclear weapons and after 1970 largely for civil nuclear power. We can still identify where nearly all of this is located today. Most (well over half) is in the form of depleted uranium, the second most plentiful form is used fuel from reactors, while the remainder is held in a variety of other forms, in many cases for potential future use. Historical uranium production therefore remains highly relevant to the nuclear fuel business today because material still containing fissile isotopes can potentially be processed for re-entry into the fuel cycle. The economics as well as the politics of recycling are the limiting factors. For example, there are acute political pressures to reduce the large quantities of military surplus highly enriched uranium (HEU) and military plutonium by using them as fuel in civil nuclear power reactors. Use of HEU presents few technical difficulties and has already become a major secondary supply. With the importance of historical production, the nuclear fuels business bears some similarity to precious commodities such as gold and diamonds being that these are rarely destroyed, so stockpiles and other secondary supplies are important.

Another notable feature is that the contractual arrangements normally used within the nuclear fuel market are peculiar when compared with trading in other energy commodities. With most reactors being refueled at intervals of 1 year or more, the demand for nuclear fuel is "lumpy" rather than continuous, as it is for the fossil fuels. Plant operators or their procurement agencies usually contract either directly or indirectly via intermediaries with uranium mining companies for the supply of uranium concentrates. They then have this uranium processed into a usable form through separate agreements with conversion, enrichment, and fuel fabrication suppliers. The obvious question is why they do not simply buy the fabricated fuel? Although there are moves today to offer a complete "cradle to grave" fuel package (maybe even taking on responsibilities for the "back end"), most buyers prefer to buy the four components—uranium, conversion, enrichment, and fuel fabrication—separately. This is for a variety of historical, economic, and (some would say) self-interested reasons. Hence four separate markets exist.

Another important feature of the nuclear fuel cycle is its international dimension. Uranium is relatively abundant throughout the earth's crust, but distinct trade specialization has occurred, due partly to the high energy density and therefore the low costs of transportation, as compared with coal, oil, and gas. For example, uranium mined in Australia can be converted in Canada, enriched in the United Kingdom, then fabricated as fuel in Sweden for a German reactor. Recycled reactor fuel may follow similar international routes, with related political as well as economic implications. With relative ease of transport and storage, inventories are an important feature of the nuclear fuel business. On the other hand, in the past there have been notable trade restrictions that have impacted the market, while today various constraints on transporting fissile materials have become an important issue.

## THE IMPORTANCE OF NUCLEAR FUEL

Ready availability of nuclear fuel is obviously important because, without it, the reactor will not run and generate electricity. So any delays and disruption to the timely arrival of the fabricated fuel at the reactor will be fatal. Yet, despite the complications of the fuel cycle outlined above, the possibilities of regulatory hindrances, and the potential for political, trade, or transport difficulties, there are very few cases where fuel has failed to reach reactors. The international nuclear fuel market is clearly somewhat imperfect, but it has always performed well in its basic function of supplying reactors. One obvious recent instance of fuel not getting to reactors is that of India, where nonproliferation restrictions and India's poor domestic uranium supply situation combined to prevent reactors from running at full capacity.

Nuclear fuel is quite a big business. Table 11-1 shows a rough calculation of the cost of 1kg of enriched uranium, present and ready to be loaded into a reactor.

| Uranium | 9.0 kg U308 | $25 per lb | 495 |
|---|---|---|---|
| Conversion | 7.6 kg U | $13 per kg | 99 |
| Enrichment | 7 SWU | $135 per SWU | 945 |
| Fabrication | 1 kg | $300 per kg | 300 |
| Total | | | $1839 |

**Table 11-1. Cost of 1kg of Nuclear Fuel.**

To refuel a large 1GWe reactor on an annual basis, about 20 tons of enriched uranium are needed, at a total cost of about $40 million. Multiplying by the 400-plus reactors in operation around the world and adjusting for their size gives a world market for nuclear fuel of $15-20 billion on an annual basis, depending of course on the contract prices. This is a small figure by comparison with the coal, oil, and gas trade, but is still a significant business, employing many thousands of people.

A significant paradox surrounds nuclear waste — it offers the biggest advantage of nuclear power, but at the same time, arguably, its greatest handicap. On one hand, the small amount of uranium required to produce a huge amount of nuclear energy leaves a correspondingly small amount of solid waste which, as far as the industry is concerned, can be safely contained and managed without environmental harm. Because nuclear fuel supplies are relatively inexpensive and highly energy-intensive (and thus small in volume), they can readily be stockpiled, affording a major buffer against energy insecurity. Finally, because fuel represents a small proportion of the generating costs of nuclear power, relative price stability for power is assured regardless of price fluctuations.

On the other hand, those opposed to nuclear power have identified the small volume of nuclear waste as its Achilles heel. As yet, there are no operating repositories for high-level waste (HLW), and there remains a very lively debate, both within and outside the industry, on the merits and demerits of reprocessing, which creates in turn additional public affairs debates. Additionally, in the oil and gas industry, the importance of fuel means that big and powerful companies like Shell, BP,

Exxon, and Total are able to devote huge resources to massaging their corporate reputations. With the exception of BP, of course, given its travails over the Gulf of Mexico oil spill, this results to some extent in a generally favorable public image of their industry. The reputation of nuclear has undoubtedly suffered because its fuel business is not so significant—the largest uranium producer, Cameco, is tiny by comparison with the oil giants. Most companies in nuclear are involved in other, sometimes mutually competitive, energy sectors too, and with the exception of Areva in France, are not as yet profitable and powerful enough to massage their image into a favorable industry reputation.

But in an economic sense, the relatively low cost of fuel (and indeed its relative stability) is nuclear's key card to play. On all the other elements of the cost structure of generating electricity, nuclear is disadvantaged—from the capital cost of the plants and the time it takes to build them, to the operating and maintenance (O&M) costs of running them, to the costs of eventually decommissioning the facilities and returning the sites to alternative use. In addition, nuclear projects are often regarded as relatively risky by investors, and the cost of securing finance may well be higher than for other energy-related ventures, too.

The relatively low cost of nuclear fuel includes, in addition to the "front end" costs outlined above, a full contribution to the cost of waste management, which is prescribed by national rules. But for nuclear plants already in operation, the fuel cost is a relatively small part of generating costs, at around a quarter (see Figure 11-2).[2] The costs of operating oil- and gas-powered electricity generating plants derive almost entirely from the fuel price while the profits of coal-powered

486

plants, too, are significantly affected by the cost of coal. Despite some movements up and down in the price of uranium, the nuclear fuel cost has remained very stable over time. However, the reactor fuel buyers fight hard to save every last cent because this is cost they feel they can influence. Where they are selling power in competitive markets, they cannot pass on increased fuel prices to customers, and higher prices will directly hit profits.

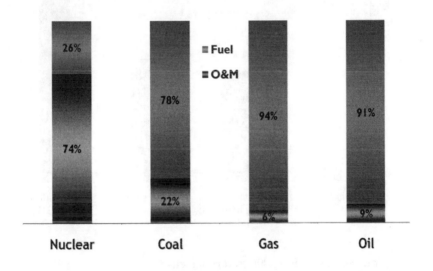

Source: Global Energy Decisions, ERI, Inc.

**Figure 11-2. Fuel as a Share of Electricity Generating Costs, Current Plants in USA.**

When it comes to new nuclear plants, their economics are even less sensitive to the fuel cost, as shown in Figure 11-3. The economics of new nuclear **DEPENDS** heavily on the capital cost of the plant and the rate of interest, with fuel costs playing only a relatively minor role. Once a nuclear plant is started up, the economics depend on it running 24 hours a day/7 days a week, with long periods (sometimes now up to 24 months) between shutdowns for maintenance and refuelling.

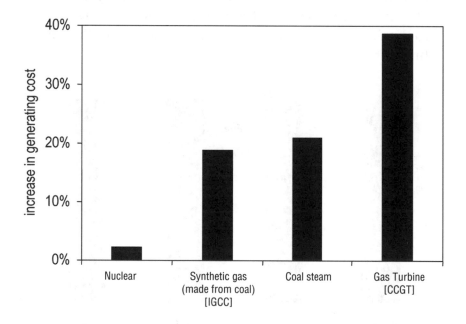

Source: IEA WEO 2006, reference case.

**Figure 11-3. Impact of 50 Percent Increase in Fuel Cost on Generating Cost, New Plants.**

## URANIUM IS NOT GEOLOGICALLY SCARCE

One of the great myths perpetuated about nuclear power is that uranium is scarce in a geological sense, on a par with diamonds, gold, and other precious metals. It is true, however, that (rather like gold) a significant amount of emotion surrounds its discovery and exploitation. Indeed, there was a uranium rush in the western United States in the 1950s, on a par with the Californian gold rush of the late 19th century, often mythologized in "B" movies depicting fathers and sons going prospecting in the badlands.

The reality is a little different.[3] Uranium is a slightly radioactive metal that occurs throughout the Earth's crust, about 500 times more abundant than gold, 40 times than silver, and about as common as tin, tungsten, and molybdenum. It occurs in most rocks in concentrations of two to four parts per million, for example, at about four parts per million (ppm) in granite, which makes up 60 percent of the earth's crust. In fertilizers, uranium concentration can be as high as 400 ppm (0.04 percent), and some coal deposits contain uranium at concentrations greater than 100 ppm (0.01 percent) (fertilizer and coal ash exploitation for uranium has been viable in the past and may conceivably be so again). It is also found in the oceans, at an average concentration of 1.3 parts per billion. The Japanese, and possibly others, have seriously studied possible extraction from seawater.

The bigger issue is one of economics. Apart from the 1950s, the late 1970s, and once again today, uranium prices have been relatively low, thus limiting usable deposits where to extraction is economically feasible. Economics is certainly related to the percentage of uranium in the ore (the grade), but that is only part of the story. The depth below the surface, geological setting, and a variety of other factors are also important. Uranium occurs in a number of different igneous, hydrothermal, and sedimentary geological environments, with deposits world-wide having been grouped into 14 major categories, based on geological setting. When mined, it yields a mixed uranium oxide product, ($U_3O_8$) which is yellow in color. Uraninite or pitchblende is the most common uranium mineral.

For many years from the 1940s, virtually all the uranium mined was used in the production of nuclear weapons, but this ceased to be the case in the 1970s. Today the only substantial use for uranium is as fuel in nuclear reactors, mostly for electricity generation. Uranium-235 is the only naturally-occurring material which can sustain a fission chain reaction, releasing large amounts of energy.

## Plenty of Uranium to Fuel Any Conceivable Nuclear Future.

There is every reason to expect that the world supply of uranium is sustainable, with adequate proven reserves being continuously replenished at costs affordable to consumers. Speculation to the contrary represents a misunderstanding of the nature of mineral resource estimates and reflects a short-term perspective overlooking continuing advances in knowledge and technology and the dynamic economic processes that drive markets.

Concerns about limitations of the Earth's resources go back more than a century. Although they appear intuitive and logical on the basis that mined mineral resources are clearly finite and physically nonrenewable, analysis in most cases shows that encountering limits to the supply of resources lies so far in the future that present-day concerns have little practical meaning. There are, however, examples such as oil, where prices may now be indicating that proven reserves are indeed beginning to run out. Concerns about resource depletion therefore deserve careful examination.

Characteristically, dire predictions of scarcity based on published proven mineral reserve figures have faltered by taking inadequate account of "resource-expanding factors," namely, gains in earth knowledge and discovery capabilities, gains in mining technology, and changes in mineral economics.

To achieve sustainability, the combined effects of mineral exploration and technology development need to discover proven recoverable reserves at least as fast as they are being used. Historical data teach this important lesson regarding most minerals. Reserve margins for metals, stated in terms of multiples of current use, have been continuously replenished or — more often — increased. On average, real prices for metals, including uranium, have tended to fall over time. It is important to recognize — with any commodity at any time — that one should never expect to see proven reserves of more than a few decades' worth because exploration will take place only if companies are confident of gaining a financial return. The prospect of a return is usually dictated by strong prices flowing from the prospect of imminent undersupply. When this happens, there tends to be a strong surge of exploration effort, yielding significant new discoveries. Weak uranium prices have held back exploration for much of the nuclear age — increased prices in recent years have led to a renewed exploration boom with the sudden appearance of over 400 "junior" uranium companies raising money through initial public offerings. These are already leading to upgrades in uranium resource estimates.

Today annual requirements to fabricate fuel for current power reactors call for about 65,000 tons of uranium. According to the authoritative Nuclear Energy Agency (NEA)-International Atomic Energy

Agency (IAEA) "Red Book,"[4] the world's present proven reserves of uranium, exploitable at below $80 per kilogram of uranium, are some 3.5 million tons. This proven reserve is therefore enough to last for 50 years at today's rate of usage—a figure higher than for many common metals. Current estimates of all expected uranium resources (including those not yet economic or properly quantified) are six times as great, representing 300 years' supply at today's rate of usage.

It cannot be overemphasized that these numbers, though providing a favorable prospect, almost surely understate future uranium availability because proven reserves of most minerals bear little relationship to what is actually in the outer part of the Earth's crust and potentially extractable for use. Proven reserves are an unrealistic indicator of what will actually be available in the long term. At most, they are useful as a guide to what is available for production in an immediate future spanning no more than a few decades. In the case of current proven reserves of uranium, the 50-year quantification is no more than a rear-view mirror perspective on supply. During future consumption of these reserves, the dynamics of supply and demand will produce price signals that inevitably trigger effects involving all three of the "resource-expanding factors" cited above. This is already evident in today's uranium market.

### Additional Supplies of Nuclear Fuel.

As noted below, up to 40 percent of recent world uranium demand has been filled by so-called secondary supplies from military and civilian stockpiles or from reprocessing of used fuel. In the period since 1985, excessive commercial inventories have been consumed as East-West arms control efforts began to dictate substantial dismantling of nuclear warheads, yielding commercially usable fissile material. These secondary supplies will remain an important part of the market for some years to come, but they are clearly limited, as their source is previously-mined uranium. As secondary supplies are depleted, primary uranium production will pick up strongly to fill their place.

It should also be noted that the element thorium, which is even more abundant in the Earth's crust than uranium, constitutes an additional potential source of nuclear fuel. Although thorium is not fissile, it is "fertile" — i.e., capable of being converted into fissile U-233 — and technologies for making this conversion are already well advanced in some places, notably India.

## LOWER URANIUM USE

Even with the current stock of operating nuclear reactors, there are ways of saving on uranium if prices rise, reflecting market scarcity due, perhaps, to production problems. It is possible to increase the amount of enrichment services in a given quantity of enriched uranium by varying the assay of the waste stream (the "tails assay" — see below), while reactor operating cycles can also be adjusted to make savings. Reactor design is, however, continuously developing. Evolutionary light-water reactor designs, which are all more fuel-efficient than their predecessors, will be the mainstay of nuclear programs over the next decade. However, in the period beyond 2030, advanced reactor designs such as those included in multinational research programs (Generation IV and INPRO) represent a further step forward in fuel efficiency.[5]

Some advanced reactor designs are fast-neutron types, which can utilize the U-238 component of natural uranium (as well as the 1.2 million tons of depleted uranium now stockpiled). When such designs are run as "breeder reactors" — with the specific purpose of converting non-fissile U-238 to fissile plutonium — they offer the prospect of multiplying uranium resources 50-fold, thereby extending them into a very far distant future. Others will be "burners" configured to utilize much of the world's used nuclear fuel inventory as future reactor fuel.

It may therefore be fairly concluded that uranium supplies will be more than adequate to fuel foreseeable expansions of nuclear power, even if the number of reactors runs into the thousands compared with the hundreds today. Indeed, in addition to its other noteworthy virtues, an abundant fuel resource will remain a crucial advantage of nuclear power. Those investors currently considering nuclear power are, of course, perfectly aware of this. It is somewhat curious why many of those opposed to nuclear power focus on the imaginary weakness of limited supply, when supply is actually plentiful. But ultimately, if investors are happy to put their money into new reactors, it is their problem, not the public's, if the reactors run out of fuel.

## Future Nuclear Generating Capacity.

The magnitude of future nuclear fuel demand depends on two factors: first, the number and size of reactors in operation (nuclear generating capacity); and second, how they are run (key operating parameters). In reality, nuclear generating capacity is by far the most important factor, and efforts to forecast the future of nuclear power concentrate heavily on it.[6]

The two main aspects to forecasting nuclear generating capacity are the outlook for the continued operation of existing plants and the prospects for the construction of new reactors. How long existing reactors will, in fact, remain in operation depends on a number of factors, which vary from country to country. The most important of these are the licensing procedures applying to life extensions and the economic attractiveness of continued operation. The latter will depend partly on the state of the electricity market in which the reactor is operating; that is, the price for which the plant's output can be sold, the types of electricity supply contracts permitted, the availability of capital for construction of replacement generating capacity, etc. Environmental (e.g., the avoidance of carbon dioxide emissions) and security of energy supply considerations may also influence reactor lifetimes in the future.

In principle, extending the lifetime of existing nuclear plants should normally be economically attractive. Nuclear power is characterized by high initial capital costs and low fuel costs, with operations and maintenance (O&M) costs varying according to operator efficiencies and regulatory practices. For well-managed plants with low O&M costs, the cost of producing electricity will be very competitive. The licensing obstacles to be overcome for life extension vary significantly from country to country. In the United States, reactor operating licenses are limited to 40 years of operation, but a procedure has been adopted by the Nuclear Regulatory Commission (NRC) to consider applications for life extensions. Most U.S. reactor operators have applied for and/or given notice that they will apply for life extensions for the operating licences. Some industry commentators have predicted that over 90 percent of the U.S. reactors could apply for and be granted life extensions to 60 years.

In some other countries, the situation regarding licenses is more flexible, with no fixed lifetime. So long as the regulatory authorities are satisfied that a reactor is safe, it can continue to operate. Of course, regulators may insist on additional checks on older plants, and may require upgrades to be carried out. But such requirements may be imposed at any time, and are not linked to a fixed nominal lifetime.

Life extensions, however, may be only one side of the coin. There is nothing which guarantees that reactors will operate even for their nominal 40-year lifetime if their operating costs are too high or if they encounter licensing or political problems. Even if operating costs are not too high, a closure decision may come because a plant requires major additional capital expenditure to keep it in operation (for example, steam generator replacement). The cost of servicing the additional capital, added to existing costs, may make the plant uneconomic.

There have already been individual instances where operable plants have been closed permanently well short of their intended lifetime, either because the utility judged that the cost of power generated was or would become too high, or because of failure to secure necessary licenses for their renewals. Politics have also unfortunately intruded here. The United States and Germany have been particularly affected by closures owing to economic factors, although no U.S. plants have closed since 1998. The Swedish government forced the premature closure of reactors in 1999 and 2005 for political reasons. Also in a political move, the German government enacted a law in April 2002 effectively limiting the operating lifetime of nuclear power plants. The highly economic nature of nuclear generation in Germany may, however, prompt a reversal of this if political change is forthcoming. The expense and possible adverse environmental effects of providing replacement power may prove significant.

A final factor to consider when discussing existing plants is the potential available for up-rating their capacity by capital expenditure on the plant, such as modifying the steam generators and/or replacing the turbine generator set. Several countries have already benefited from this, notably Finland, Germany, Spain, Sweden, Switzerland, and the United States, and it may represent a highly economic way of generating more power in many others. For example, some U.S. reactors are now up-rating their power output by up to 20 percent as part of plans to seek extensions for total operating lives of 60 years. Power up-rates in boiling water reactors (BWRs) tend to be much larger than in pressurized water reactors (PWRs), owing to the greater ease of changing the size of the fuel array.

Estimating the likely number of new reactors is particularly challenging, given the wide range of important factors to consider. It is reasonable to divide the likely new reactors over the next 25 years or so into three groups:

1. Those currently under construction around the world, which currently amounts to about 40;

2. Those for which a significant amount of planning, financing, and approval activity has already taken place, currently about 100; and,

3. Those which have been proposed, but without any commitment of significant funds towards financing and approval, currently up to 300.

The degree of uncertainty as to completion of reactors obviously increases in the third category. The usual approach to projecting numbers is to build scenarios based on different mixes. This is the approach of the World Nuclear Association (WNA), which offers three country-level scenarios to 2030:

1. A reduced-scope scenario in which many existing reactors do not operate beyond currently licensed lives and there are very few new reactors—indeed, some of those under construction today are never completed.

2. A reference scenario, where most existing reactors get some extensions to their operating licenses and there are increasing numbers of new reactors, particularly after 2020, comprising those under construction and planned, plus a few of those merely proposed.

3. An increased-scope scenario, in which many reactors run for 60 years and there are large numbers of new reactors, including all those planned and many of those currently merely proposed.

In reality, the picture for overall world nuclear generating capacity (and effectively the demand for nuclear fuel) depends on a few major countries. Despite the possibility of many new countries getting nuclear power, by 2020 there are unlikely to be more than five to add to the 30 countries which currently do. By 2030, there could conceivably be a much larger additional number,[7] but nuclear generating capacity will be driven by what happens in the United States; some major European countries like the United Kingdom, Germany, and Russia; and the big developing countries, China and India.

Figure 11-4 shows the WNA world nuclear generating capacity scenarios to 2030. Up to 2020, there is not a major difference between the scenarios as there are relatively few reactor closures in even the lower scenario. The number of new reactors which can come into operation by 2020 is somewhat limited by the time it takes to license and construct new reactors (an allowance of 4 years for each of these stages is customary, meaning 8 years in total). After then, significant numbers of reactors go out of service in the lower scenario (there were over 200 current reactors completed in the 1980s), while the reference and upper scenarios show large numbers of new reactors. By 2030, the scenarios diverge markedly, with nuclear generating capacity in the upper scenario roughly double today's level at 720 gigawatts (GWe), but less than 300 GWe in the lower case. However, because world electricity generation is also expected by the International Energy Agency (IEA) to double by 2030, even the upper scenario will not increase the share of nuclear from the current 15 percent.

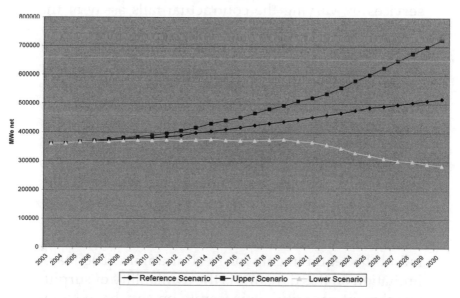

Figure 11-4. WNA World Nuclear Generating
Capacity Scenarios.

# FUTURE NUCLEAR FUEL DEMAND

The generating capacity scenarios can form the basis of similar ones for complete fuel demand (uranium, conversion, enrichment, and fuel fabrication). These require a computer-based model for the calculations, using the key parameters (such as the reactor load factor, the enrichment level, the fuel burn-up, and the tails assay at the enrichment plant). Perhaps the most important of these is the tails assay, that is, the measure of the amount of fissile uranium (U-235) remaining in the waste stream from the uranium enrichment process. There is a link between uranium and enrichment services, to the extent that they are at least partial substitutes. To obtain supplies of enriched uranium, required for 90 percent of all commercial nuclear reactors, fuel buyers can alter the quantities of uranium and enrichment services by varying the contractual tails assay at the enrichment plant. When uranium becomes relatively more expensive, there is an incentive to supply less of it and use more enrichment, thus "extracting" more U-235 from each pound. When uranium prices were around U.S.$10 per pound, the optimum tails assay was about 0.35 percent, but with the quadrupling of uranium prices since 2003 and a much smaller upward movement of enrichment prices, the optimum is now around 0.25 percent. Assuming such price relativities are sustained into the long term (which is arguable), there could be a substantial (20 percent or more) increase in enrichment demand and a corresponding fall in the requirements for fresh uranium. The major limitation on this dynamic is the availability of surplus enrichment capacity — constraints on this have so far limited the possibility of buyers to take full advantage.

Nevertheless, higher uranium prices are undoubtedly a positive inducement for future enrichment demand and will no doubt be taken into account in the coming major plant investment decisions.

Figure 11-5 shows the WNA world uranium requirements scenarios to 2030. The shape of the scenarios is, of course, very similar to those for generating capacity, with the lower scenario very robust until 2020, after which demand begins to diminish with reactor closures. This consistency of uranium demand is unusual among metal commodities, which usually suffer from significant demand cycles — with nuclear, however, once a reactor starts up, it tends to run for many years. The reference and upper scenarios both show rapidly rising uranium demand beyond 2015. The growth rates are actually slightly ahead of the growth of generating capacity because the fuel enrichment levels and the load factors of the reactors (essentially the percentage of time they are on-line) are both expected to rise from the levels of today.

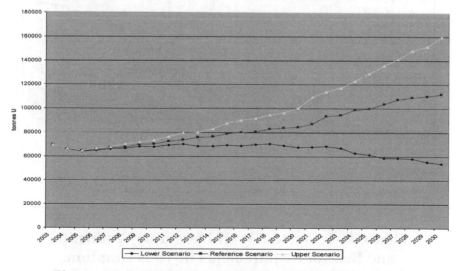

**Figure 11-5. WNA Scenarios for World Uranium Requirements.**

501

## HISTORICAL URANIUM PRODUCTION

Figure 11-6 shows the peaks and troughs of uranium production in the western world since 1945 and also plots the level of demand to feed commercial reactors. It is clear that supply and demand are not always in sync. The difference can be explained by there being essentially "four ages of uranium":

1. A military age, from 1945 to the late 1960s. Uranium demand from this source fell sharply from 1960 onwards and, in response, production halved by the mid 1960s.

2. An age of rapidly expanding civil nuclear power, lasting from the late 1960s to the mid 1980s. Production peaked in 1980 and stayed above annual reactor requirements until 1985.

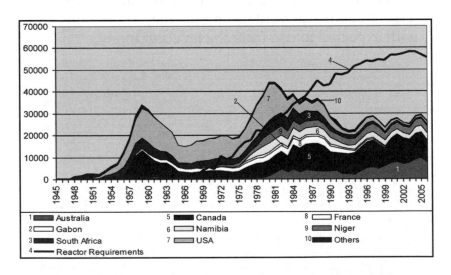

**Figure 11-6. Western World Uranium Production and Reactor Requirements in Tons Uranium.**

3. An age dominated by an inventory over-hang, extended by supply from the former Soviet Union, lasting from the mid-1980s up to 2003.

4. From 2003, a strong market reaction to the perception that additional primary production is needed to support accelerating nuclear growth and to offset declining and finite secondary supplies.

The gap between production and demand is still apparent today, but it is beginning to close as the so-called "secondary supplies begin to diminish in significance. The third age, "inventory overhang," led to a long depression in the uranium price, shown in Figure 11-7. This led to production becoming concentrated in a small number of major mines in a limited number of countries, with Canada and Australia producing around half of the world total by the early years of this century. The significant price reaction since 2003 (the fourth age) is discussed in more detail below, but has had the effect of stimulating exploration and plans for new mine development. Kazakhstan is the rising world producer and is set to overtake Canada as the leader by 2010. Production is also now rising in Africa, with increases in Namibia, Niger, and Malawi, with Malawi now expecting its first mine opening. Plotting future production against the demand scenarios for uranium has to take into account the secondary supplies of uranium as shown in Figure 11-8.

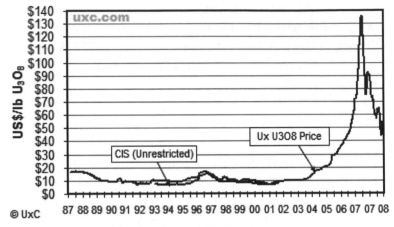

\* Commonwealth of Independent States

**Figure 11-7. Spot Uranium Prices.**

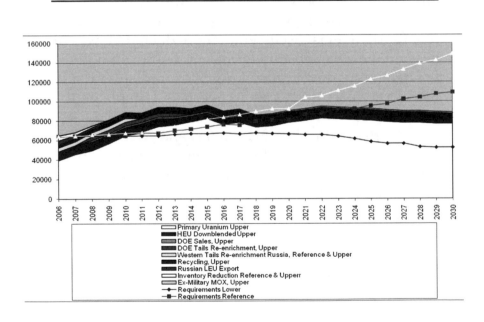

**Figure 11-8. Reference Case Supply and Uranium Demand Scenarios.**

Primary uranium production must now rise from around 40,000 tons worldwide to 60,000 tons to satisfy market demand. Beyond 2020, however, it is currently hard to predict where and when new mines will open, but with the reference and upper demand cases, world production will have to rise to 80,000 tons and beyond, double today's level.

It is believed that there are now over 400 junior uranium companies, the overwhelming majority still at the exploration stage. Few are yet moving towards mine development, but the front-runners, such as Paladin and Uranium One, are already producing and growing rapidly. Moreover, a high degree of consolidation is beginning to take place amongst these companies. Some are being acquired by the established producers (such as UraMin by Areva) but the better-established juniors are also acquiring each other — Uranium One's successive acquisitions of Southern Cross, UrAsia, and Energy Metals are particularly notable.[8]

## MINING TECHNIQUES AND THE ENVIRONMENT[9]

The decision as to which mining method to use for a particular deposit is governed by the nature of the ore body, safety, and economic considerations. Excavation may be either underground or open pit mining. In the case of underground uranium mines, special precautions, consisting primarily of increased ventilation, are required to protect against airborne radiation exposure. But in many respects uranium mining is much the same as any other mining. Projects must have environmental approvals prior to commencing, and must comply with environmental, safety, and occupational health conditions applicable.

Increasingly, these are governed by international standards, with external audits.

Milling, which is generally carried out close to a uranium mine, extracts the uranium from the ore. Most mining facilities include a mill, although where mines are close together, one mill may process the ore from several mines. Milling produces a uranium oxide concentrate which is shipped from the mill, usually referred to as "yellowcake," and generally contains more than 80 percent uranium. The original ore may contain as little as 0.01 percent uranium. The residue, containing most of the radioactivity and nearly all the rock material, becomes tailings, which are deposited in engineered facilities near the mine (often in mined-out pits). Tailings contain long-lived radioactive materials in low concentrations and toxic materials such as heavy metals; however, the total quantity of radioactive elements is less than in the original ore, and their collective radioactivity will be much shorter-lived. These materials need to be isolated from the environment.

Conventional mining will remain important (for example, the huge Olympic Dam deposit in South Australia is currently an underground mine, but the owner, BHP Billiton, is investigating a four-fold expansion as an open pit from about 2015). But an increasing proportion of the world's uranium now comes from in situ leaching (ISL).[10] This technique involves leaving the ore where it is in the ground, and using liquids which are pumped through it to recover the minerals from the ore by leaching (i.e., dissolving out soluable target constituents by percolation). If there is significant calcium in the ore body (as limestone or gypsum), alkaline (carbonate) leaching must be used, otherwise, acid (sulfate) leaching is generally better.

There is little surface disturbance, and no tailings or waste rock are generated. However, the orebody needs to be permeable to the liquids used, and located so that they do not contaminate groundwater. About a quarter of world uranium production is now by ISL (including nearly all the rapidly-rising Kazakh output). Techniques for ISL have evolved to the point where it is a controllable, safe, and environmentally benign method of mining which can operate under strict controls.

## SECONDARY SUPPLIES STILL IMPORTANT

Secondary supplies may be defined as all materials other than original, out-of-earth products sourced to satisfy reactor requirements. They include inventories, the draw-down of surplus military stockpiles, and other recycled materials of various types. In the widest sense, secondary supplies may be regarded as previous uranium production, returned to the commercial nuclear fuel market. Uranium production historically has not been closely correlated with actual reactor fuel requirements, leading to cycles of substantial inventory buildup and then disposal. In particular, there was a substantial buildup of commercial inventories in the late 1970s and early 1980s, when production rose sharply at a time when many reactor projects were being cancelled. The subsequent gradual exhausting of these inventories depressed the uranium market for many years.

Much of the secondary supply reaching the market in recent years has been down-blended highly enriched uranium (HEU) from military stockpiles declared surplus by arms limitation treaties.[11] A deal between Russia and the United States involving Russian stockpiles has satisfied roughly half of the U.S. nuclear fuel requirements since the deal's commencement in the mid-1990s and has also substantially contributed to important nonproliferation goals. The commercial terms, however, are now judged by the Russians to be unfavorable, as they were signed at a time when the Russians needed hard currency (whereas today they have lucrative oil and gas export earnings). They have now announced that there will be no renewal after the current deal expires in 2013. There will, however, be substantial quantities of surplus Russian HEU available for down-blending in the period beyond 2013, so it is reasonable to expect that it will be mostly employed to meet internal needs such as fueling Russian-origin reactors both at home and in export markets such as China and India. The United States also has some quantities of HEU which are surplus to military requirements, which will likely enter the commercial nuclear fuel market at some point in the future.

Finally, the reprocessing of used nuclear fuel is one fuel cycle option which can allow the recycling of plutonium and uranium to displace fresh uranium.[12] Programs for the recycling of plutonium were developed in the 1970s when it appeared that uranium would be in scarce supply and would become increasingly expensive. It was originally proposed that plutonium would be recycled through fast breeder reactors, that is, reactors with a uranium "blanket" but which would produce slightly more plutonium than they consume. Thus it was envisaged that the world's "low cost" uranium resources, then estimated to be sufficient for only 50 years' consumption, could be extended for hundreds of years.

As things transpired, the pressure on uranium resources was very much less than expected, and prices remained low in the period up to 2003. This was caused by the discovery of several new extensive and low-cost uranium deposits, the entry onto the world market of large quantities of uranium from the dismantling of nuclear weapons, and the slower growth of nuclear power than was expected back in the 1970s. Thus there was little incentive to develop fast breeder reactors, particularly as they present major engineering challenges which could prove expensive to resolve. Nevertheless, since the late 1970s, around 30 percent of used fuel arising from commercial nuclear reactors outside the former Soviet Union and its satellite states have been covered by breeder reprocessing contracts with plants in France and the UK.

Mixed Oxide (MOX) fuel was introduced mainly to reduce the stockpiles of plutonium, which were building up as spent fuel reprocessing contracts were fulfilled. MOX was therefore an expedient solution to a perceived problem, which had been created by changed circumstances. The MOX programs have demonstrated that plutonium has some advantages as a nuclear fuel and so the stockpiles have economic value.

Currently 12 of the countries with nuclear energy programs are committed to a closed nuclear fuel cycle, but there are signs that the number may soon increase. In particular, the United States is reassessing its previous policy, set strongly against reprocessing with subsequent recycling of recovered materials. The decision to introduce MOX fuel from ex-weapons plutonium in civil reactors was an important element in this and the first assemblies are now in use in reactors operated by Duke Power.

The "once through" cycle uses only part of the potential energy in the fuel, while effectively wasting substantial amounts of usable energy that could be tapped through recycling. In the United States, this question is pressing since significant amounts of used nuclear fuel are stored in different locations around the country awaiting shipment to the planned geological repository at Yucca Mountain in Nevada. This project is much-delayed, and, in any case, will fill very rapidly if it is used simply for used fuel rather than the separated wastes after its reprocessing.

The strong upward movement in uranium prices suggests that utilities owning inventories of reprocessed uranium (RepU) will look once again at utilizing these. The greater expense during the conversion and enrichment stages may now be outweighed by the substantially increased prices for fresh fuel. EDF, the operator of all the French nuclear plants, is at center stage here, owning significant quantities of RepU as a strategic asset. A few years ago, these could fairly be viewed on the other side of the balance sheet, as a long-term liability, but such an assessment is now outdated. Certainly many European utilities (and maybe also some in the United States) are looking at RepU in a new light and will possibly seek to add to those plants which have already gone down this road (albeit in relatively small quantities).

## THE URANIUM MARKET

Most uranium is traded on the basis of multi-annual contracts, based on perceived utility requirements. The spot market in uranium is driven by shorter-term adjustments to utility procurements and by uranium production plans rather than by annual reactor requirements, with price quotes provided by traders and brokers. Unlike the case of many other commodities, there is no terminal clearing market place such as the London Metal Exchange (LME) or its equivalents, though a market for financially settled futures, involving very small quantities, has been established at NYMEX. In addition, mutual funds have been created to allow investors to buy directly in and own uranium inventories.

The market has now moved up from a long period of oversupply in the 20 years up to 2003, where hopes for new demand from additional reactors were frustrated and abundant secondary supplies pushed the price down to around $10 per pound. Although there was plenty of industry speculation about this period's inevitable end (secondary supplies can clearly not last forever), there were few price signals until the market suddenly tightened during 2003, and a sharp price spike began. Financial speculators became interested in uranium (indeed, the price became an easy one-way bet for a time), while hundreds of small mining exploration companies added uranium to their portfolio and raised substantial sums on the stock markets.

The spot price peaked at $137 per pound in the middle of 2007 but has since slipped back sharply, in a series of stages, to end 2008 at around $50.[13] While volatility is a characteristic of most commodity prices, with tendencies to both over- and under-shoot deeper market fundamentals, the extent of the price decline now raises worry that projects will not go ahead and potential supply shortages could appear in the future (together with another and possibly more dramatic price spike). Everyone knows there are plenty of proven uranium resources in the ground — the question is how to get these to market in a timely manner and at prices which balance the interests of both producers and consumers in an equitable way.

This balance should really not be too difficult to achieve, as both uranium producers and reactor investors/operators have similar time horizons, with new projects going through lengthy approval stages and then taking several years in the construction stage, before running for 40 years and beyond. Reactors are generally fuelled only once per year (or longer), so that demand is discontinuous (contrast this with a coal-fired generating station). This pattern lends itself to long-term contracts, negotiated between buyer and seller, which may last for up to 20 years. These are highly confidential, and while they may reference quoted industry spot prices, they also contain escalation clauses, caps, and floors. This has been the traditional approach to selling nuclear fuel, with producers using the security of long-term contracts as collateral for raising project capital.

## URANIUM CONVERSION

This enrichment process requires uranium in gaseous form, which is achieved by converting it to uranium hexafluoride ($UF_6$) gas at relatively low temperatures. At a conversion facility, uranium is first refined to uranium dioxide, which can be used as the fuel for those types of reactors that do not require enriched uranium. Light water reactors (LWRs) require enriched uranium as do the UK's gas-cooled reactors. Heavy water reactors (HWRs), which are mainly of the CANDU design, require conversion from natural uranium concentrates directly to $UO_2$.

Worldwide requirements for $UF_6$ conversion services, averaged over an extended period, will be equal to aggregate demand for uranium requirements after allowing for the small number of reactors which do not require conversion. Countries operating CANDUs or other HWRs with requirements for $UO_2$ conversion are Argentina, Canada, China, India, Korea, Pakistan, and Romania. The key to future growth in demand is the magnitude of the Indian nuclear program, which so far has relied heavily on HWRs.

Worldwide, five major suppliers meet the majority of the demand for $UF_6$ conversion services, namely Cameco in Canada, Converdyn in the United States, Areva in France, Westinghouse in the United Kingdom, and Rosatom in Russia. The market is therefore quite concentrated, but there is sufficient competition to avoid monopolistic abuse. With regard to $UO_2$ conversion supply, Cameco's plant in Canada is by far the largest supplier, with a licensed annual capacity of 2,800 tU. In addition, smaller plants exist to meet the local needs in India, Argentina, and Romania.

## URANIUM ENRICHMENT

The enrichment of uranium constitutes a necessary step in the nuclear fuel cycle to fuel more than 90 percent of operating reactors worldwide.[14] The process involves increasing the isotopic level of the uranium-235 contained in natural uranium (0.711 percent) relative to the level of uranium-238 (99.3 percent). The majority of nuclear power reactors use low enriched uranium with up to 5 percent U-235. This enables greater technical efficiency in reactor design and operation, particularly in larger reactors, and allows the use of ordinary water as a moderator. The process of enriching the U-235 content to up to 5 percent is currently carried out utilizing two proven enrichment technologies, gaseous diffusion, and centrifugation. The first of these to be developed was gaseous diffusion, in which $UF_6$ gas is pumped through a series of diffusion membranes. The lighter U-235 passes through the porous walls of the diffusion vessels slightly faster than U-238, resulting in a higher concentration of U-235 in the product. Centrifugation is a more recent technique in which $UF_6$ gas is spun at high speed in a series of centrifuges. This tends to force the heavier U-238 isotope closer to the outer wall of the centrifuges, leaving a higher concentration of U-235 in the center.

The enrichment stage has traditionally represented the largest single front-end fuel cycle expense for utilities, but with the uranium price increases since 2003, the relative uranium cost has risen. The process is measured in terms of the separative work completed, defined as the amount of enrichment effort expended upon a quantity of uranium in order to increase the contained assay of U-235 by a given amount relative to that of U-238. This is measured in separative work units (SWU).

On the enrichment supply side, the most obvious feature is the gradual replacement of the old gas diffusion facilities of the U.S. Enrichment Corporation (USEC) in the United States and Areva in France with more modern and economical centrifuge plants. Even with favorable supply contracts, the huge amount of power required by the diffusion process renders it uneconomic against the centrifuges, as currently used by Urenco in Europe and by the Russian plants. Areva will gradually replace diffusion equipment with centrifuges derived from a technology-sharing agreement with Urenco, while USEC has decided to develop its American centrifuge technology, based on U.S. Department of Energy (DoE) programs in the 1970s and 1980s. Urenco and Areva are also building

U.S. plants in New Mexico and Idaho, respectively. Assuming USEC can overcome the financing and technical issues surrounding its plans, the last gas diffusion capacity should disappear around 2015 and the whole of the enrichment market should then be covered by centrifuges. The only likely alternative is the Australian SILEX laser enrichment technology, which has the support of GE-Hitachi for its possible commercial development. This latter may yet turn out to be the technology of the future, as was thought 10 years ago when USEC and others were investing significant amounts in laser technology, but its widespread commercialization (if it turns out to be technically and economically viable) may have to await the next generation of heavy investment in capacity, in the period after 2015. For the near future at least, centrifuges will be the technology of choice. The Russian centrifuge capacity is not known with any degree of accuracy, but is believed to be in the range of 25 million SWUs per year. This is believed to be rising slowly, as old centrifuges are replaced by new.

The enrichment stage in the fuel cycle creates much interest because of the possible weapons proliferation issues — the enrichment plants could be used to enrich uranium up to the levels required for a nuclear bomb, over 90 percent U-235. This topic will be considered below, but the large quantity (about 1.3 million tons worldwide) of depleted uranium (DU) from enrichment plants is also a live issue. Every ton of natural uranium produced and enriched for use in a nuclear reactor provides about 130 kilograms (kg) of enriched fuel (3.5 percent or more U-235). The balance is DU (U-238, with 0.25-0.30 percent U-235). It is stored either as $UF_6$ or converted back to $U_3O_8$, which is less toxic and more benign chemically, and thus more suited for long-term storage. Every year over 50,000 tons of depleted uranium join already substantial stockpiles in the United States, Europe, and Russia.

Some DU is drawn from these stockpiles to dilute high-enriched (>90 percent) uranium released from weapons programs, particularly in Russia, and destined for use in civil reactors. Other uses are more mundane, and depend on the metal's very high density (1.7 times that of lead). Hence, where maximum mass must fit in minimum space, such as aircraft control surfaces and helicopter counterweights, yacht keels, etc., DU has been found to be well-suited. It has also been used for radiation shielding, being some five times more effective than lead. Also because of its density, it is used as solid slugs or penetrators in armor-piercing projectiles, alloyed with about 0.75 percent titanium. This final use has caused much controversy, with the allegation that there are radiation risks when such shells explode.

## FUEL FABRICATION

Little similarity exists between the workings of the uranium, conversion, and enrichment markets and that of fuel fabrication. Nuclear fuel assemblies are highly engineered products, made especially to each customer's individual specifications. These are determined by the physical characteristics of the reactor, by the fuel cycle management strategy of the utility, and national, or even regional, licensing requirements.

Many fuel fabrication companies are also reactor vendors, and they usually supplied the initial cores and early reloads for reactors built to their own designs. As the market developed, however, each fabricator began to offer reloads for its competitors' reactor designs. This has led to an increasingly competitive market for fuel. Moreover, with several suppliers competing to supply different fuel designs, a trend of continuous fuel design improvements has emerged focusing on improving performance.

Currently, fuel fabrication capacity for all types of light water reactor (LWR) fuel throughout the world exceeds the demand by a considerable amount. Outside the LWR fuel market, fuel fabrication requirements tend to be filled by facilities dedicated to one specific fuel design, usually operated by a domestic supplier. For example, all fabrication requirements for AGR and Magnox reactors in the UK are supplied by dedicated domestic facilities. CANDU fuel is also produced almost exclusively within the country where the reactor is located, by $UO_2$ conversion and fabrication facilities dedicated to such supply. Fuel fabrication supply is therefore less concentrated than that of conversion and enrichment.

Given the very competitive nature of the LWR fabrication business and overcapacity in supply, the industry has reorganized and now seen some mergers, possibly driven by the expectation of the apparent nuclear renaissance. For example, British Nuclear Fuels (BNFL) sold Westinghouse Electric to Toshiba, and General Electric has, as a consequence, formed a joint nuclear company with its Global Nuclear Fuels partner, Hitachi.

The mergers a few years ago were expected to result in reduction of existing over-capacities, but only production consolidation has happened so far. Some plants have even increased their capacity along with modernization and relicensing projects.

## NONPROLIFERATION CONCERNS

A web of licensing, surveillance, and national and multinational regulations is in place throughout the nuclear fuel cycle to ensure that safety and nonproliferation objectives are met. This is administered by governments, by regional organizations such as Euratom Supply Agency (ESA) in the European Union (EU), and by the IAEA. Despite the evident success (as international treaties go) of the Treaty on the Non-Proliferation of Nuclear Weapons (NPT) in preventing many more countries from developing nuclear bombs, the expected expansion of nuclear power has brought forth new concerns.[15]

These concerns essentially started with the announcement from North Korea claiming it has an operating centrifuge enrichment program. There remain substantial doubts about this claim, but it was followed by further revelations from Iran and Libya showing that they had similar programs. Centrifuge enrichment technology is very difficult to master and needs high-quality plant components, but it appears that in each case, substantial progress has been made towards achieving facilities which could enrich uranium to weapons-level assays.

The common link in each of these countries has been technology transfer from the enrichment program in Pakistan, which uses old Urenco-derived centrifuge technology. This has clearly worried those concerned with weapons proliferation, although the quantities of enriched material produced and its assays remain unknown. These revelations have led to proposals for strengthening the nonproliferation regime. A big concern is that countries may develop various sensitive nuclear fuel cycle facilities and research reactors under full safeguards and then subsequently opt out of the NPT, as North Korea has done. This suggests that moving to some kind of intrinsic proliferation resistance in the fuel cycle itself is timely. There are several ideas, floated many years ago, which have been dug out and revamped. One key principle is that the assurance of nonproliferation must be linked with assurance of supply and services within the nuclear fuel cycle to any country embracing nuclear power. In addition to the need to accelerate adherence to the IAEA Additional Protocol, which ensures a stricter inspection regime, the IAEA, the United States, and Russia have proposed that enrichment facilities should be confined to the small number of countries already involved in the business. These will then offer full and fair trade to only those who accept full scope safeguards, perhaps with the provision of fuel banks and possibilities of fuel leasing. A similar regime has been proposed for spent fuel reprocessing, which also carries proliferation risks.

Those opposed to such measures see them as essentially a solution looking for a problem. The number of new nuclear countries is likely to be very limited for many years, and few countries that have moved to civil nuclear power have shown any desire to get involved in weapons. The commercial nuclear fuel market arguably works very well in securing regular supplies for any potential customer, and restrictions on supply may be deemed anti-competitive and potentially lead to higher prices.

## TRADE AND TRANSPORT RESTRICTIONS

Few countries possess the full range of facilities required to carry out all steps of the nuclear fuel cycle. The degree of specialization in the nuclear fuel industry clearly contributes to the overall economic efficiency of the nuclear fuel markets, as it would be prohibitively expensive for a country with a small or fledgling nuclear power program to develop all the necessary fuel cycle facilities. Hence those that attempt to do so (for example, Brazil) naturally arouse suspicions on grounds of possible proliferation risk. They may argue, in return, that they are concerned by possible trade and transport restrictions and want to develop local natural and labor resources.

Nevertheless, it is the case today that international nuclear commerce does not face particularly onerous barriers, provided that nations fit in with the obligations imposed by the NPT. Indeed, by comparison with the trade in agricultural commodities, it can be argued that the rules and regulations in force today are not particularly onerous and should not prevent new countries from acquiring power reactors, if they wish to do so. With the general easing of governmental restrictions on nuclear material flows for political or protectionist reasons, it is concerns about transport that are now threatening the future of nuclear commerce.[16] At the very least, they impose substantial cost increases, but also threaten security of supply. They are being addressed by establishing a better dialogue between government, industry, and the contractors themselves. Both port and carrier shipments need to be freed up in order to provide the confidence that is needed for a sound industry future.

## SUMMARY AND CONCLUSIONS

There is clearly sufficient uranium in reserve to fuel any conceivable expansion of nuclear power over the next few decades, and the costs of nuclear fuel are unlikely to be material in the decision whether to go ahead with new reactor plans. The key feature of the nuclear fuel market over the coming period is likely to be the ability of primary uranium production to expand rapidly, despite the continued important part which secondary supplies will play. With firmer world uranium prices, it has now become easier for primary producers to compete with the remaining secondary supplies, the production costs of which are largely sunk. Much consolidation has already taken place within the uranium production industry, and new uranium projects nearly always face various delays and frustrations before getting into production.

Within the conversion, enrichment, and fuel fabrication sectors, there are interesting market developments, but capacities appear likely to be sufficient to cope with demand. The enrichment sector is facing a technology shift in the period to 2015, by when it is generally expected that the older gas diffusion technology will have been replaced by centrifuges. During the years of poor fuel prices, the supply infrastructure in the industry was badly neglected, and this damage is at last being repaired so as to cope with escalating demand.

Looking to the very long term, beyond 2030, there is the promise of new reactor designs making fundamental changes to the nuclear fuel business. In particular, they may act as an effective solution to disposing of the substantial quantities of used nuclear fuel around the world, as many designs are characterized as "burners." Uranium, conversion, and enrichment requirements, as we currently know them, may gradually pass into history.

# ENDNOTES - CHAPTER 11

1. For further detail on the fuel cycle, see World Nuclear Association, "The Nuclear Fuel Cycle," available from *www.world-nuclear.org/info/inf03.html*.

2. This point is made well in the chapter on nuclear power in *World Energy Outlook 2006*, Washington, DC: International Energy Agency, 2006.

3. This and the following sections draw heavily on a WNA paper on uranium sustainability, World Nuclear Association, "Uranium: sustaining the Global Nuclear Renaissance?" September 2005, available from *www.world-nuclear.org/reference/position_statements/uranium.html*.

4. *Uranium 2007: Resources, Production & Demand*, Washington, DC: Organization for Economic Cooperation and Development (OECD)-Nuclear Energy Agency (NEA) and International Atomic Energy Agency (IAEA), 2008.

5. World Nuclear Association, "Generation IV Nuclear Reactors," updated June 2010, available from *www .world-nuclear.org/info/inf77.html*.

6. World Nuclear Association, *The Global Nuclear Fuel Market: Supply and Demand 2009-2030*, London, UK: World Nuclear Association, 2009.

7. World Nuclear Association, "Emerging Nuclear Energy Countries," updated August 16, 2010, available from *www.world-nuclear.org/info/inf102.html*.

8. World Nuclear Association, "World Uranium Mining," updated May 2010, available from *www.world-nuclear.org/info/inf23.html*.

9. World Nuclear Association, "Environmental Aspects of Uranium Mining," September 2009, available from *www.world-nuclear.org/info/inf25.html*.

10. World Nuclear Association, "In Situ Leach (ISL) Mining of Uranium," June 2009, available from *www.world-nuclear.org/info/inf27.html*.

11. World Nuclear Association, "Military Warheads as a Source of Nuclear Fuel," October 2009, available from *www.world-nuclear.org/info/inf13.html*.

12. World Nuclear Association, "Mixed Oxide (MOX) Fuel," updated March 2009, available from *www.world-nuclear.org/info/inf29.html*.

13. The Ux Consulting Company, "UxC Nuclear Fuel Price Indicators (Delayed)," available from *www.uxc.com/review/uxc_Prices.aspx*.

14. World Nuclear Association, "Uranium Enrichment," updated June 2010, available from *www.world-nuclear.org/info/inf28.html*.

15. World Nuclear Association, "Safeguards to Prevent Nuclear Proliferation," updated May 2010, available from *www.world-nuclear.org/info/inf12.html*.

16. World Nuclear Association, "Transport of Radioactive Materials," January 2010, available from *www.world-nuclear.org/info/inf20.html*.

# CHAPTER 12

# THE COSTS AND BENEFITS OF REPROCESSING

## Frank von Hippel

Since 1974, when India tested a nuclear bomb made with plutonium that it separated with U.S. assistance under the Atoms for Peace Program, there has been a debate within the global nuclear-power community about the desirability of reprocessing spent power reactor fuel. Today, about one-quarter of the world's spent fuel is reprocessed (see Table 12-1). Seven of the 31 countries with nuclear power reactors are having at least some of their spent power-reactor fuel reprocessed. A dozen more countries that had been sending their spent fuel to one of the three merchant reprocessing countries (France, Russia and the United Kingdom (UK) have decided, however, not to renew their contracts.

In the 1960s and 1970s, reprocessing of spent light water reactor fuel was justified by the need to obtain plutonium for startup cores for liquid-sodium-cooled plutonium-breeder reactors. The concern was that the world's resources of high-grade uranium ore would not be able to support the thousands of gigawatts (GWe) of nuclear capacity then projected by the year 2000. Current generation reactors can efficiently exploit only the fission energy stored in chain-reacting U-235, which makes up 0.7 percent of natural uranium. Breeder reactors would turn the U-238 that constitutes virtually all the remainder of natural uranium into chain-reacting plutonium.

| Countries that reprocess or plan to (GWe, [109 Watts]) | Customer Countries that have quit or plan to quit (GWe) | Countries that have not reprocessed (GWe) |
|---|---|---|
| China (30%) 8.4 | Armenia (in Russia) 0.4 | Argentina 0.9 |
| France (80%) 63.3 | Belgium (France) 5.8 | Brazil 1.8 |
| India (≈50%) 3.8 | Bulgaria (Russia) 1.9 | Canada 12.6 |
| Japan (90% planned) 46.1 | Czech Republic (Russia) 3.6 | Lithuania 1.2 |
| Netherlands (in France) 0.5 | Finland (Russia) 2.7 | Mexico 1.3 |
| Russia (15%) 21.7 | Germany(France/UK) 20.5 | Pakistan 0.4 |
| United Kingdom 10.1 | Hungary (Russia) 1.9 | Romania 1.3 |
| | Slovak Republic (Russia) 1.7 | Slovenia 0.7 |
| | Spain (France, UK) 7.5 | South Africa 1.8 |
| | Sweden (France/UK) 9.0 | South Korea 17.6 |
| | Switzerland (France/UK) 3.2 | Taiwan 4.9 |
| | Ukraine (Russia) 13.1 | U.S. (since 1972) 100.6 |
| Total (65%) 153.9 | Total 71.3 | Total 145.1 |

**Table 12-1. Status of Reprocessing in the Countries with Nuclear-Power Reactors.[1]**

However, the commercialization of breeder reactors has therefore not happened because:

1. global nuclear capacity is still below 400 GWe;

2. rich deposits of uranium were found in Australia, Canada, and Kazakhstan;

3. it was learned from demonstration breeder-reactor projects that liquid sodium brings with it many reliability and safety problems; and.

4. that breeder reactors would be *much* more costly than light water reactors.[2]

Nevertheless, a commitment to reprocessing persists in seven countries (see Table 12-1), with France recycling the separated plutonium into the fuel for the light water reactors from which it came, Japan is about to start doing so, and others are simply stockpiling their separated plutonium. The result is a global stockpile of about 250 tons of separated civilian plutonium—about as much as was separated for nuclear weapons by Russia and the United States during the Cold War—i.e., enough to make tens of thousands of nuclear weapons.[3] Most of this separated plutonium is stored at the reprocessing plants where it was separated, with some also at France's Melox mixed-oxide (MOX, plutonium-uranium) fuel fabrication plant.

As discussed below, both France and Japan have published analyses comparing the costs of reprocessing and plutonium recycling from their light-water reactors with the costs of simply storing the spent fuel—i.e., the "once-through" fuel cycle. Both nations have found that the once-through fuel cycle is lower in cost. However, they continue to be committed to reprocessing. Why?

At the same time, as noted above, a dozen countries that sent their spent fuel abroad for reprocessing have not renewed their contracts. Why did these countries find reprocessing attractive in the first place, and why did they change their minds?

The UK has lost its foreign reprocessing customers and had its government-owned reprocessing company go bankrupt. The reprocessing site has been taken over by a Nuclear Decommissioning Authority that has not yet decided whether or not to continue to reprocess UK domestic spent fuel. Russia and India continue to justify their reprocessing programs by expectations of the imminent commercialization of plutonium breeder reactors. Finally, China is patterning its nuclear-energy program on those of France and Japan and has completed the construction of a pilot reprocessing plant and plans to build a commercial-scale plant.

The "once-through" fuel cycle as currently practiced in the United States and many other countries is shown above the dotted horizontal line in Figure 12-1. Low-enriched uranium (LEU) fuel is irradiated in a light-water reactor and then stored. The reprocessing and recycle system that is in operation in France and soon will be in Japan is shown below the line. It involves the separation and recycle of the plutonium in "mixed-oxide" (MOX, uranium-plutonium) fuel. The spent MOX fuel is then stored. Because of the high cost of reprocessing, the cost of this MOX fuel is much higher than the cost of LEU fuel and most countries have decided that it is not worthwhile.

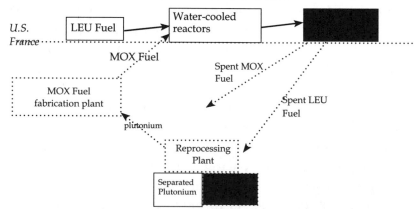

**Figure 12-1. Currently Competing Spent Fuel Management Strategies.**

Figure 12-2 depicts a U.S. nuclear power plant dry cask storage facility. Each cask weighs over 100 tons and typically holds about 10 tons of heavy metal (mostly uranium) in spent fuel that was discharged from a reactor 20 or more years earlier. A 1-GWe light-water reactor discharges about 20 tons per year. Each cask costs $1-2 million. Reprocessing of 10 tons of spent fuel would cost on the order of $20 million.

Source is available from *www.connyankee.com/html/fuel_storage1.html*.

## Figure 12-2. Dry-Cask Storage of Older Spent Fuel at a U.S. Nuclear Power Plant.

## THE FUEL CYCLES

Figure 12-1 shows the two different fuel cycles in use in the world today. Above the horizontal dotted line is the once-through fuel cycle in which low-enriched uranium (LEU) fuel is used in a reactor and the spent fuel stored. The United States has the largest group of nuclear power plants operating in this mode, with the spent fuel accumulating on the reactor sites because of the lack of a central site to ship to (see Figure 12-2). Utility dissatisfaction with this situation led the G.W. Bush administration to advocate reprocessing and plutonium recycling, but Congress was skeptical. The Obama administration shares this skepticism and is likely to limit U.S. reprocessing activities to research and development (R&D).

Shown below the horizontal dotted line in Figure 12-1 is the light-water reactor fuel cycle as practiced in France, the country that has gone the furthest in recycling plutonium. There, spent LEU fuel is reprocessed and the plutonium recovered from about seven tons of spent LEU fuel is mixed with depleted uranium to make a ton of mixed-oxide (MOX) fuel, which replaces about one-seventh of the LEU fuel that otherwise would have been used. The spent MOX fuel is then shipped back to the reprocessing plant but is not reprocessed again, despite the fact that it still contains about five times as much as plutonium as spent LEU fuel.[4] The reason given is that the mix of plutonium isotopes in the spent MOX fuel contains a lower fraction of chain-reacting Pu-239 and a larger fraction of even atomic number isotopes (Pu-238, Pu-240, and Pu-242) that are not as effectively fissioned as the odd isotopes by the slowed neutrons in light-water reactors.[5]

France therefore proposes to leave this plutonium in the spent MOX fuel until the commercialization of liquid sodium-cooled fast-neutron plutonium-burner reactors—the same reactors previously designed to be plutonium breeder reactors—is achieved. Reconfigured as plutonium-burners, they could fission the even plutonium isotopes more effectively than can light-water reactors. The only problem with this strategy is that liquid-sodium-cooled reactors are so much more costly than light-water reactors, so there is little prospect that they will be commercialized in the foreseeable future. In that case, France will only have complicated its radioactive-waste disposal problem by creating multiple waste streams—some of them quite voluminous—where previously there was only one waste form.[6]

France, like all other countries with nuclear power plants, does not yet have an operating geological repository for its high-level radioactive waste. The net effect of its reprocessing and plutonium recycle therefore is to shift the storage of spent fuel from France's reactor sites to its reprocessing facility. The plutonium is stored both in separated form (about 55 tons, enough for about 7,000 nuclear weapons, as of the end of 2007[7]) and in spent MOX fuel, while the uranium recovered from the spent LEU fuel is stored separately.[8] The fission products and the transuranic elements other than plutonium are stored in liquid form and then mixed into glass and the resulting "vitrified" high-level waste is stored on site. Long-lived medium and low-level radioactive wastes produced by reprocessing and MOX-fuel fabrication are also stored on site pending identification of one or more ultimate disposal sites.[9] France has also turned La Hague into a central storage site for LEU spent-fuel, holding in its pools about 60 percent as much French spent fuel as it has reprocessed. As of the end of 2008, only about 10 percent of that stored fuel was spent MOX fuel.[10]

## ECONOMICS OF DOMESTIC REPROCESSING IN FRANCE

Through 2005, almost half of the spent fuel reprocessed in France was of foreign origin—about 10,000 metric tons.[11] At perhaps $2 million per ton,[12] those reprocessing contracts were a significant source of

531

foreign exchange and France's policy of reprocessing its own spent fuel may have been in part a way to help support this important industry. Reprocessing has not gone completely unquestioned, however. In 2000, Socialist Prime Minister Jospin requested an analysis of the costs and benefits of continuing to reprocess most of France's spent fuel. Three scenarios were considered:

1. Continue reprocessing about 70 percent of France's low-enriched uranium (LEU) spent fuel with the separated plutonium being recycled in mixed oxide (MOX, plutonium-uranium) fuel;

2. Increase reprocessing to 100 percent of LEU spent fuel but stop when the separated plutonium could no longer be recycled because of the approaching end-of-life of the reactors — in effect, reprocess about two thirds of the LEU fuel discharged during the reactors' lifetimes); and,

3. End reprocessing in 2010 (corresponding to reprocessing 27 percent of the LEU fuel discharged during the reactors' lifetimes).

The panel also constructed a counterfactual scenario in which France had never embarked on reprocessing at all. Finally, from scenarios 1 and 2, one also can derive a second counterfactual scenario in which all of the LEU fuel is reprocessed and the plutonium recycled — one-third of it in a successor generation of light-water reactors. Table 12-2 shows the front and back-end costs of the fuel cycles for these four scenarios along with the inputs of materials and separative work and outputs of spent fuel and various radioactive wastes.

| | Percentage of Spent LEU Fuel Reprocessed | | | |
|---|---|---|---|---|
| | 67% (S6) | 27% (Reprocessing Ends in 2010, S4) | 100% (Derived Scenario) | No Reprocessing (S7) |
| Fuel cycle costs (billions of 2006 $ undiscounted, assuming $0.2 per 1999 French Franc) | | | | |
| Front End | 116 | 120 | 112 | 122 |
| Back End | 74 | 61 | 84 | 41 |
| Total | 190 | 182 | 196 | 162 |
| Cost/kWhr (cents) | 0.94 | 0.91 | 0.97 | 0.80 |
| INPUTS | | | | |
| Natural Uranium Mined ($10^3$ metric tons) | 437 | 460 | 418 | 475 |
| Separative Work (millions SWUs) | 313 | 330 | 299 | 341 |
| LEU Fuel Fabricated ($10^3$ tons uranium) | 54 | 56 | 52 | 58 |
| MOX Fuel Fabricated ($10^3$ tons) | 4.8 | 2 | 7.1 | 0 |
| LEU Fuel Roprocessed ($10^3$ tons) | 36 | 15 | 52 | 0 |
| WASTES | | | | |
| Depleted Uranium ($10^3$ tons) | 379 | 401 | 360 | 417 |
| LEU Spent Fuel ($10^3$ tons) | 18 | 41 | 0 | 58 |
| MOX Spent Fuel ($10^3$ tons) | 4.8 | 2 | 7.1 | 0 |
| Transuranic Waste ($10^3$ cubic meters) | 18 | 12 | 23 | 0 |
| High-level Waste ($10^3$ cubic meters) | 4.8 | 1.6 | 7.5 | 0 |
| Plutonium/Americium in Spent Fuel (tons) | 514 | 602 | 441 | 667 |
| Reprocessed Uranium ($10^3$ tons) | 34 | 14 | 50 | 0 |

Table 12-2. Costs, Inputs, and Outputs
for Different Scenarios for the Future of France's
Nuclear Fuel Cycle Based on a Study Done for the
Prime Minister in 2000 ($20.2 \times 10^{12}$ nuclear kWh).[13]

With regard to the inputs, it will be seen that by dividing the front-end costs and the quantities of natural uranium and separative-work used in the no-reprocessing scenario by the tonnage of LEU fuel produced, that the average cost of the LEU fuel was estimated to be about $2,000, with inputs of 8.2 kg of natural uranium and 5.9 SWU per kilogram of LEU fuel.

It was assumed in the French government's analysis that the price of uranium would climb slowly from $60/kg in 2000 to $80/kg in 2050.[14] As Figure 12-3 shows, uranium prices have been volatile — especially the spot market — but this still seems a reasonable average. The price spike in the late 1970s was due to the expectation that global nuclear power capacity — and therefore demand for natural uranium — would grow rapidly. In fact, new orders stopped, and it took decades for the utilities to use and sell off the uranium that they had contracted for. Hence the slump in prices. The more recent spike reflects a temporary panic over the future availability of uranium when it was realized that the selling off of the utility stockpiles and the blending down of excess Cold War weapons HEU to LEU for use in civilian power-reactor fuel had resulted in the global output from uranium mines shrinking to about half the size required to sustain the world's current fleet of nuclear power reactors.

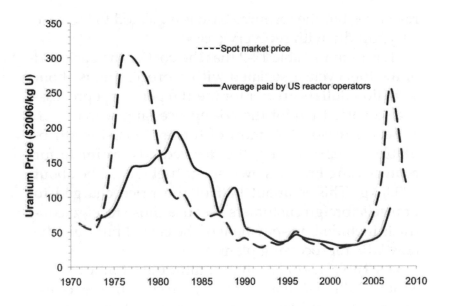

Spot market price is the broken line.

**Figure 12-3. Price of Uranium, 1970-2008.[15]**

As Table 12-2 also shows, the estimated cost of the 100-percent-reprocessing scenario was $34 billion higher than that of the no-reprocessing scenario—despite the fact that the consumption of uranium would be 57 million kilograms less. If the price of uranium increased by $600/kg while all other prices were unchanged, the cost of the once-through would be the same as the closed fuel cycle. Such a price increase is highly unlikely, however. There are over 5 million tons of identified resources of natural uranium—more than 70 years of consumption at the current rate—recoverable at an estimated cost of less than $130/kg and, despite mining and inflation, identified resources at less than this cost continue to increase.[16] Understandably, there has been little exploration of higher-cost resources but the resource base is expected to increase very rapidly with recovery cost.[17]

It is seen in Table 12-2 that the cost of the back end of the fuel cycle associated with reprocessing is about 43 billion dollars greater for the 100 percent reprocessing scenario than for the no-reprocessing scenario. If, as seems reasonable, most of this extra cost is attributable to reprocessing, the derived cost estimate for reprocessing France's own spent fuel would be about $800/kg. This is about half of the price charged to France's foreign customers because those foreign contracts included pre-payment of the cost of building the new UP3 reprocessing plant.

As Table 12-2 also shows, although it amounts to about a billion dollars per year, the cost difference estimated by the French Government in 2000 between no reprocessing and all reprocessing amounts to only about 0.2 cents per kilowatt-hour or perhaps 5 percent the total cost of generating nuclear power. In the past, France's national utility *Électricité de France* (EDF) has been able to pass this extra cost on to its customers. As Europe's electric power market has been deregulated, however, foreign competition has become more of a concern. As the reprocessing contract drew to an end in 2007, EDF tried hard to get a lower price for reprocessing, while Areva, the government-owned company that provides reprocessing services, lost virtually all of its foreign customers and insisted on a higher price. It took a year after the old contract had expired for a new one to be agreed upon, and with the old contract extended to bridge the gap.[18]

So why has the French government decided to continue its commitment to reprocessing despite the higher cost to the economy and the loss of almost all of its foreign reprocessing business? Probably part of the answer is that so much of the extra cost is now sunk cost, which was spent building the reprocessing and MOX-fuel fabrication plants.[19] Another part is the political weight of 6,000 jobs in rural Normandy.[20] The ability to move spent fuel off the nuclear power plant sites to a central location and thereby delay confronting the problem of siting a radioactive waste repository may also have been a consideration, as is suggested by the case of Japan.

## REPROCESSING IN JAPAN[21]

Japan's continued commitment to reprocessing is in large part a result of the unwillingness of local governments to allow increased storage of spent fuel on-site. This is in contrast to the situation at almost all U.S. power-reactor sites, where, when storage pools fill up, the oldest spent fuel is removed to make way for newly discharged spent fuel. The old fuel is stored on-site in dry casks (see Figure 12-2).

Japan's utilities were unable to interest any prefect in hosting a central spent-fuel storage facility. They therefore took the only option open to them at the time, which was to ship the spent fuel abroad to France and the UK to be reprocessed. This only bought time, however, because public opinion in France and the UK — and hence their reprocessing contracts — required Japan to take back the high-level waste resulting from the reprocessing of its spent fuel. Therefore, when Japan built a domestic reprocessing plant, it obtained an agreement from the local host government of Aomori Prefecture that the site would also accommodate the high-level waste coming back from Europe.

Reprocessing — like all things nuclear — is controversial in Japan, and the government periodically feels obliged to justify its policies as prudent. In 2004, the

Planning Committee of Japan's Atomic Energy Commission (JAEC) published, as a backup to the Japan Atomic Energy Commission's (JAEC) Long-Term Nuclear Plan, an evaluation of the costs of four scenarios for spent-fuel management in Japan:

1. Full reprocessing of all spent fuel;

2. Reprocessing only of the spent fuel that could be accommodated by the new Rokkasho Reprocessing Plant operating at nominal capacity (800 metric tons/year);

3. Direct disposal of all spent fuel; and,

4. Interim storage of all spent fuel.

The resulting cost estimates, shown as costs in cents per nuclear kilowatt hour (approximating one 2004 yen = one cent), are given in Table 12-3.

| | Full Reprocessing | Direct Disposal | Partial Reprocessing | Interim Storage |
|---|---|---|---|---|
| Front-end cost | 0.63 | 0.61 | 0.63 | 0.61 |
| Back-end cost | 0.93 | 0.32-0.46 | 0.77-0.85 | 0.48-55 |
| Total fuel-cycle cost | 1.56 | 0.93-1.07 | 1.4-1.48 | 1.09-1.16 |

**Table 12-3. Estimated Cost of Different Back-End Fuel-Cycle Options in Japan (cents/kWh).**[22]

As in France, it was found that reprocessing and plutonium recycling are more costly than the once-through fuel cycle. The cost difference between full reprocessing and direct disposal was found to be about 0.6 cents/kWh. This is more than twice as large as the corresponding cost difference found by France based on Table 12-1 and reflects the fact that Japan spent about as much to build its French-designed reprocessing plant as Areva claims to have spent for its UP2 and UP3 reprocessing plants, which together have more than twice the capacity. Also, Japan appears to be incurring about twice the annual operating cost as France—or about four times as much per ton of reprocessing capacity.[23]

The Planning Committee concluded that, nevertheless, reprocessing would be the less costly option for Japan for two reasons:

1. The Rokkasho Reprocessing Plant was already built and the $20 billion for its construction plus the projected $13 billion decommissioning cost would have to be paid in any case. These costs, divided by the nuclear kWhrs expected to be generated from the spent fuel reprocessed during the plant's 40-year planned life come to about 0.24 yens/kWh.

2. If Rokkasho became unavailable as an off-site destination for the spent fuel from Japan's nuclear power plants, they would have to shut down as soon as their spent-fuel storage pools filled up and replacement electricity would have to be generated by fossil-fueled plants. The JAEC estimated that the replacement electricity would cost 0.7-1.3 Yen/kWh. This cost seems remarkably low,[24] but it is large enough to tip the balance in favor of reprocessing.

Thus, this analysis clearly bases the rationale for the reprocessing of Japan's spent fuel on the need to have an off-site destination for this spent fuel or shut down all of Japan's power reactors.

## THE DOZEN COUNTRIES THAT DID NOT RENEW THEIR REPROCESSING CONTRACTS

What about the dozen countries listed in Table 12-1 that did not renew their reprocessing contracts? Here the situation is different for the seven countries that sent their spent fuel to Russia (Armenia, Bulgaria, the Czech Republic, Finland, Hungary, the Slovak Republic, and Ukraine) and the five that were customers of France and the UK (Belgium, Germany, Spain, Sweden, and Switzerland).

For the seven countries that sent their spent fuel to Russia, the cost was low, $300-620 per kg of heavy metal,[25] and nothing came back! In fact, only the fuel that was sent to Russia from first-generation VVER-440 light-water reactors was actually reprocessed at Russia's small RT-1 reprocessing plant in the Urals.[26] The spent fuel from the VVER-1000s is sent to a large spent-fuel storage pool at the never-completed RT2 reprocessing plant near Krasnoyarsk.

In the post-Soviet era, however, Russia began to raise its prices. Also, the leadership of Russia's nuclear-energy establishment came under public pressure not to make Russia a dumping ground for foreign radioactive waste and began to put clauses into its contracts that would allow it to ship high-level waste or unreprocessed spent fuel back to the country of origin. At the same time, most of Russia's former reprocessing customers had become members of the European Union (EU), and the EU has rules against transferring spent fuel to any country that cannot guarantee the same level of safety as is required in the EU. Finally, all of Russia's customers found that, like the United States, they were politically able to site and build adequate interim domestic storage for their spent fuel — either centrally or at the reactor sites.[27]

With regard to Belgium, Germany, Spain, Sweden and Switzerland, the story is different for each country. Because of domestic political opposition, Sweden decided not to have its spent fuel reprocessed after all and sold its contracts to other countries. Spain only sent spent fuel for reprocessing to France that came from its French-supplied gas-cooled reactor, which ended operations in 1990.[28] It also had a small (145 ton) reprocessing contract with the UK, equivalent to only about 1 year of discharges from its 7.5 GWe of light water reactor (LWR) capacity.[29]

Belgium, Germany, and Switzerland all have had significant quantities of spent fuel reprocessed in France,[30] and Germany and Switzerland have substantial reprocessing contracts in the UK that have not yet been completed because of the plant's poor operation and prolonged shutdown after a major pipe-break accident in 2005.[31] However, nuclear power and reprocessing became a contentious issue in all three countries. Belgium and Germany passed laws to end reprocessing and phase out nuclear power in the longer term. Switzerland's voters rejected a phase-out of nuclear power but voted for a 10-year reprocessing moratorium (2006-2016).[32]

## THE CASE OF THE UK

Reprocessing in the UK started with its first-generation Magnox gas-cooled, graphite-moderated power reactors. The design of these reactors was based on the Calder Hall and Chapelcross dual-purpose reactors that produced most of the plutonium for the UK's nuclear weapons as well as electric power. The fuel of the Magnox reactors is designed for easy reprocessing and not storage. The fuel "meat" is uranium metal, which, unlike the uranium oxide used in LWR fuel, oxidizes rapidly in water, and the cladding is a magnesium alloy, which also corrodes easily in water. Although the UK could have converted to a storable fuel form after its needs for weapon plutonium were satisfied, it did not do so and all of the Magnox fuel has been reprocessed. The last Magnox reactor will be shut down in 2010, however, and the associated B-205 reprocessing plant will be decommissioned after it has reprocessed the spent fuel.

The UK has a second reprocessing plant, the THermal Oxide Reprocessing Plant (THORP), which was built primarily with prepaid contracts to reprocess foreign LWR fuel. One third of the base-load tonnage to be reprocessed in THORP is from second-generation UK Advanced Gas-cooled Reactors (AGRs) that are fueled with oxide fuel.[33] British Nuclear Fuels Limited, which operated the plant, went bankrupt when the foreign contracts were not renewed. The UK government therefore established a Nuclear Decommissioning Authority (NDA) to take over and decommission the reprocessing plant and the Magnox reactors. The NDA's first priority has been to fulfill the base-load contracts for reprocessing foreign spent fuel that paid for the construction of the plant and were to have been fulfilled by 2004, but this date keeps slipping.

The situation with regard to the UK's domestic reprocessing customers is that they have contracts under which the reprocessing plant simply takes the AGR spent fuel and can either store or reprocess it. The cheapest option, of course, would be to store the spent fuel, but the chemistry in the spent-fuel storage pools is so poorly controlled that corrosion appears to be forcing reprocessing.[34] Although NDA's 2006 strategy document discussed the option of shutting down the reprocessing plant and storing the AGR fuel,[35] the 2008 NDA plan discussed only plans to reprocess.[36] Its most recent statement, with no detail offered, is that "Thorp is currently programmed to operate until 2016."[37] In the meantime, the NDA is also beginning to

grapple with the challenge of disposing of the approximately 100 tons of separated UK plutonium that will
have accumulated in storage at its reprocessing plants
by the time the current contracts are completed.[38]

## THE CASES OF RUSSIA AND INDIA

Reprocessing in Russia and India continues to
be driven by the expectation of the near-term commercialization of plutonium breeder reactors. Russia has only a small reprocessing plant, RT-1, in the
Urals. It reprocesses naval and other fuels containing
highly enriched uranium (HEU) to recover the HEU
for blend-down to LEU for recycle into power reactors. It also reprocesses about 50 tons per year of spent
fuel from Russia's first six first-generation VVER-440
LWRs. As of the end of 2007, Russia had 43.6 tons of
separated plutonium stored at the RT1 reprocessing
plant.[39] Russia is also storing tens of thousands of tons
of spent fuel from VVER-1000 reactors and RBMK-
1000 (Chernobyl-type) reactors at a never-completed
reprocessing plant near Krasnoyarsk, Siberia.

Russia has ambitious plans to shift to building
plutonium breeder reactors during the next decade
(see Figure 12-4) as a way to conserve its uranium resources for export. It would use its separated plutonium — first excess weapon plutonium and then civilian
plutonium — to start up these reactors. Whether these
plans will be realized remains to be seen. As a result of
the global recession, Russia's program to bring one or
two light-water reactors online every year during the
next decade will take longer than planned.[40]

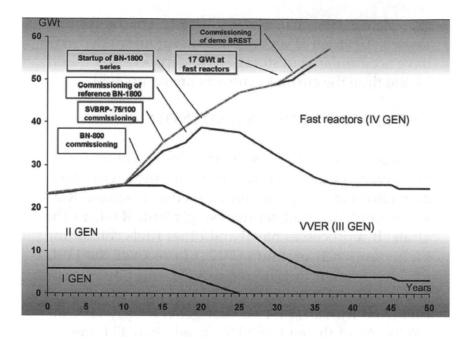

Zero is the year 2000.[41]

**Figure 12-4. Recent Plans for Russian Nuclear Power Expansion.**

Because of its limited resources of high-grade uranium ore, India has, for the past 50 years, premised its plan for nuclear power on breeder reactors.[42] It is currently reprocessing the spent fuel from 3.5 GWe of unsafeguarded heavy-water reactors to provide startup plutonium for a fleet of plutonium-breeder reactors. One 0.5-GWe prototype fast breeder reactor is under construction. India's Department of Atomic Energy (DAE) projects 43 GWe of breeder capacity by 2032,[43] however an insufficient amount of plutonium would be produced to support anywhere near this rate of growth.[44] For this and other reasons, this projection is likely to continue to retreat into the future, as have all past projections of imminent breeder commercialization by DAE and its counterpart nuclear-energy R&D establishments worldwide.

## THE CASE OF CHINA

China has plans underway for a huge expansion of its nuclear generating capacity from 9 GWe in 2009 to 120-160 GWe by 2030.[45] The Chinese nuclear energy establishment has been heavily influenced by that of France and emulates that of Japan. It has just completed a pilot reprocessing plant (50-100 tons/year) and is discussing with France the acquisition of reprocessing plant on the same scale as Japan's Rokkasho Reprocessing plant (800 tons/year).[46]

## CONCLUSIONS

There is no debate over the fact that the economic cost of reprocessing is significantly higher than that for interim spent-fuel storage. This is why the international trend continues to move away from reprocessing.

There must therefore be special explanations for the policies of the countries that continue to reprocess—and there are. In Japan, it is the unwillingness of local governments to allow expanded onsite spent-fuel storage. In India and Russia, politically powerful nuclear establishments continue to dream of a massive buildup of plutonium breeder reactors just over the planning horizon. In France, reprocessing is sustained by sunk costs, the political power of France's nuclear conglomerate Areva and its associated nuclear union, and Areva's hopes of building $20-40 billion worth of reprocessing plants in the United States and China. In

China, the nuclear establishment is emulating France and Japan but may still decide to postpone a major commitment to reprocessing.

In the longer term, these decisions are too important to remain the province of nuclear bureaucracies. Utilities are becoming increasingly unwilling to carry the economic burden of reprocessing and governments are becoming increasingly sensitive to the security and proliferation issues. It is therefore likely that the trend will continue whereby, one-by-one, utilities that reprocess find ways to implement the less costly and less controversial option of interim spent-fuel storage.

## ENDNOTES - CHAPTER 12

1. Nuclear capacities as of the end of 2008, based on the International Atomic Energy Agency's (IAEA) Power Reactor Information System, available from *www.iaea.org/programmes/a2/*.

2. A review of the history of breeder reactor commercialization efforts is forthcoming from the International Panel on Fissile Materials, available from *www.fissilematerials.org*.

3. International Panel on Fissile Materials, *Global Fissile Material Report 2008*, Chap. 1, available from *www.fissilematerials.org*.

4. *Plutonium Fuel: An Assessment*, Paris, France: Nuclear Energy Agency, OECD Publishing, September 30, 1989, Tables 9 and 11. Numbers cited are for spent fuel with a burnup of 53 Megawatt-days per kilogram of original heavy metal in the fuel and assuming that the spent LEU fuel was 10 years post discharge at the time it was reprocessed. The plutonium from the spent LEU fuel contains 50 percent Pu-239 and 15 percent Pu-241. The latter is less valuable, however, because it decays to nonchain-reacting Am-241 with a half-life of 14 years. Spent MOX fuel contains 37 percent Pu-239 and 17 percent Pu-241.

5. The fission cross-sections in the thermal neutron energy range of about 0.1 eV is about 1 barn for Pu-238, almost 1000 barns for Pu-239, less than 0.1 barns for Pu-240, several hundred barns for Pu-241, and less than 0.001 barns for Pu-242. See National Nuclear Data Center, Evaluated Nuclear Data File, available from *www.nndc.bnl.gov/exfor/endf00.jsp*.

6. See, e.g., Mycle Schneider and Yves Marignac, "Spent Fuel Reprocessing in France"; and Martin Forwood, "The Legacy of Reprocessing in the United Kingdom," International Panel on Fissile Materials, 2008, both available at *www.fissilematerials.org*.

7. "Communication Received from France Concerning Its Policies Regarding the Management of Plutonium," INFCIRC/549/Add.5/12, IAEA, October 30, 2008, available from *www.iaea.org/Publications/Documents/Infcircs/2008/infcirc549a5-12.pdf*. The IAEA assumes that 8 kg of plutonium is sufficient to make a first-generation Nagasaki-type plutonium bomb. More modern designs require less.

8. This irradiated uranium, although it contains about the same percentage of U-235 as natural uranium, is less valuable because it also contains a comparable amount of U-236 created by non-fission neutron absorption in U-235. Since U-236 absorbs neutrons without fission (i.e., is a neutron poison), it reduces the reactivity of the fuel. This effect must be offset by enriching the U-235 to higher levels than in LEU produced from natural uranium. France has re-enriched less than one-third of its reprocessed uranium. See *Management of Reprocessed Uranium: Current Status and Future Prospects*, IAEA-TECDOC-1529, IAEA, February 2007, p. 57.

9. "Spent Fuel Reprocessing in France."

10. As of the end of 2005, the spent-fuel pools at La Hague held about 8,100 tons of spent fuel and about 11,700 tons of spent fuel had been reprocessed at La Hague. See "Spent Fuel Reprocessing in France," p. 22. At a discharge rate of 1,200 tons a year from France's reactors and a contracted rate with Électricité de France for reprocessing of 850 tons/year, these totals would have climbed to about 9,000 and 14,000 tons by the end of 2008. As of the end of 2005, 543 tons of the spent fuel stored at La Hague was

French spent MOX fuel. See "Spent Fuel Reprocessing in France," Table 5. By the end of 2008, this would have increased to about 850 tons,

11. "Spent Fuel Reprocessing in France," Figure 7.

12. Matthew Bunn, Steve Fetter, John Holdren, and Bob van der Zwaan, "The economics of reprocessing versus direct disposal of spent nuclear fuel," *Nuclear Technology*, Vol. 150, June 2005, p. 209.

13. Table adapted from Frank von Hippel, "Managing Spent Fuel in the United States: The Illogic of Reprocessing," International Panel on Fissile Materials, 2007, Appendix. All scenarios discussed here assume that France's LWRs operate for an average of 45 years.

14. J. M. Charpin, B. Dessus, and R. Pellat, "Report to the Prime Minister: Economic Forecast Study of the Nuclear Power Option," 2000, p. 60, English translation available at *www.fissile-materials.org/ipfm/site_down/cha00.pdf* (CDP Report).

15. This figure is based on Figure 5 of "The economics of reprocessing versus direct disposal of spent nuclear fuel," updated by Steve Fetter through 2006 and the author through 2007 (average U.S. price) and 2008 (spot price) based on U.S. Energy Information Administration, "Average Price and Quantity for Uranium Purchased by Owners and Operators of U.S. Civilian Nuclear Power Reactors by Pricing Mechanisms and Delivery Year," and *Uranium Intelligence Weekly*, respectively.

16. "Uranium 2007: Resources, Production and Demand," OECD Nuclear Energy Agency and International Atomic Energy Agency, 2008, Table1 and Figure 8.

17. K. S. Deffeyes and I. D. MacGregor, "World Uranium Resources," *Scientific American*, January 1980, pp. 50-60; and Erich Schneider and William Sailor, "Long-term uranium supply estimates," *Nuclear Technology*, Vol. 162, June 2008, p. 379.

18. Ann MacLachlan, "EDF, Areva in tug-of-war over reprocessing price," *Nuclear Fuel*, February 25, 2008; "EDF-Areva pact ensures reprocessing, recycle," *Nuclear Fuel*, December 29, 2008.

19. According to a study funded by Areva and based on Areva proprietary data, the "overnight" cost of these plants, not including interest on the investment during construction, was about $18 billion. *Economic Assessment of Used Nuclear Fuel Management in the United States*, Boston, MA: Boston Consulting Group, 2006, Figure 8.

20. AREVA, "All about operations: La Hague," available from *www.lahague.areva-nc.com/scripts/areva-nc/publigen/content/templates/show.asp?P=50&L=EN*.

21. See also Tadahiro Katsuta and Tatsujiro Suzuki, "Japan's Spent Fuel and Plutonium Management Challenges," International Panel on Fissile Materials, 2006.

22. "Long-Term Nuclear Program Planning Committee publishes costs of nuclear fuel cycle," Citizens Nuclear Information Center, *Nuke Info*, Tokyo, Vol. 103, November/December 2004.

23. The nominal capacity of the Rokkasho plant is 800 tons per year vs. 1,000 tons/year for each of the two French reprocessing plants, UP2 and UP3. The Rokkasho plant has a design life of 40 years. As of the end of 2004, the costs incurred in its construction were 2.14 trillion yen (about $18 billion at a an exchange rate of 120 yen/$). In November 2003, the Federation of Electric Power Companies of Japan gave the total cost of the Rokkasho plant, including operations for 40 years, vitrification of the high-level waste, low-level-waste disposal, and decommissioning the plant as 11 trillion yen ($92 billion), Masako Sawai, "Japanese Nuclear Industry's Back End Costs," Citizen's Nuclear Information Center, *Nuke Info Tokyo*, No 98, November 2003-February 2004. This implies an operating cost of a little less than $2 billion per year vs. AREVA's claimed cost of $1 billion a year for operating both its reprocessing plants and its MOX fuel-fabrication facility. See *Economic Assessment of Used Nuclear Fuel Management in the United States*.

24. The only way to get a cost this low is to assume that Japan has enough coal-fired capacity fueled with low-cost coal to replace the output of all its nuclear-power plants.

25. Alexei Breus, "Russian Imports of Spent Fuel from Europe to Resume by Year-End," *Nuclear Fuel*, Vol. 28, No. 8, April 14, 2003.

26. Armenia has one VVER-440, Bulgaria has two VVER-1000, the Czech Republic has 4 VVER-440s and 2 VVER-1000s, Finland has two VVER-440s, Hungary has four VVER-440s, the Slovak Republic has two VVER-1000s, and Ukraine has two VVER-440s and 13 VVER-1000s.

27. Alexei Breus and Ann MacLachlan, "Russia and Hungary sign protocol for fresh and spent fuel trade," *Nuclear Fuel*, May 10, 2004; Ann MacLachlan and Daniel Horner, "Russia drops plans for taking in foreign spent fuel, citing other priorities," *Nuclear Fuel*, July 31, 2006.

28. "The curious case of Vandellós-1," *Plutonium Investigation*, No. 16, July 1999, p. 4.

29. David Albright, Frans Berkhout, and William Walker, *Plutonium and Highly Enriched Uranium, 1996*, Oxford, UK: Oxford University Press, 1997, Table 6.4.

30. In France: Germany, 5672 metric tons; Switzerland, 766 tons; and Belgium, 672 tons. See "Spent Fuels Reprocessing in France," Figure 7.

31. Contracted in the UK: Germany, 969 tons on base-load contracts and 787 tons on post-2003 contracts; and Switzerland, 422 tons base-load contracts. See M. Forwood, *The Legacy of Reprocessing in the United Kingdom*, Research Report No. 5, Table 1, Princeton, NJ: The International Panel on Fissile Material, July 2008.

32. Bruno Pellaud, personal communication with the author, March 1, 2009.

33. *The Legacy of Reprocessing in the United Kingdom*, Table 1.

34. Martin Forwood, "Quotes from CoRWM docs Re: Problems with Management of Spent AGR Fuel," January 29, 2009.

35. *Strategy*, Cumbria, UK: Nuclear Decommissioning Authority (NDA), March 2006, p. 46.

36. *Ibid.*, Draft Business Plan 2009/2012, 2008, p. 32.

37. Pearl Marshall, "Thorp expected to soon restart normal reprocessing operations," *Nuclear Fuel*, March 9, 2009.

38. NDA, NDA Plutonium Topic Strategy: Credible Options Technical Analysis, January 30, 2009. The UK had 81.2 tons of separated UK civilian plutonium in storage at its Sellafield reprocessing site as of the end of 2007. "Communication Received from the United Kingdom of Great Britain and Northern Ireland Concerning Its Policies Regarding the Management of Plutonium," INFCIRC/549/Add.8/11, IAEA, July 2, 2008.

39. "Communication received from the Russian Federation Concerning Its Policies Regarding the Management of Plutonium," INFCIRC/549/Add.9/10, IAEA, October 30, 2008.

40. "Russian Nuclear Program Slowed on Weak Energy Demand," *Uranium Intelligence Weekly*, March 9, 2009.

41. O. Saraev, Rosatom, "Prospects of Establishing a New Technology Platform for Nuclear Industry Development in Russia," International Congress on Advances in Nuclear Power Plants, Nice, France, May 13-18, 2007.

42. H. J. Bhabha and N. B. Prasad, "A study of the contribution of atomic energy to a power programme in India," Proceedings of the Second United Nations International Conference on the Peaceful Uses of Atomic Energy, Geneva, Switzerland, 1958, pp. 89-101.

43. R. B. Grover and Subhash Chandra, Department of Atomic Energy, India, "Scenario for growth of electricity in India," *Energy Policy*, Vol. 34, 2006, p. 2834.

44. M. V. Ramana and J. Y. Suchitra, "Plutonium Accounting and the Growth of Fast Breeder Reactors," forthcoming.

45. China National Development and Reform Commission, May 2007.

46. Mark Hibbs, "CNNC favors remote site for future reprocessing plant," *Nuclear Fuel*, April 7, 2008.

# PART IV:

# USING MARKET ECONOMICS TO PRICE
# NUCLEAR POWER
# AND ITS RISKS

# CHAPTER 13

## THIRD PARTY INSURANCE:
## THE NUCLEAR SECTOR'S "SILENT" SUBSIDY
## IN EUROPE

### Antony Froggatt
### Simon Carroll

## OVERVIEW OF NUCLEAR POWER
## IN THE EUROPEAN UNION

Within the European Union (EU), nuclear power is a divisive issue on a public and political level. Of the 27 Member States, 15 have nuclear power, with a total of 145 nuclear reactors providing 30 percent of the EU's electricity. France has by far the largest nuclear fleet, operating with 45 percent of the EU's total capacity.

Since the Chernobyl accident in 1986, there has been a downturn in the fortunes of the nuclear industry, and the absolute number of reactors in operation is expected to decline from 172 reactors in 1987 down to 135 reactors by 2010, and in 2006 eight reactors were shut. However, there is renewed interest in nuclear power and reactors are under construction in Bulgaria (Belene), Finland (Olkiluto 3), and France (Flamanville 3), the first new reactors orders in a Member State for over a decade.

Proposals are being developed in a number of countries in the EU to order new nuclear power plants. This includes a proposal for a reactor in Lithuania that would be jointly owned by Estonia, Latvia, and Poland. The proposals in other new Member States (Romania and Slovakia) also involve considerable cooperation with international utilities or constructors. In the United Kingdom (UK), the Government has stated its desire to see the continual use of nuclear but says the decision rests with the utilities. A number of vendors (Areva and AECL) have submitted designs for approval.

In some countries, there are no plans to build new reactors, although the existing reactors are being subject to plant life extensions, which simultaneously expands the output from each unit and prepares to extend their operating lives.

A number of countries have politically agreed to phase out plans. The most active is in Germany where a number of reactors have been closed. In Sweden, the original timetable for closure of reactors has slipped significantly. In January 2003, in Belgium, an agreement to limit the operating life of the reactors to 40 years and to stop building nuclear power plants was reached.

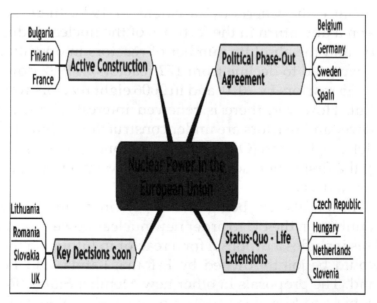

**Figure 13-1. Status of Nuclear Power in Europe.**

The other countries in the EU do not have nuclear power, and their view on it varies considerably. Austria has been outspoken in its criticisms of nuclear power and has been actively engaged in the nuclear debate in neighboring countries. Similarly, Ireland has actively engaged in the UK nuclear debate, particularly as it relates to the Sellafield reprocessing plant.

Under *Business As Usual* scenarios, the number of reactors being built will not even replace those due to be closed at the end of their working lives. Both the International Energy Agency (IEA)[1] and the European Commission[2] anticipate a drop of installed nuclear capacity, no later than 2030, by 44 percent and 25 percent, respectively.

## STATUS OF NUCLEAR INSURANCE REGIMES IN MEMBER STATES OF THE EU

There are two basic international legal frameworks contributing to an international regime on nuclear liability: The International Atomic Energy Agency's (IAEA) 1963 Convention on Civil Liability for Nuclear Damage (Vienna Convention), the Organization for Economic Cooperation and Development's (OECD) 1960 Convention on Third Party Liability in the Field of Nuclear Energy (Paris Convention), and the associated "Brussels Supplementary Convention"[3] of 1963. The Vienna and Paris liability conventions are also linked by a Joint Protocol adopted in 1988.[4]

## THE ORIGINAL LIABILITY AND COMPENSATION REGIMES

Negotiated at a time when the nuclear power industry was in its infancy, the Vienna and Paris Conventions had two primary goals: first, to create an economic environment where the nascent nuclear industry could flourish; and, second, to ensure that clear procedures and some compensation would be available in the event of an accident. The first aim would be achieved by removing legal and financial uncertainties over potentially enormous liability claims that could arise in the event of an accident. From the industry's development, it was clear that nuclear power could only be exploited as an efficient and independent source of energy if a reasonable amount of financial protection were available for private investors who were placing their financial resources in an unknown and potentially dangerous sector.[5]

While there are some differences in detail, the Vienna and Paris Conventions have some important features in common. In particular they:
- Allow limitations to be placed on the amount, duration, and types of damage for which nuclear operators are liable;[6]
- Require insurance or other surety to be obtained by the operator;
- Channel liability exclusively to the operator of the nuclear installation;
- Impose strict liability on the nuclear operator, regardless of fault, but subject to exceptions; and,
- Grant exclusive jurisdiction to the courts of one country for any given incident, normally the country in whose territory the incident occurs.

Chernobyl clearly revealed a number of deficiencies in the regimes established by both the Vienna and Paris Conventions.[7] Compared with the damage caused by the Chernobyl accident, it was obvious that the liability amounts were woefully low. Many countries were not party to either Convention.[8] Not all of the damage, or even the most serious damage, caused by Chernobyl was covered by the definition of damage applicable under either Convention. There were also problems with the limits on the time in which claims for compensation could be brought, the claims procedures, and the limitations on which courts had jurisdiction to hear claims.

## POST-CHERNOBYL REVISIONS TO THE LIABILITY AND COMPENSATION REGIMES

Following the 1986 accident at Chernobyl, significant effort was made by the international nuclear community to modernize a number of conventions. This eventually led to the revision of the international regime and the adoption of a number of new conventions, including:

- The International Nuclear Safety Convention, June 1994.
- Joint Convention on the Safety of Spent Fuel Management and on the Safety of Radioactive Waste Management, June 2001.
- Convention on the Early Notification of a Nuclear Accident, October 1986.
- Convention on Assistance in the Case of a Nuclear Accident, September 1986.

On nuclear liability, as an interim step intended primarily to address the lack of membership in both the IAEA and OECD liability regimes, the parties to both the Vienna and Paris Conventions adopted the 1988 Joint Protocol.[9] The Joint Protocol generally extends to states adhering to it the coverage that is provided under the convention (either Paris or Vienna) to which it is not already a Contracting Party.[10] It thus creates a "bridge" between the two conventions, effectively expanding their geographical scope. In doing so, it ensures that only one of the two conventions will be exclusively applicable to a nuclear incident.[11] At the time, it was believed that the link established by the Joint Protocol would induce a greater number of Central and Eastern European countries to join the Vienna Convention, particular those that had formed part of the former Soviet Union, a hope only partially realized.[12]

The international community soon recognized, however, that the Joint Protocol was not enough to redress the liability and compensation problems brought to harsh light by the Chernobyl accident. To attract broad adherence to the international nuclear liability conventions and to make them really effective, reform had to be more far reaching. In short, it had to ensure that in the case of a nuclear accident, much greater financial compensation would be made available to a much larger number of victims, in respect of a much broader scope of nuclear damage, than ever before.

The process of negotiating amendments to the Vienna Convention began in 1990 and concluded in 1997. Work then began officially in 1997 on revisions to the Paris Convention and in 1999 for the Brussels Supplementary Convention. Amending protocols to the Vienna, Paris, and Brussels Conventions have been adopted[13] as well as the new Convention on Supplementary Compensation (CSC), a completely new convention intended to establish a global regime of liability and compensation.[14]

The revisions to the Vienna and Paris/Brussels Conventions do increase the amount of compensation available, expand the time periods during which claims might be made, and expand the range of damage that is covered by the Conventions. It is also worth noting that in the formula to be used for State contributions to the combined fund under the revised Brussels Supplementary Convention, the proportion to be raised is more closely related to the actual generation of nuclear power by the participating states.[15]

The new liability and compensation amounts would be higher than before, with operator liability under the revised Paris Convention required to be at least €700 million and total compensation available under the revised Brussels Supplementary Convention would be €1500 million. Nonetheless, the overall amounts remain worryingly low when compared with the costs of the Chernobyl accident, currently estimated to be in the order of tens and hundreds of **billions** of euros.[16] Further, setting fixed compensation sums is not only arbitrary (in the absence of genuinely robust estimates of probable damage) but it is also unlikely to be valid over the longer term (unless they can be continually adjusted to take into account changes in the economic profile of accident consequences).[17] Table 13-1 depicts compensation and liability.

| Convention | Operator Liability + Installation State | Total Combined Contributions from Other States Party | Total Compensation Available |
|---|---|---|---|
| Paris, 1960 | €6 to €18 | - | €6 to €18 |
| Brussels, 1963 | Up to €202 | €149 | €357 |
| Paris, 2004 | At least €700 | - | At least €700 |
| Brussels, 2004 | Up to €1200 | €300 | €1500 |
| Vienna, 1963 | €50 | - | €50 |
| Vienna,1997 | Up to €357 | - | €357 |
| CSC, 1997 | At least €357 | Depends | At least €713 |

All figures are rounded and for **millions** of euros (€) — See Annex 1 for details.

**Table 13-1. Summary Table Showing Liability and Compensation Amounts for Different Conventions.[18]**

## CURRENT STATUS

Less than half the world's nuclear reactors are covered by any of the existing international agreements.[19] Moreover, although there are unifying features, the nuclear liability conventions do **not** provide one comprehensive and unified international legal regime for nuclear accidents. In fact, there is a labyrinth of intertwined international agreements on nuclear liability, the interrelations of which have become increasingly complicated.[20] Currently, there are at least eight such agreements, including the 1960 Paris Convention, the 1963 Vienna Convention, the 1963 Brussels Supplementary Convention, the 1988 Joint Protocol, the 1997 Protocol to Amend the Vienna Convention, the 1997 Convention on Supplementary Compensation, the 2004 revised Paris Convention, and the 2004 revised Brussels Supplementary Convention. The complications arise because the earlier and revised versions of some of these instruments may coexist, and states may become party to more than one instrument.[21] Figure 13-2 is depicts reactor insurance only.

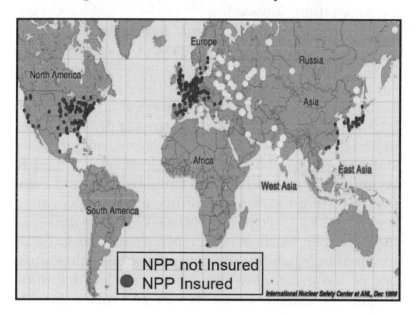

**Figure 13-2. Insurance of Nuclear Reactors.[22]**

The goal to ensure broad participation in the new regimes has not been achieved. At this point, only five countries have ratified the 1997 Vienna Convention. This was enough to bring the Protocol to amend the Vienna Convention into force in 2003, but the lack of wide adherence remains problematic.[23] There has also been a delay in the ratification of the revised Paris Convention and the revised Brussels Supplementary Convention. [24] In order for the Protocol amending the Paris Convention to enter into force, it must be ratified by **two-thirds** of the Contracting Parties. For EU Member states, this was supposed to have taken place by the end of 2006, but it has not yet been done.[25] For the Protocol amending the Brussels Convention, ratification by **all** Contracting Parties is required. Only three countries have ratified the new Supplementary Compensation Convention.[26]

The revisions of the original liability and compensation Conventions may not be supportive of ensuring broad adherence by a large number of states. To ensure a favorable environment for those considering investing in nuclear programs, it is necessary for installation states, states involved in the supply of nuclear materials or services for these programs, and **all** other states that might be affected by a nuclear accident to be under the umbrella of the same liability and compensation regime. For a liability and compensation regime to be attractive to states seeking to maintain or increase their nuclear power programs, the burdens imposed by a liability and compensation regime must not be too great. However, the expanded definition of damage, extended time frames, and raised liability and compensation amounts are proving problematic for some countries.

Conversely, to be attractive for a state without nuclear power plants, the liability and compensation conventions must offer sufficient compensation, and the regime must not introduce unacceptable restrictions or burdens when seeking to obtain compensation for losses incurred. For such states, becoming party to one of the nuclear liability conventions is not necessarily an attractive proposition, even if the revisions are taken into consideration. This is not surprising as the Paris and Vienna Conventions were essentially developed to nurture nascent nuclear industries, and the recent revisions have done little to alter this fundamental characteristic of the instruments and protecting and promoting nuclear power remains a central feature. Even as revised, the levels of compensation are relatively low when compared to the likely costs of a serious accident. By becoming a party, a non-nuclear power generating state might actually restrict its possibilities for obtaining legal remedies in the event of an accident.[27]

Until recently most EU Member States were party to the Paris/Brussels regime of nuclear liability and compensation, and this was considered a sufficiently uniform situation for the European Commission not to consider specific EU measures in this field.[28] Since the 2004 EU enlargement, this is no longer the case (see Table 13-2). EU States variously are party to the original Vienna convention; the revised Vienna convention; the Paris Convention; some have signalled their intention to adhere to the revised Paris Convention; and some are party to both the Paris and Brussels Conventions. The current range of operator liability in Member States goes from the low of €50 million in Bulgaria and Lithuania to unlimited liability in Germany. A full list of the different liability and compensation requirements in the EU Member States can be found in Appendix 2.[29]

| Not Party to Any Nuclear Liability Convention | Paris Convention, 1960 | Paris Convention + Brussels Supplementary Convention | Vienna Convention | | Convention on Supplementary Compensation, 1997 |
|---|---|---|---|---|---|
| | | | Original (1963) | Revised (1997) | |
| Austria, Cyprus, Ireland, Luxembourg, Malta | Greece, Portugal | Belgium, Denmark, Finland, France, Germany, Italy, Netherlands, Slovenia, Spain, Sweden, United Kingdom | Bulgaria, Czech Republic, Estonia, Hungary, Lithuania, Poland, Slovakia | Latvia, Romania | Romania |

See Annex 2 for details.

## Table 13-2. Summary of EU Country Participation in International Nuclear Liability Regimes.

Some EU Member States are not party to any of the international nuclear liability Conventions. Indeed, for EU countries like Ireland, Luxembourg, and Austria[30] — gravely concerned about the risks of nuclear power in neighboring countries, but with no nuclear power plants of their own — it would be difficult, indeed, to identify many, if any, reasons why they should accede to the current nuclear liability conventions.[31]

From the discussion above, it can be seen that there are widely divergent nuclear liability and compensation arrangements currently in place across the various EU Member States. These have profound implications for reactor safety, compensation of victims in the event of an accident, and for competition in the EU electricity market.

The problem created by this current situation has been recognized by the European Commission, which was to undertake an impact assessment in 2007 to explore the range of possible solutions and prepare a proposal to the Council.[32]

## IMPACTS ON THE PRIVATE NUCLEAR INSURANCE MARKET

The capacity of the private nuclear insurance market is also a major factor in determining the amount and extent of liability imposed on nuclear operators. According to the concept of the international nuclear liability conventions, coverage and liability amount are interlinked. The problems which insurers might have with the revised conventions could therefore put the results of the revision exercises at risk. In general terms, the shortcomings in the size and extent of coverage have a direct impact on the size and extent of the operator's liability. As a consequence, liability amounts exist worldwide which largely correspond to the insurance capacity but which do not match the nuclear risk.[33] The expanded scope of operator liability and the raised liability limits introduced by the 2004 amendments to the Paris Convention need to be seen in this context.

During the negotiations to revise the Vienna and Paris Conventions, representatives of the nuclear insurance industry stated that some of the proposed amendments would be problematical.[34] In particular, the nuclear insurance industry was concerned that there was:

- Insufficient private insurance market capacity to insure nuclear operators against raised liability amounts;
- An unwillingness of the market to cover extended /extinction periods during which an operator would be liable;

- A difficulty in that private insurance could not cover all the categories included in the expanded definition of damage. [35]

Effectively, as a consequence of the revisions introduced into the Vienna and Paris Conventions, nuclear operators might no longer be able to obtain private insurance coverage to cover their full liabilities under the revised Conventions. Tetley argues that, if no insurance cover is available, then the liability for the revised scope of cover must fall either on the operator directly or on the national government. The gap which has opened up between what the liability risks the operators are required to assume under the revised convention and the coverage available from private insurers, is causing problems and is delaying ratification of the revised liability Conventions.[36] Tetley summarized the concern thus: "The financial uncertainties introduced by the new heads of cover under the revised conventions will cause a reduction in insurance cover unless a consistent approach is found to deal with the unquantifiable risks imposed upon the nuclear operators."[37]

Another problem has to do with the new perception of the possibilities of terror attacks against nuclear installations. Under the Vienna Convention (both the original Convention and as amended by the 1997 Protocol) and the original Paris Convention, terrorism is not a ground for exoneration. This is because acts of terror are not explicitly given as a basis for exoneration of operator liability and the kind of terrorism like the events of September 11, 2001 (9/11) cannot be considered as an armed conflict, hostilities, civil war, or insurrection.[38] Consequently, the operator of a nuclear installation is liable for damage due to acts of terrorism. After the events of 9/11, the insurance pools reappraised the risks associated with acts of terror, concluding that the probability of a nuclear reactor becoming the target of such an attack was significantly higher than had been previously considered to be the case. During the negotiations on the revision of the Paris Convention, there was a call from the nuclear insurance industry for a review of the provisions of Article 9, with a view to exonerating an operator from liability for damage arising as a consequence of an act of terror.[39] This was not accepted by the parties, and consequently damage resulting from terrorism will still be covered by the revised Conventions also.[40] Nevertheless, some insurers may be able to limit their coverage to operators for damage caused by a nuclear incident resulting from a terrorist act — requiring state intervention to insure this risk.[41] This means that a new gap has opened up between the obligations on operators under the Conventions and what the private nuclear insurance market is prepared to cover.[42]

The problems with private insurance can be seen to be, at least partly, a financial question. It is not that insurance is unavailable, it is just that "few can be purchased at reasonable cost or at least at costs that are competitive with rates offered by the nuclear insurance pools."[43] The UK government laid out the current difficulties in its recent consultation paper when it said:[44]

> When the revised Conventions are implemented in the UK there will be an increase in the liability amount and the cost of insurance for UK nuclear operators (present ones and any future ones). To the extent that commercial cover cannot be secured for all aspects of the new operator liabilities, the Government will explore the alternative options available — including providing cover from public funds in return for a charge.

However, it is also, at least partly, a political decision. Simply because the private insurance industry is not able or willing to make cover available at the appropriate price to the industry does not mean that the risks are not there. As Pelzer has commented:[45]

> Tetley's conclusion clearly confirms the old school of thinking that liability means insurability. Legislators cannot agree to that view nor is it in the best interest of operators — not to mention the interest of possible victims — to be tied to the insurance industry without alternatives. For good reasons and after long difficult negotiations, States agreed on the revised conventions with a view to establishing a more risk adequate liability regime and to better protecting victims. There is no "inconsistent approach" which would warrant a change or an insurance adequate streamlining of the new liability concept only for the reason that the insurance industry is unable to cover the liability.[46]

The only conclusion which can be drawn from the insurers' reluctant position is to look for coverage other than insurance.

From the perspective of potential victims, there is a pressing need to ensure full and effective compensation for the full risks of nuclear accidents, and it is less of an issue what the specific modalities are. In accordance with the conventions, gaps in insurance coverage have to be covered by the installation state that has to step in to the extent that insurance or other financial security is not available or not sufficient to satisfy claims.[47] Pelzer argues that it would send the wrong signal if the advantages of the revised nuclear liability law could only be implemented with the help of state money. It derogates from the "polluter pays" principle, unless of course a nonsubsidized fee or premium is paid for that security. He argues that operators, in their own best interest, would therefore be well advised if they look for solutions to cover the insurance gaps by means of their own.[48]

From the perspective of the efficient functioning of the energy markets (for example, avoiding subsidies to nuclear power by failing to internalize the full costs of nuclear generation), whatever modalities are chosen must be reflected in the price of electricity from nuclear generation. Instituting some form of operators' pooling (rather than pooling of State funds) could be one way of realizing this objective.

## Europe's Changing Energy Market.

The energy sector in Europe will undergo considerable change over the coming decades. A combination of aging infrastructure, a growing awareness of climate change, and the dwindling of European fossil fuel reserves will result in considerable investment in noncarbon or low carbon emitting energy sources.

The scale of the investment anticipated in the EU over the next decade is unprecedented, the International Energy Agency (IEA) estimates that between 2005 and 2030, the EU will need 862 GW in total new capacity to replace aging conventional and nuclear power plants and meet increases in demand.[49] Of this additional installed capacity, 465 gigawatts (GWe) or 61 percent of the current total of 395 GWe will be replaced during this period.

The EU is a global driving force on climate change and has set targets to reduce its Greenhouse Gas emission by 20 percent by 2020. The European Commission has placed at the heart of its attempt to reduce emissions the European Emissions Trading Scheme

(ETS). This was finally adopted by the EU in October 2003 and was intended to introduce a "cap and trade" system for stationary $CO_2$ emissions. This was to cover around 40 percent of all Greenhouse Gas emissions from EU 27. The scheme became operational in January 2005, with the first trading scheme running until the end of 2007. The second period runs until the end of 2012.

The ETS covers only large industrial $CO_2$ producers, including: power stations over 20 MW; oil refineries; coke, iron, and steel; lime and cement production; glass production; ceramics; and paper and pulp production. The methodology of the scheme is for a set number amount of emission allowances to be allocated to each Member State, based on the existing emissions. In the first phase, a minimum of 95 percent of the allowance has to be allocated for free (or grandfathered). In the second phase, this is 90 percent. The rest are supposed to be auctioned.

The key issue, therefore, is the level of allocations. Member States made applications to the European Commission, who then offered revised allocations. The allocations for phase II are to be announced by the Commission in December 2007.

It has been retrospectively shown that in phase I, there was a major over-allocation of emission permits to such an extent that Member States handed out permits for 1829 million tons of $CO_2$ in 2005, while the actual emissions were 1,785 million tons.[50] With such overcapacity, it is hardly surprising that, despite a buoyant start, the price of carbon has dropped to close to zero from a high of over €30/ton. Despite the fact that virtually all of the permits were given to the utilities for free, the introduction of the ETS has had a measurable impact on the price of electricity in Europe. This has been the advantage of the large electricity generating sources, and it has been said that the main economic winners of the current scheme have been the coal and nuclear utilities.[51] This is all the more remarkable as the nuclear industry is currently excluded from the scheme.

It is anticipated that the second phase of the ETS will introduce lower permits for Member States, however, the impact that this will have on the price of $CO_2$ is still to be determined. In particular, the phase II allows, through a linking directive, the use of carbon credits gained through the Joint Implementation and Clean Development Mechanism of the Kyoto protocol. Depending on the volumes involved, which could be significant, this may have a considerable impact on the carbon price in the ETS.

The lack of price consistence in the carbon market has led the nuclear industry to call for a guaranteed floor price. The industry argues that its long investment cycle means that it needs some certainty over the market fluctuations and the chief executive of Electricité de France (EdF) has stated, "To make a commitment of billions of pounds to a project with a time-scale of half a century, investors above all need predictability about price. They must know the value society will place on carbon reduction not just tomorrow, but 10, 20, 30, 40 years from now."[52]

Despite the uncertain start for the ETS, it is clear that it remains a central part of the EU's policy on climate change. Over the coming years, further measures will be introduced to enlarge and refine the ETS. For it to be successful, it will require a significant and relatively certain price for carbon. Given the importance that the EU has placed on ETS in its fight against climate change, it should be assumed that a long-term carbon price, conducive to the nuclear industry, will be introduced.

### Costs of a Nuclear Accident and the Challenges of Externalities.

Limits on the liability of nuclear operators for off-site damage caused by a severe nuclear accident amount to an implicit subsidy of nuclear electricity. While there is some disagreement as to the exact degree of the subsidy, several assessments have confirmed that limits on the liability exposure of operators below the anticipated costs of a nuclear accident act as a significant subsidy to nuclear power generation.[53]

One study suggested that if EdF, the main French electric utility, was required to fully insure their power plants with private insurance but using the current internationally agreed limit on liabilities of approximately €420m, it would increase EdF's insurance premiums from €0.017/MWh, to €0.19/MWh, thus adding around 8 percent to the cost of generation. However, if there was no ceiling in place and an operator had to cover the full cost of a worst-case scenario accident, it would increase the insurance premiums to €5/MWh, thus increasing the cost of generation by around 300 percent.[54]

### THE COSTS OF SEVERE REACTOR ACCIDENTS

One reason for the lack of consensus on the precise extent of the subsidy resulting from limited liability of operators is that, while it is acknowledged that the consequences of serious nuclear accidents are very large and widespread, estimates of the likelihood and off-site consequences of a severe nuclear reactor accident vary widely.[55]

It is not unusual for different risk analyses carried out at the same reactor or different reactors of the same type to produce central value estimates that differ from one another by several orders of magnitude, and upper- and lower-bound estimates of damage can vary similarly, with no secure criteria for selecting among the conflicting expert assessments.[56] In the literature various accident scenarios are sketched, whereby the damages typically range from €100 million to €10 billion,[57] although some cost estimates are dramatically higher.[58]

An early estimate put the minimum **near-term costs** of the Chernobyl accident to be in the neighborhood of $15 billion, with longer-term costs of $75-150 billion.[59] A 1990 report prepared by Yuri Koryakin, the then-chief economist of the Research and Development Institute of Power Engineering of the Soviet Union, estimated that the costs from 1986 through to 2000 for the former Soviet Republics of Belarus, Russia, and Ukraine, would be 170-215 billion rubles (at the then official exchange rate this would be equivalent to $283-358 billion).[60] The Belarus Government estimate the total economic damage caused between 1986-2015 will be $235 billion (June 1992 prices).[61] Another estimate suggests overall economic costs **in the Ukraine alone** of **$130 billion**.[62]

Following the Chernobyl accident, the U.S. General Accounting Office (GAO) conducted an analysis of the off-site financial consequences of a major nuclear accident for all 119 nuclear power plants then operating in the United States. The estimates ranged between a low of $67 million to a high of $15,536 million.[63]

Four reactor accident scenarios considered by the EU ExternE project, yielded cost estimates for damage ranging from €431 million to €83,252 million.[64] It should be noted that these cost estimates exclude decontamination, although it is acknowledged that these costs "can rapidly be very high," and that there are major limitations to the economic evaluation,[65] arising from:

- Uncertainties on the impact (evaluation of source term, difficulties to estimate the environmental impacts due to the long-term contamination, uncertainties on the radiation health effects, etc);
- Uncertainties on the efficiency of countermeasures; and,
- Economic evaluation of some social consequences is nearly impossible.

At the same time, the often-cited expert opinion is that the type of reactors used in Western Europe have a very low probability of the kind of failure that would produce a severe accident. The exact values associated with an event in which there is a failure of containment and hence potentially significant damage vary from one set of experts to another but, in general, experts consider that the probabilities in the order of $10^{-6}$ and lower. Normalized to the probability of the event and to the electricity generation over a power plant's lifetime, the expected value of formal risk (i.e., probability x consequences) from an accident appears low, even against uncertainties in the accident probability.

However, the applicability of such tools is at least questionable, as it is also widely accepted that it is not only the expected value of risk (i.e., probability x consequence) that is important for the valuation of major accidents. Moreover, it appears that the estimates of the externalized costs of nuclear electricity are much more sensitive to changes in expert assessment of the expected off-site consequences of a worst-case accident, than of its likelihood of occurrence.[66]

## EXTERNALITIES

In general, comparing externalities between different energy sources and processes remains problematic. When comparing external costs of energy options, the same standard of environmental effects should be applied to all the options. However, the classification of environmental effects is inconsistent. Different valuation studies address different stages of the fuel cycles and different phases in the life cycles of the associated facilities.[67] Other difficulties arise owing to ignorance or damages that are effectively valued at zero. They are likely to be ignored in the pricing of electricity, selection of resources, and for any other policymaking purpose. Our knowledge of the environmental damages and the future is too uncertain to allow reliable estimates of damages. The consequence is that there can be little confidence that efforts to quantify and aggregate environmental externalities will yield systematic, comprehensive, or perhaps most importantly, comparable results.[68]

The ExternE Project set out to be the first systematic approach to the evaluation of external costs of a range of different fuel cycles.[69] The study's principal objectives to the end of 1995, when the first series of reports was published, were:

- To develop a unified methodology for quantifying the environmental impacts and social costs associated with production and consumption of energy;
- To use this methodology to evaluate the external costs of incremental use of different fuel cycles in different locations in the EU; and,
- To identify critical methodological issues and research requirements.

The 1995 ExternE report sought to quantify impacts and their associated externalities using an approach that accounted for the latest developments in environmental research. It reported external fuel cycle costs spanning three orders of magnitude.[70] Among the main contributing factors leading to this large range of results were the differing methodologies and assumptions used for the assessment of severe nuclear accidents.[71]

It has been subsequently noted that the boundaries and limitations of the estimation of the economic consequences, including the remaining uncertainties and nonquantifiable effects, "show the limitations of the economic modeling of the costs of accidents **which cannot integrate the complexity of a post-accidental situation**."[72] Despite further work on refining the analyses, the treatment of severe nuclear reactor accidents by ExternE remains problematic. Indeed, in a subsequent review and after considerable additional work, the ExternE team concluded: "The subject is one of the most difficult to be faced in the project: indeed despite earlier extensive research a clear solution to the problem is still to be identified."[73] The portion of the external costs that might be internalized by nuclear accident insurance was **not** addressed by ExternE.[74]

According to ExternE there remains wide divergence in opinion on what consequences should be looked at and hence what probabilities should be attached to those consequences. For the analysis of nuclear accidents from a PWR reactor, different source terms for release have been used as base data in France, Germany, and the UK. The significant differences in the release categories analyzed and in the probabilities attached to those releases, leads to considerable variation in assessments making cross-country comparisons difficult. Even more importantly, it makes it difficult to accept that there is a unique expert view of the accident probabilities that can be defined as objective. It shows that the accident scenarios and their associated probabilities are determined partly by judgement and partly by more "objective" considerations.[75] This implies that expert opinion should not be seen as single-valued and objective, and policymakers have to choose between different sets of consequences and probabilities.[76]

Related to the discussion above is the issue of how one treats public estimates of probabilities versus expert estimates in the assessment of accidents. Clearly both matter; one cannot ignore the careful analysis carried out by the experts, but at the same time one cannot overlook the opinions of the public. In the case of accidents which occur with reasonable frequency, this problem is resolved by looking at the relative frequencies of different accidents and basing the probabilities of such accidents on the relative frequencies. For nuclear accidents, there is no such history to draw on.[77] There have been hardly any major incidents with serious consequences in the history of nuclear power; for some experts, the one at Chernobyl is not considered relevant to the reactors deployed in Western Europe. Hence the divide between public and expert opinion has not narrowed appreciably over time. The expected value of damages is not enough. The public is willing to pay something for the reduction in risk per se, which is not captured in the expected value.[78]

The potential consequences from a single incident are also recognized as an important key criteria on its own.[79] While the approach of explicitly suggesting acceptable risk levels is partly established in environmental policy, up to the present ExternE failed to consistently integrate the level of potential consequences as an individual parameter into the valuation framework.[80]

The 2005 ExternE Methodology Update concluded, with respect to severe accidents in the nuclear sector, that:

> It is sometimes argued that, for so-called Damocles risks, i.e., risks with a very high damage and a low probability, the risk assessment of the public is not proportional to the risk. The occurrence of a very high damage should be avoided, even if the costs for the avoidance are much higher than the expectation value of the damage. However past attempts to quantify this effect have not been successful or accepted, so there is currently no accepted method on how to include risk aversion in such an analysis. Consequently, it is currently not taken into account within the ExternE methodology. Research on how to assess this, for example with participatory approaches, is clearly needed. [81]

### Proposals for New Nuclear Legislation.

From an economic perspective, the basic rule which should underlay a nuclear liability regime is rather straightforward: the legal regime should provide for incentives to nuclear operators to internalize their risk costs in order to maximize prevention. The basic idea is that by exposing nuclear operators to the full risk costs they are generating, an efficient internalization of the nuclear risk can take place. Of course, this internalization can be reached through a variety of legal and economic tools. For the nuclear sector, safety regulation plays a crucial role (i.e., so that nuclear reactors are designed, built, and operated in such a way as to minimize the risk of accidents). Liability rules have an important function in complementing safety regulation.

However, on the basis of straightforward economic analysis of nuclear liability law, it is clear that a nuclear operator should be exposed to the full costs his activity generates in order to provide optimal incentives for prevention.[82] From this simple rule a few equally simple rules of thumb follow as far as the structure of the regime of nuclear liability is concerned: nuclear operators should in principle be fully liable for the potential damage caused by their activity and, to the extent that compensation is provided through another source (government or insurance), mechanisms should be put in place as a result of which the nuclear operators' preventive efforts are taken into account. In insurance, these are the well-known techniques of risk differentiation as a remedy to moral hazard;[83] in case of government provided compensation the financing should in principal also be risk related whereby, a government fund is financed by risk-based premiums paid by operators.[84] The international nuclear compensation regime has been heavily criticized in the law and economics literature for not respecting these rules of thumb.[85]

It has been pointed out that the international regime of the conventions and the U.S. national nuclear compensation schemes were originally very similar, but they have since evolved along different lines to be quite markedly different today. The discussions on the international conventions and the American compensation scheme beginning in the 1950s started from the idea that nuclear energy development had to be supported. This entailed limiting the nuclear operator's liability and making public funding available to compensate for victims of a nuclear accident. In the United States, it was accepted much faster than in Europe, that this justification cannot be upheld forever. As a result, already in 1982, the United States completely abandoned the public funding of nuclear damage, whereas the international regime today still to a large extent relies on public funding.[86]

Faure and Vanden Borre have concluded that the economic goal of cost internalization cannot be reached in the current international conventions regime for two main reasons: the individual liability of the nuclear operator[87] is only a small fraction of the potential costs of a nuclear accident (looking at the damage estimated between €10 billion and €100 billion). Moreover, the second layer of compensation in the international regime is entirely provided for through public funds (the installation state in a second layer and a collective state fund in the third layer), whereby no risk-related financing takes place whatsoever. The second and third layer of public funds are a pure subsidy to the nuclear industry and fail to make any contribution to cost internalization.[88] Faure and Vanden Borre argue that the U.S. model shows that if a compensation regime were to be organized as a collective responsibility of the nuclear industry (thus excluding public funding), much higher amounts of compensation can be provided to victims and a better internalization of the nuclear risk can be promoted. They point to the operators' pooling systems established in the United States and Germany as having demonstrated the capacity to deploy many times the amounts required under the revised nuclear liability conventions and in particular the amounts offered by the insurance industry.[89]

Pelzer has also looked at alternatives to the current system and identified international operators' pooling is an interesting option. Operators' pooling is meant to provide financial security if and to the extent insurance coverage is not available and state intervention is regarded as being an inappropriate means to cover private liabilities because it would conflict with the polluter-pays principle and would interfere with principles of market economy. Under these circumstances, the pooling could serve two purposes. First, it could be used to fill gaps in coverage due to specific exclusions from insurance coverage. Second, it could be used to increase the total amount of compensation beyond the capacity of the insurance industry. Using the pooling for both purposes is desirable. The principal advantage of an operator pooling system such as that adopted in Germany or the United States is that large sums of private money, as opposed to public funds, can be made readily available to compensate victims of a nuclear accident. Pelzer also argues that there are advantages for the liable operator, as this option could be an attractive supplement and alternative to other forms of financial security provided pooling can be organized appropriately.[90]

## ADAPTING THE U.S. MODEL TO EUROPE?

Faure and Vanden Borre have suggested the creation of an international nuclear liability system modeled on that currently in place in the United States. In their approach, a key issue is to phase-out all state funding in the international (and national) nuclear compensation scheme, i.e., by replacing the current collective state funding, by a collective tier funded by the nuclear operators.[91]

The last change to the Price-Anderson Act so far was made in 2005.[92] The liability of the individual operator amounts to $300 million. However, the amount available in the second (collective) tier, is set at $95.8 million, plus an extra 5 percent for legal costs, with a maximum of $15 million per reactor per year. Given the fact that in 2005 a total of 104 reactors had a licence, the total available amount in the United States is $10.76 billion.[93] If the nuclear power industry grows, the funds available in the event of a serious accident will increase. **It should be noted that a pooling system at the U.S. national level requires that premiums or shares to be paid by an individual operator are only due after a nuclear incident has occurred causing damage in excess of a defined size.**[94]

In nine Western European countries alone, there are 135 nuclear reactors in operation — this is more than the current 104 reactors in the second tier of the U.S. compensation system. If all these operators should contribute, e.g., €10 million in the second tier (one-tenth of the current amount of the U.S. second tier), an amount of €1.35 billion of private funding would be immediately available in the second tier. Applying the same level currently in place in the United States would raise more than €10 billion.[95]

Faure and Vanden Borre identify two barriers for pursuing a U.S.-type approach: (1) in the EU, every Member State has its own regulatory structure (on nuclear safety, but also concerning the approval of the form of financial security to be presented by the nuclear operator); and, (2) differences in the way nuclear power plants are being operated throughout Europe, despite several EU Directives on nuclear safety (focussing more on issues concerning radiation protection and less on operational safety issues).[96] Furthermore, the EU has not developed common safety standards and currently relies on the international safety guidelines of the International Atomic Energy Agency and the safety requirements instituted by the individual Member States.[97]

There are several advantages to considering action at EU level when developing a pooling approach. Pooling is easier to agree upon if it takes place among operators of like-minded states that preferably cooperate already in other fields. States that are contracting parties to an organization of regional integration or other nature provide a good basis for operators' pooling. This applies particularly to EU Member States. Limitation of the system to a certain geographical region makes pooling more reasonable because only in a geographically limited area, may a natural transboundary risk community exist. To minimize the described problems and to prevent discrimination against operators that join a pooling regime, installation states should ensure harmonized economic and legal conditions. That requires common arrangements among the concerned states and in this context, the EU could possibly play a supportive role regarding pooling among EU operators.[98]

While the U.S. system is based on a statutory obligation or duty of the individual operator to contribute, Pelzer suggests that this is not the model to follow at the international level. Instead, he considers that it should be left up to the industry to decide if, and to what extent, and under what conditions they are prepared to embark on international pooling of financial means to cover their mandatory nuclear liability. How and to what extent they do so should not, in his view, be a business of states.[99] Leaving the mandatory or voluntary nature of a pool to one side, for now at least, it is nonetheless useful to consider how an EU-wide pool might be designed to reflect any specific characteristics of the EU nuclear electricity generating sector. Previously nuclear reactors in Western Europe tended to be operated by state agencies or national companies. This is no longer the case. Reactor ownership is also shared among private companies in an increasingly privatised electricity sector often operating at EU (and wider) rather than national levels of organization. Individual reactors may have multiple owners, in some cases there are multiple "part" owners of reactors, with large multinationals like Vattenfall, EON, etc., who have interests in nuclear reactors located in several EU Member States. Pelzer believes that the organization and structure of any international nuclear operators' pooling should be left to the discretion of operators and their respective parent companies— consideration should be given not just to operators, but owners too. It is worth noting that this is the situation today with the German national nuclear pooling system described earlier. The four parent companies owning the 19 German nuclear power plants have established a joint arrangement for nuclear pooling.[100] Based on, and corresponding to, the shares a partner holds in an individual power plant, the percentage of this plant will be attributed to the partner; the sum of all percentages for all power plants forms the total size of the guarantee of that partner.[101]

As noted above, the pooling approach is attractive because of the potentially much higher amounts of compensation and the improved internalization of the risks of nuclear power in the costs of generation of nuclear electricity. However, the extent to which these potential benefits can be realized will depend much on the details and implementation of any planned new scheme. Pooling per se is no panacea—a flawed and inadequate pooling system will not improve the current situation. At this stage, there are two principal issues of concern: the extent to which the full costs of a Chernobyl-scale accident would be covered; and the potential for unscrupulous operators to spread their risk through the pool.

While the current pooling arrangements in Germany and United States offer considerably greater compensation amounts than the current system of liability conventions, including the revised Paris/Brussels conventions, they still do not come close to matching the actual costs of an accident on the scale of Chernobyl. Obviously, the total amount of funding that could be realized by a pooling arrangement is a function of the design of the pool (especially its financial obligations and the levels of contributions) and the number of contributors. However, it needs to be considered that a severe accident may exhaust even the large financial resources provided through a pooling mechanism. To address this concern requires maintaining options to supplement the finances made through the pool to en sure additional compensation is available for victims and to remedy damage in the event that the pool funds are insufficient. It should be recalled that the pooling itself is a funding mechanism designed to facilitate availability of funds up to a certain preferably high amount. However, the creation of such a pool should not affect the ultimate liability of the operator, which should be unlimited. This is the current situation with the German nuclear liability pool arrangement.[102] Accordingly, such a pooling system would be designed to create an EU-wide international pool to provide a large fund (in the order of tens of billions of euros, at least an order of magnitude larger than the German national pool). This would be coupled with unlimited liability of individual nuclear operators.

Any pooling arrangement spreads the risk among its members, with the result that: (1) for any individual operator, the internalization of the nuclear risk is less than complete; and (2) the risk per reactor is averaged, so that a "risky" operator transfers a part of its risk to the pool, whereas a "safer" operator accepts a portion of the extra risk. One virtue of the pooling system is that there will be an element of self-policing by the pool members, in their own self-interest. Pool members themselves will have at least minimum requirements concerning the level of nuclear safety and security of the nuclear installations with which the risk will be shared. Operators will only be prepared to pool if the safety and security standards of other installations are up to the standards of their own installations.[103] There also has to be an adequate nuclear regulatory legal framework in all states whose operators wish to cooperate in the pool and, as noted above, there may well be a need for an EU-wide approach to safety regulation and standards.

## An Opportunity for Intervention.

On a relatively ad hoc basis the European Commission publishes a background paper on the state of nuclear power in Europe (Nuclear Illustrative Program, also called the PINC paper). The most recent was published in January 2007 and stated "The Commission is aiming at harmonising the nuclear liability rules within the Community. An impact assessment will be started to this end in 2007."[104]

This was officially proposed, because, as noted above, some Member States are parties to different versions of the Vienna Convention and Paris Convention, some are not party to the Brussels Convention, and some Member States are not party to any nuclear liability instrument. The Commission was therefore hoping to introduce measures to harmonize this current situation. In addition, there are other issues that may be included within this harmonization process.

1. In 2004, the limits and other provisions of the Paris Convention were revised. For these changes to enter into force, two-thirds of the signatory states must ratify the 2004 Protocol. This will take place when the EU Member States complete their procedure of simultaneous ratification required by the Council Decision of March 8, 2004. The deadline for this was by the end of 2006, but the EU Member States did not meet that deadline so it was (informally) reset for the end of 2007. Similarly, the 1997 Vienna Convention was only ratified by two EU Member States, and only by five countries in total worldwide. This issue is said to be causing increasing concern among legal experts.

As indicated earlier, the gap between what the nuclear insurers are willing to insure and what the operators are liable for is causing problems for the nuclear operators and governments and is delaying ratification. As Pelzer has noted, the current difficulties of the insurance industry to cover certain nuclear risks offers a chance to break new ground in providing financial security. The still-pending ratification and entry into force of the improved international nuclear liability regime creates some time pressure. All stakeholders are responsible for making those enhancements effective in a timely fashion. In his view and despite the inherent challenges, operators' pooling is a means to speed up the process and the time is ripe to explore the option more closely.[105]

2. Some Member States are not party to either the Vienna or Paris conventions and therefore do not recognize the limiting factors that these impose upon potential victims in the event of an accident. This is especially true for Austria, which is not only a nonparty to the conventions, but has domestic legislation that enables unlimited liability. Given the transboundary nature of large scale nuclear accidents, this undermines the effectiveness of the regimes to limit liability.

3. The EU is not party to either convention.

Formally, a number of bodies will now be asked to provide their opinion on the PINC paper. So far, only the European Economic and Social Committee (EESC) have done so. Concerning nuclear insurance the EESC stated:

> A harmonised liability scheme, including a mechanism to ensure the availability of funds in the event of damage caused by a nuclear accident without calling on public funds, is in the view of the EESC also essential for greater acceptability of nuclear power. The current system (liability insurance of $700 million) is inadequate for this purpose. [106]

The Commission is also trying to put its message across on this issue to a wider audience. At a recent meeting of the Nuclear Inter Jura Conference organized by the International Nuclear Law Association (INLA) the Commissioner in charge of Energy, Andris Piebalgs, stated that a "harmonized liability scheme, including a mechanism to ensure the availability of funds in the event of damage caused by a nuclear accident, is essential to the long-term acceptability of nuclear power." The Commissioner then went onto say that: "Therefore, before the end of the year, the

Commission will undertake an impact assessment to explore the range of possible solutions and prepare a proposal to the Council." [107]

Governments have signed up to the revised arrangements for nuclear liability and compensation that the nuclear insurance industry finds difficult to implement and which the nuclear industry is not comfortable with. The current nuclear insurance pooling system does not give adequate cover and the private insurance market is more expensive. Consequently, operators are putting pressure on governments not to ratify the revised conventions without having first guaranteed that their additional exposure risks will be met with Government assistance. Consequently, the public is not yet being given the fairly modest (when compared to the actual likely costs of a major nuclear accident) increase in compensation levels developed over 2 decades following the Chernobyl disaster.

Even with the proposed increase in operator liability and the new compensations arrangements — should they enter into force — only a small fraction of the potential costs of a nuclear accident will be covered. Any limitation in operator liability below the likely costs of a major nuclear accident constitutes a subsidy to the nuclear industry. Existing compensation arrangements allowing for state funds to be provided in lieu of industry responsibility for the economic consequences of an accident also are a pure subsidy to the nuclear industry and fail to make any contribution to cost internalization of the risks of nuclear power in electricity pricing.

It is nonsensical to persist with a system that:

- reduces the incentives for the nuclear industry to pursue the highest possible levels of safety by shielding the nuclear industry from the economic consequences of a nuclear accident;

- provides at best partial compensation for the damage caused by a major nuclear accident; and,
- adds an additional market distortion to the electricity market at a time when the EU is seeking to internalize environmental and other costs.

The recognition by the European Commission of the need to address the disparities and incongruities in nuclear third party liability currently existing in the EU has opened the door on this issue anew. There is now a real opportunity to develop and implement a fairer more efficient and effective nuclear liability and compensation scheme to the benefit of all.

## ENDNOTES - CHAPTER 13

1. *IEA 2006: World Energy Outlook*, Paris, France: International Energy Agency, OECD/IEA, 2006.

2. *European Energy and Transport: Trends to 2030*, Luxembourg: Office for Official Publications of the European Communities, 2008.

3. Convention Supplementary to the Paris Convention of July 29, 1960, on Third Party Liability in the Field of Nuclear Energy.

4. The Joint Protocol Relating to the Application of the Vienna Convention and the Paris Convention, September 1988. The Joint Protocol entered into force on April 27, 1992.

5. The limitation of liability was considered to be necessary in order to not obstruct the development of the nuclear industry. The Exposé des Motifs for the 1960 Paris Convenion notes that "unlimited liability could easily lead to the ruin of the operator without affording any substantial contribution to compensation for the damage caused." See Exposé des Motifs, Motif 45. The reason for this limitation was therefore purely economical: the li-

ability of the operator was limited to the amount for which the insurance market was able to provide coverage. OECD Council, *Exposé des Motifs of the 1960 Paris Convention,* approved by the OECD Council on November 16, 1982, available from *www.nea.fr/html/law/nlparis_motif.html.*

6. The Vienna Convention allows liability to be extended by national legislation to cover additional types of damage.

7. See for example, Michael G. Faure and Tom Vanden Borre, "Economic Analysis of the Externalities in Nuclear Electricity Production: The U.S. versus the International Nuclear Liability Scheme," paper presented to Nuclear Inter Jura 2007, Brussels, Belgium, October 2, 2007. The major criticisms of the international compensation regime can be summarized as follows:
- the low financial cap may provide insufficient deterrence for the prevention of accidents by the nuclear operator;
- the financial cap (combined with the public compensation scheme) constitutes a subsidy to the nuclear power since full costs are not internalized;
- the monopoly of the national nuclear insurance pools creates problems for the operators (paying high premiums) and for the authorities (who are confronted with a problem of information asymmetry, not knowing what the exact capacity of the insurance market is);
- despite the public compensation scheme (and notwithstanding recent increases), victims will not be fully compensated in case of a major (or even an average) nuclear accident. Recently, the international conventions have been amended to (in some cases rather substantially) increase the amounts of liability and to increase the amounts available in the public compensation scheme, but the same criticisms still apply.

8. The Vienna Convention was intended to be a global instrument governing civil liability for nuclear damage. However by the time of the 1986 Chernobyl accident, only 10 states had ratified it and not one of these had a major nuclear program (Argentina, Bolivia, Cameroon, Cuba, Egypt, Niger, Peru, Philippines, Trinidad & Tobago, and Yugoslavia). The Paris (and Brussels) Conventions were originally negotiated to provide a regional liability and compensation regime for nuclear damage for West-

ern Europe. They had achieved wide-spread, but not universal, participation of Western European countries by the time of the Chernobyl accident.

9. The Joint Protocol entered into force in April 1992.

10. For example, where a nuclear incident occurs for which an operator in a Paris Convention/Joint Protocol state is liable and damage is suffered by victims in a Vienna Convention/Joint Protocol state, those victims will be able to claim compensation for damage suffered against the liable operator as if they were victims in a Paris Convention state.

11. The exclusive application to a nuclear incident of only one of the two conventions is accomplished by means of the conflict rule contained in Article III of the Joint Protocol.

12. Some 18 countries from those parts of Europe have ratified or acceded to that convention, more than half the total number of Contracting Parties thereto. Yet only 11 of those 18 countries have ratified or acceded to the Joint Protocol, the instrument which would link them to the regime established by the Paris Convention, a disappointing development for those who had hoped to link all of Europe with one single nuclear liability and compensation regime. See Julia A. Schwartz, "International Nuclear Third Party Liability Law: The Response to Chernobyl," in *International Nuclear Law in the Post-Chernobyl Period*, Paris, France: OECD, 2006, pp. 37-72.

13. The 1997 Protocol to Amend the 1963 Vienna Convention; the 2004 Protocol to Amend the 1960 Paris Convention; and the 2004 Protocol to amend the 1963 Brussels Supplementary Supplementary Convention.

14. The new Convention on Supplementary Compensation for Nuclear Damage was adopted in 1997. It is intended to be a free-standing instrument which may be adhered to by all states irrespective of whether or not they are party to any of the existing nuclear liability conventions. Its objective is to provide additional compensation for nuclear damage beyond that established by the existing conventions and national legislation. It would do this through additional financial contributions from states which become parties.

15. Following the example of the Supplementary Compensation Convention which imposes greater responsibility upon nuclear power generating states to provide compensation, the formula for calculating contributions to the international tier under the Brussels Supplementary Convention Protocol moves from one based equally on gross national product and installed nuclear capacity to one based on 35 percent on gross domestic product and 65 percent on installed nuclear capacity. See Schwartz, "International Nuclear Third Party Liability Law."

16. The total costs of the 1986 Chernobyl accident remain uncertain, but typical estimates place the costs in the order of € tens of billions and € hundreds of billions. See the discussion below on the estimates of costs of severe nuclear reactor accidents.

17. An illustration of this can be seen in the context of natural disasters. In the United States, until recently, the number of lives lost to natural hazards each year has declined. However, the economic cost of response to, and recovery from, major disasters continues to rise. Each decade, the cost in constant dollars, of property damage from natural hazards, doubles or triples. See "Facing Tomorrow's Challenges — U.S. Geological Survey Science in the Decade 2007-2017," Circular 1309, Washington, DC: U.S. Department of the Interior/U.S. Geological Survey, 2007, p. 30. A similar inflation would be expected for the costs of man-made disasters also.

18. This table summarizes the detailed information on the financial requirements and limits of the different Conventions, including the original and revised Conventions, set out in Appendix 1.

19. McRae has calculated that of the 10 countries with the largest installed nuclear capacity, one half are members of the international scheme. Overall, the nuclear power generating countries that operate outside the international compensation regimes account for more than half of worldwide installed capacity. See Ben McRae, "Overview of the Convention on Supplementary Compensation," in *Reform of Civil Nuclear Liability*, Paris, France: OECD, 2000, p. 175.

20. For a comprehensive discussion of the interrelationship of the various conventions, see N. L. J. T. Horbach, ed., *Contemporary Developments in Nuclear Energy Law: Harmonising Legislation in CEEC/NIS*, Cambridge, UK: Kluwer Law International (ISBN 90-411-9719-2), 1999, pp. 43-85. See also O. F. Brown and N. L. J. T. Horbach, "Liability for International Nuclear Transport: An Overview," International Symposium on Reform of Civil Nuclear Liability, Budapest, Hungary, June 1999.

21. A further complication is introduced by transitional measures introduced in the various new instruments, designed to facilitate adherence by new States (see Appendix 1).

22. Note that the figure illustrates reactor insurance only, not whether any of the liability conventions apply. For example, although they are insured, reactors in the USA, Canada, Japan, and Republic of Korea are not covered by any international liability convention. See Mark Tetley, "Underwriting the Nuclear Risk," Presentation for the West Minister Energy Forum, 2005, *Nuclear Risk Insurers Ltd*, London, UK: YJL, April 24, 2005.

23. Five countries have ratified the 1997 Vienna Convention: Argentina, Belarus, Latvia, Morocco, and Romania. Only Argentina and Romania have nuclear power generating capacity, and according to the IAEA's Power Reactor Information System, as of August 30, 2007, those capacities were 935 MWe and 1310 MWe, respectively (available from www.iaea.org/programmes/a2/). As Schwartz has noted:

> The adoption of the VC Protocol was one of the most significant developments to have taken place in nuclear liability law for several decades. It was hoped that this new instrument would attract broad adherence by both nuclear power generating states and non-nuclear power generating states, whether Party to the Vienna Convention or not. Despite the many years of difficult negotiations required to reach agreement on this instrument, the keen interest it elicited from a broad range of interested states, and the many provisions it contains to encourage and facilitate adherence to it, the VC Protocol has not drawn the wide support originally hoped for or expected. Some 80 states participated in its negotiation

and in the Diplomatic Conference which culminated in its adoption. Yet only 15 countries have actually signed the Protocol, and 14 of those did so within one year of its adoption, when motivation and impetus were both still strong. The Protocol entered into force on 4 October 2003, some six years after it had been adopted, having been ratified by the number of states required for that purpose.

Schwartz, "International Nuclear Third Party Liability Law."

24. The Protocol to the Paris Convention and the Protocol to the Brussels Supplementary Convention were opened for signature on February 12, 2004, but in October 2007, neither of these instruments had entered into force.

25. According to the Council Decision of March 8, 2004, Member States which are party to the Paris Convention shall take necessary steps to deposit simultaneously their instruments of ratification of the Protocol with the Secretary General of the OECD "within a reasonable time and, if possible, before 31 December 2006." See "Council Decision of March 8, 2004, authorizing the Member States which are Party to the Paris Convention of July 29, 1960, on Third Party Liability in the Field of Nuclear Energy to ratify, in the interest of the European Community, the Protocol amending that Convention, or to accede to it," Brussels, Belgium, *Official Journal of the European Union* (OJ), L 97/53, April 1, 2004.

26. The three states which have ratified the CSC are Argentina, Morocco, and Romania. Entry into force requires the ratification of at least five states with a combined minimum of 400,000 installed units (MWthermal) of nuclear capacity. On August 3, 2006, the U.S. Senate took the most important constitutional step of consenting to ratification of the treaty by adopting an Act to implement the CSC. The only remaining step is enactment of implementing legislation that will set forth the mechanism the United States will use to fund any contribution it might have to make in the future to the international fund established by the CSC. The U.S. Congress was considering this implementing legislation, but the biennial Congress adjourned before this Act was approved by the House of Representatives. The implementing legislation is expected to be enacted by the new Congress that was elected in November 2006. According to McRae, the United States expects

to deposit its instrument of ratification with the IAEA in the very near future. See McRae, "Overview of the Convention on Supplementary Compensation." The aim of the (draft) Act is to establish a funding mechanism under the Price-Anderson Act for the U.S. contribution to the international nuclear liability compensation system. It is also said that "CSC benefits US suppliers who face potentially unlimited liability for nuclear accidents outside the coverage of the Price-Anderson Act by replacing potentially open-ended liability with a predictable regime." If, as seems quite likely, the United States will eventually ratify the CSC, it will change U.S. policy by entering the international nuclear liability regime (only CSC). The fact that the United States ratifies the Convention will in itself not be enough for the entry into force of the convention; it is, however, a very important step and might motivate other states to join the convention as well. If another state with a large nuclear capacity such as Japan were to become a member, this would trigger the entry into force of the CSC. Second, it is important because the United States immediately shifts the financial burden it will have under the collective tier of the convention to the private sector (the nuclear suppliers).

27. See for example, Philippe Sands and Paolo Galizzi, "The 1968 Brussels Convention and Liability for Nuclear Damage," *Nuclear Law Bulletin*, No. 64, December 1999, pp. 7-27; and Paolo Galizzi, "Questions of Jurisdiction in the Event of a Nuclear Accident in a Member State of the European Union," *Journal of Environmental Law*, Vol. 8, No. 1, 1998, pp. 71-97.

28. Answer of Commissioner Matutes to Written Question E-2489/93 (S. Kostopolous), September 1, 1993 (94/C 240/45), in which it is stated, *inter alia*, that:

> All the Member States are parties to the 1960 Paris Convention save Luxembourg and Ireland, which have no nuclear installations on their territory. There is thus no need for the Commission to take the initiative suggested by the Honourable Member [to lay down provisions in insurance law relating to the civil liability of operators of nuclear installations for any damage to persons, property and the environment].

OJ, C240/24, August 29, 1994.

29. Appendix 2 contains a table setting out the different liability and compensation regimes applicable in the individual EU Member States and the varying operator liability amounts and financial security limits.

30. Although not a party to any of the conventions, Austria has enacted specific legislation covering liability for nuclear accidents. Austria's nuclear liability legislation rejects many of the fundamental principles underlying the current nuclear liability regimes. Under its legislation, for example, the operator of a nuclear installation may not be exclusively liable. Victims may even assert a claim against a nuclear operator or supplier pursuant to other liability legislation in force, for example, product liability legislation. Nor are victims precluded from pursuing claims against more than one defendant. The liability imposed is in all cases unlimited. There are no time limits during which claims may be brought. Prescription periods are determined by the general law of civil procedure of Austria. Austrian courts have jurisdiction to determine claims and Austrian law is applicable, regardless of where the incident causing damage took place, subject only to certain limited exceptions. See *Bundesgesetz über die zivilrechtliche Haftung für Schäden durch Radioaktivität* (Federal Law on Civil Liability for Damages Caused by Radioactivity) *Atomhaftungsgestz,* 1999, BGB1, Vol. I, No. 170, 1998. A description is given in M. Hinterregger, "The New Austrian Act on Third Party Liability for Nuclear Damage," *Nuclear Law Bulletin,* No. 62, 1998, pp. 27-34.

31. It should be noted, in this respect, that Commissioner Matutes' response to the Parliamentary question described above, is deficient. Although neither Ireland nor Luxembourg have nuclear installations, they may be affected by a nuclear accident at a reactor located in one of the other EU Member States. In such circumstances, the fact that they are not party to the Paris Convention would pose problems in that **none** of the provisions of the Paris Convention would apply with respect to them. This creates the possibility of claims being pursued through other mechanisms, without the limitations on type of damage, time periods and amounts of liability of the operator, or the channeling, exclusivity, and other special requirements favorable to the nuclear operator, which are established by the Paris Convention. Plaintiffs in such countries might seek compensation through the courts in their

own country, i.e., where the damage occurred (or, at the plaintiff's discretion, in the country where the incident occurred), relying on the general conflict of law rules relating to international jurisdiction, including, for example, the 1968 Brussels Convention on the Jurisdiction and Enforcement of Judgements in Civil and Commercial Matters. See Galizzi. While the outcome of such a proceeding is by no means certain, it might be considered to offer certain advantages not found in pursuing claims pursuant to the limitations of the Paris Convention. See, also: Sands and Galizzi.

32. In January 2007, the Commission stated: "The Commission is aiming at harmonising the nuclear liability rules within the Community. An impact assessment will be started to this end in 2007." See Illustrative Nuclear Programme, Presented Under Article 40 of the Euratom Treaty for the Opinion of the European Economic and Social Committee, *Communication from the Commission to the Council and the European Parliament*, COM(2006) 844 Final, European Commission, July 12, 2007. This aim was restated recently by Energy Commissioner Piebalgs at the *Nuclear Inter Jura Conference 2007* in Brussels, Belgium, on October 2, 2007.

33. See, for example, Norbert, Pelzer, "International Pooling of Operators' Funds: An Option to Increase the Amount of Financial Security to Cover Nuclear Liability?" Discussion Paper for the IAEA INLEX Group Meeting on June 21-22, 2007, pp. 37-55.

34. *Ibid.*, p. 9. The nuclear insurance industry made its concerns known at an early stage in the discussion of amendments of the Paris Convention, see Letter of the Comité Européen des Assurances of December 8, 2000.

35. The nuclear insurance industry concerns with regard to the full insurability of these various risks stems from a variety of issues. In some cases, particularly for "reinstating a significantly impaired environment," insurers take the view that there is no insurable interest to be protected, or that there is no quantifiable economic interest. They maintain that it will be difficult to establish the type and extent of damage caused by the accident and at what stage of progression that damage occurred; they point out that it is not always easy to relate decreases in land values to a particular source. They have expressed concerns over uncertainty as to how courts may define or interpret a significant impairment

of the environment. Finally, they have indicated their opposition to extended prescription periods both on the basis of problems related to causality, but as well, the difficulty of quantifying exposure, the need to defend against speculative claims and the questioned value of legally authorised exposure limits. See M. Tetley, "Revised Paris and Vienna Nuclear Liability Conventions – Challenges for Nuclear Insurers," *Nuclear Law Bulletin*, No. 77, June 2006, pp. 27-39; and Faure and Vanden Borre.

36. Julia A. Schwartz, "Alternative Financial Security for the Coverage of Nuclear Third Party Liability Risks," OECD-NEA, paper presented to Nuclear Inter Jura 2007, Brussels, Belgium, October 2, 2007.

37. Tetley, "Revised Paris and Vienna Nuclear Liability Conventions," p. 39.

38. Important in answering the question whether the nuclear operator is liable in case of a nuclear incident caused by an act of terrorism is Article IV.3 of the 1963 Vienna Convention and Article 9 of the Paris Convention. These provide essentially the same exoneration from liability. Article IV.3 of the 1963 Vienna Convention states:

a. No liability under this Convention shall attach to an operator for nuclear damage caused by a nuclear incident directly due to an act of armed conflict, hostilities, civil war or insurrection.

b. Except in so far as the law of the Installation State may provide to the contrary, the operator shall not be liable for nuclear damage caused by a nuclear incident directly due to a grave natural disaster of an exceptional character.

Article 9 of the Paris Convention states that:

The operator shall not be liable for damage caused by a nuclear incident directly due to an act of armed conflict, hostilities, civil war, insurrection or, except in so far as the legislation of the Contracting Party in whose territory his nuclear installation is situated may provide to the contrary, a grave natural disaster of an exceptional character.

Article 6.1 of the Protocol amending the provisions of the Vienna Convention repeals only the exoneration for a grave natural disaster of an exceptional character.

39. Roland Dussart Desart, "The Reform of the Paris Convention on Third Party Liability in the Field of Nuclear Energy and of the Brussels Supplementary Convention: An Overview of the Main Features of the Modernisation of the two Conventions," *Nuclear Law Bulletin*, No. 75, 2005, pp. 7-33.

40. The 2004 Protocol to the Paris Convention would amend Article 9 of the Paris Convention to read: "The operator shall not be liable for damage caused by a nuclear incident directly due to an act of armed conflict, hostilities, civil war, or insurrection." Thus, for the purposes of this analysis, the two revised Conventions may be treated as equivalent in this respect, exonerating an operator from liability **only** for nuclear incidents directly due to an act of armed conflict, hostilities, civil war, and insurrection. See Tom Vanden Borre, "Are Nuclear Operators Liable and Insured in Case of an Act or Terrorism on a Nuclear Installation or Shipment?" paper presented to the Symposium Rethinking Nuclear Energy and Democracy after 09/11, PSR/IPPNW/Switzerland, April 26-27, 2002.

41. Dussart Desart, "The Reform of the Paris Convention."

42. See, for example, Schwartz, "Alternative Financial Security." In the United States, the U.S. Congress passed (and has subsequently renewed) the Terrorism Risk Insurance Act under which the U.S. Government will contribute funds in the event of significant industry needs, and pay terrorism claims; in return, the Act prohibits insurers from discontinuing their terrorism risk insurance. These provisions are temporary, allowing the insurance industry to gain enough experience to properly price its continuing coverage for terrorism risks.

43. In Europe there are two mutual insurance arrangements which supplement commercial insurance pool cover for operators of nuclear plants. The European Mutual Assurance for the Nuclear Industry (EMANI) was founded in 1978, and the European Liability Insurance for the Nuclear Industry (ELINI) was created

in 2002. ELINI plans to make €100 million available as third party cover, and its 28 members have contributed half that as late as 2007 for a special capital fund. ELINI's members comprise most EU nuclear plant operators. EMANI's funds are also only about €500 million. See UIC 2007: "Civil Liability for Nuclear Damage," Uranium Information Centre, Issues Briefing Paper # 70, October 2007, available from *www.uic.com.au/nip70.htm* (updated October 2010, renamed "Liability for Nuclear Damage").

44. Her Majesty's Government, "The Role of Nuclear Power," Consultation Paper May 2007, available from *webarchive.national archives.gov.uk/+/http://www.berr.gov.uk/files/file39197.pdf.*

45. Pelzer, "International Pooling," p. 47.

46. In his article, Tetley identifies a number of problematic issues in the revised conventions, and, in particular, stresses in bold that "almost all forms of environmental liability are currently uninsurable," p. 36. That may be correct generally. But a closer look into the heads of environmental damage in the nuclear conventions show that the definitions contain qualifiers which enable judges to restrict and define an individual damage quite clearly in terms of money. (This footnote is part of the above quote, taken directly from Pelzer's article.)

47. New legislation proposed in Sweden shows one way of dealing with the shortfall. In Sweden, at least for the moment, private nuclear insurance will not be available to fully cover the €700 million of liability to be imposed upon a nuclear operator under the 2004 Protocol to Amend the Paris Convention. Not only will insurance capacity be unavailable for that amount, but it will equally be unavailable (in whole or in part) for certain types of risks which nuclear operators will be required to assume once the Protocol has come into force, such as claims made more than 10 years following the date of the incident; or the costs of reinstating a significantly impaired environment. The Swedish Government's inquiry into an appropriate nuclear liability regime for that country concluded that the Government (should) be authorized by the Swedish Parliament to provide alternative financial security to supplement the amount of (currently) available insurance, subject to charges that are calculated on the basis of standard commercial terms and that conform to EU regulations regarding restrictions

against competition, within the framework of a state guarantee. This self-financed commitment should preferably take the form of a reinsurance commitment so that financial coverage of the operator's liability may be available for up to €1200 million, the amount required to be paid by operators and by their governments under the first two tiers of the Brussels Supplementary Convention as amended by the 2004 Protocol. Summary of the Report of the Swedish Government Inquiry in the Swedish Government Official Report Series (SOU) 2006:43, p. 27 *et seq.*, available from *www. sweden.gov.se/content/1/c6/06/25/90/25aa61e8.pdf.*

48. Pelzer, "International Pooling," p. 48.

49. *IEA 2006: World Energy Outlook*, see Table 6, p. 148.

50. "The High Price of Hot Air," *Open Europe*, August 2007.

51. Peter Atherton, head of European Utility Research, Citigroup, "Citigroup Analysis of the Impact of the EU Carbon Market on European Utilities," Powerpoint presentation, 2006.

52. Vincent De Rivaz, Chief Executive EDF Energy, "Can we make nuclear energy a reality in the UK?" Westminster Energy Forum Speech, London, UK, November 16, 2006.

53. For reactors in Canada, it has been estimated that the limit on the liability of operators, which currently excludes about 80 percent of expected off-site damage, amounts to an implicit subsidy of between 1 and 4 cents per kWh, depending on the risk assessments used. See Anthony Heyes, and Catherine Heyes, "An empirical analysis of the Nuclear Liability Act (1970) in Canada," *Resource and Energy Economics*, Vol. 22, 2000, pp. 91-101. In the United States, it has been estimated that the value of the Price-Anderson subsidy was $60 million per reactor year before 1982, but it then dropped to $22 million per reactor year following the 1988 amendments. See Jeffrey A. Dubin and Geoffrey S. Rothwell, "Subsidy to Nuclear Power through Price-Anderson Liability Limit," *Contemporary Economic Policy*, Vol. 8, No. 3, 1990, pp. 73-79; and A. Heyes and C. Liston-Heyes, "Subsidy to Nuclear Power Through Price-Anderson Liability Limit: Comment," *Contemporary Economic Policy*, Vol. 16, No. 1, 1998, pp. 122-124. More generally, see R. L. Ottinger *et al.*, *Environmental Costs of Electricity*,

New York: Oceana Publications, 1991. While this paper focuses on externalities resulting from damage caused to third parties after a nuclear accident, it is clear that nuclear power generation may well generate other externalities as well (e.g., including those related to the costs of decommissioning and the management of nuclear waste and spent fuel) which should be taken into account in a more comprehensive economic analysis.

54. "Solutions for environment, economy and technology," Report for DG Environment, Environmentally harmful support measures in EU Member States, European Commission, January 2003, p. 132.

55. A severe nuclear accident means one where there is a breach of the containment, loss of integrity of the core, and uncontrolled emission of core substances into the environment. The 1986 Chernobyl accident was in this category. In the Chernobyl accident, it is believed that about 4 percent of the reactor core material was released.

56. There no single, internationally accepted, methodology for assessing and valuing damage incurred as a result of a nuclear accident, particularly for damage arising in different countries. In addition, different assessments include or exclude particular categories of damage to a greater or lesser extent, sometimes even entirely excluding particular types of damage from consideration. An illustration of the complexities involved can be seen by considering the Chernobyl accident. Most of the population of the Northern hemisphere was exposed, to various degrees, to radiation from the Chernobyl accident. Even now it is possible to arrive only at a reasonable, but not highly accurate, assessment of the ranges of doses received by the various groups of population affected by the accident. Within the former Soviet Union, large areas of agricultural land are still excluded from use and are expected to continue to be so for a long time. In a much larger area, although agricultural and dairy production activities are carried out, the food produced is subjected to strict controls and restrictions of distribution and use. The progressive spread of contamination at large distances from the accident site caused considerable concern in many countries outside the former Soviet Union, and the reactions of the national authorities to this situation were extremely varied, ranging from a simple intensification of the

normal environmental monitoring programs, without adoption of specific countermeasures, to compulsory restrictions concerning the marketing and consumption of foodstuffs. Some of these restrictions remain in place today. To date we are aware of no comprehensive overall assessment of the total costs of the Chernobyl accident which compiles and integrates the costs of these different damages, preventive responses, and related actions in all affected countries.

57. See, for example, Dubin and Rothwell, "Subsidy to Nuclear Power"; Heyes and Liston-Heyes, "Subsidy to Nuclear Power Through Price-Anderson Liability Limit"; Ottinger *et al.*, "Environmental Costs of Electricity"; Heyes and Heyes, "An empirical analysis of the Nuclear Liability Act"; Anthony Heyes, "PRA in the nuclear sector: quantifying human error and human malice," *Energy Policy*, Vol. 23, No. 12, 1995, pp. 1-8; Greenpeace International, "Review of Estimates of the Costs of Major Nuclear Accidents," prepared for the 9th Session of the IAEA Standing Committee on Nuclear Committee, Vienna, Austria, February 7-11, 1994.

58. Woolley Report 17, citing Greenpeace International, Review of Estimates of the Costs of Major Nuclear Accidents, prepared for the 9th Session of the IAEA Standing Committee on Nuclear Committee, 1994. The so-called "Sandia siting report" (1982) concluded that a very large accident could cause damages in the order of $695,000 million. Cited in Michael Faure, "Economic Models of Compensation for Damage Caused by Nuclear Accident: Some Lessons for the Revision of the Paris and Vienna Conventions," *European Journal of Law and Economics*, Vol. 2, 1995, pp. 21-43.

59. Report to the Congress from the Presidential Commission on Catastrophic Nuclear Accidents (Volume One), August 1990, p. 73, footnote 10.

60. Richard Hudson, "Study Says Chernobyl Might Cost 20 times more than Prior Estimates," *Wall Street Journal Europe*, March 29, 1990.

61. *The Republic of Belarus: 9 Years after Chernobyl. Situation, Problems, Actions*, National Report, Ministry for Emergencies and

Population Protection from the Chernobyl NPP Catastrophe Consequences, 1995.

62. See G. J. Vargo, ed., *The Chornobyl Accident: A Comprehensive Risk Assessment*, Columbus, OH: Batelle Press, 2000, cited in M. C. Thorne, *Annals of Nuclear Energy*, Vol. 28, 2001, pp. 89-91.

63. Nuclear Regulation, "A Perspective on Liability Protection for a Nuclear Power Plant Accident," GAO/RCED-87-124, 1987, p. 20 and Appendix II.

64. Report Number 5, Nuclear Fuel Cycle, Externalities of Fuel Cycles 'ExternE' Project, European Commission, DGXII, Science, Research and Development (JOULE), Brussels, Belgium, 1995, p. 5.

65. Alain Sohier, ed., *A European Manual for "Off-site Emergency Planning and Response to Nuclear Accidents,"* prepared for the European Commission Directorate-General Environment (Contract SUBV/00/277065), SCK-CEN Report R-3594, December 2002, Chap. 13, "Economic Impact," in particular, pp. 245-248.

66. Heyes and Heyes, "An empirical analysis of the Nuclear Liability Act."

67. For example, while the generation of nuclear electricity does not emit $CO_2$ and other "classical" pollutants associated with the combustion of fossil fuels, there a number of environmental problems associated with nuclear technology. In addition to the risks of major accidents, in particular, it is still argued that the mechanisms to safely manage nuclear waste over the timescales necessary have not been developed or deployed. It is expected that governments will be required, at some time in the future, to pay for full cost of the safe disposal of nuclear waste, and this should somehow be factored into the present-day pricing of nuclear-generated electricity. However, there is no agreement on exactly how these costs can be factored into analyses.

68. See, for example, R. L. Ottinger, "Have recent studies rendered environmental externality valuation irrelevant?" in O. Hohmeyer, ed., *Social Costs and Sustainability: Valuation and Implementation in the Energy and Transport Sector*, Berlin, Germany: Springer, 1997, pp. 29–43; and Sang-Hoon Kim, "Evaluation of

negative environmental impacts of electricity generation: Neo-classical and institutional approaches," *Energy Policy*, Vol. 35, 2007, pp 413–423.

69. ExternE Report No. 1, "Summary, Externalities of Fuel Cycles 'ExternE' Project," European Commission, DGXII, Science, Research and Development, JOULE, 1995.

70. The results ranged from 0.1 to close to 100 mECU/kWh. ExternE Report No. 5, "Nuclear Fuel Cycle," European Commission, DGXII, Science, Research and Development, JOULE, 1995, p. 7.

71. *Ibid.* The priority impacts of the nuclear fuel cycle to the general public are radiological and nonradiological health impacts due to routine and accidental releases to the environment. The source of these impacts are the releases of materials through atmospheric, liquid, and solid waste pathways. The most important impacts to the natural environment that could be expected would be the result of major accidental releases. This type of impact has been included in the economic damage estimates as the loss of land-use and agricultural products after a potential severe reactor accident. Possible long-term ecological impacts have not yet been considered. Within the framework of the ExternE project, the cost associated with a nuclear accident was derived on the basis of the economic module available in the COSYMA code (based on the "direct" economic loss associated with health and environmental consequences), including further considerations on the probability of occurrence of the different accidental scenarios as well as specific values for health effects. At that period of time, it was clearly pointed out that this approach was limited and that there was a need for further investigation in order to deal with the risk perception.

72. Sohier, p. 248 (emphasis added).

73. 1998 Update to the ExternE Methodology Report, 1999, citing: A. Markandya and N. Dale, eds., *Improvement of the Assessment of Severe Accidents*, Unpublished report from the ExternE Core Project for European Commission, DGXII, under contract number JOS3-CT95-0002, 1998.

74. ExternE - Report No. 5, p. 5.

75. Peter Bickel and Rainer Friedrich, eds., "Methodology 2005 Update," ExternE — Externalities of Energy, European Commission, 2005.

76. In this respect, a telling slide was presented at a 2005 ExternE meeting. In response to the Stakeholders' concern: "Is the risk of a nuclear accident evaluated correctly?" the response from ExternE was "Who can agree on 'correctly'?" with the additional explanation that "Someone who does not agree can modify the assumptions of ExternE or do an alternative analysis." See *External costs of energy and their internalization in Europe - Dialogue with industry, NGO, and policy-makers*, Brussels, Belgium, December 9, 2005, available from *www.externe.info/index.html*.

77. As Tetley has pointed out, with fewer than 500 reactors worldwide — not all of which are insured — the nuclear insurance industry does not have a large database on which to base premiums and loss assessments. Much of the modeling and premium assessment is therefore done on an actuarial and theoretical basis, rather than using real data. The inherent uncertainty of this methodology makes many insurers even more reluctant to commit their capital to nuclear risks. See Tetley, "Revised Paris and Vienna Nuclear Liability Conventions."

78. 1998 Update to the ExternE Methodology Report.

79. Welt im Wandel: Strategien zur Bewältigung globaler Umweltrisiken (*World in Transition: Strategies for Managing Global Environmental Risks*), Wissenschaftlicher Beirat der Bundesregierung Globale Umweltveränderung (WGBU), Jahresgutachten 1998, Berlin, Heidelberg, Germany, and New York: Springer, 1999.

80. Wolfram Krewitt, "External costs of energy - do the answers match the questions? Looking back at 10 years of ExternE," *Energy Policy*, Vol. 30, 2002, pp. 839–848.

81. Bickel and Friedrich, eds., (emphasis added).

82. This follows from the standard economic analysis of tort. See for example S. Shavell, "Strict Liability Versus Negligence,"

*Journal of Legal Studies*, 1980, pp. 1-25; and S. Shavell, *Economic Analysis of Accident Law*, Cambridge, MA: Harvard University Press, 1987.

83. S. Shavell, "On Moral Hazard and Insurance," *Quarterly Journal of Economics*, 1979, pp. 541-562.

84. For conditions for efficient functioning of compensation funds, see M. Faure, "Financial Compensation for Victims of Catastrophes: An Economic Perspective," *Law & Policy*, Vol. 29, No. 3, July 2007, pp. 339-366.

85. See in this respect, more particularly, M. Faure and R. Van den Bergh, "Liability for Nuclear Accidents in Belgium from an Interest Group Perspective," *International Review of Law & Economics*, 1990, pp. 241-254; M. Faure and G. Skogh, "Compensation for Damages Caused by Nuclear Accidents: A Convention as Insurance," *The Geneva Papers on Risk and Insurance*, Vol. 17, 1992, pp. 499-513; Faure, "Economic Models of Compensation"; and M. Trebilcock, and R. A. Winter, "The Economics of Nuclear Accident Law," *International Review of Law & Economics*, Vol. 17, 1997, pp. 215-243.

86. The result of the changes of the Price-Anderson Act has been that the costs of a nuclear accident were increasingly internalized, that the externalities decreased to an important degree. Although there are still issues to consider under the Price-Anderson Act, in particular that operator liability remains limited, Faure and Vanden Borre conclude that there are substantially less remaining externalities under the U.S. than under the international nuclear compensation scheme. Faure and Vanden Borre, "Economic Analysis of the Externalities," pp. 34-35.

87. As far as the existing Paris Convention regime is concerned: €150 million; once the 2004 Protocol to the Paris Convention will enter into force: €700 million.

88. Faure and Vanden Borre, "Economic Analysis of the Externalities." Faure and Vanden Borre note that this last criticism can be met (partially) if the Contracting Parties charge the operators for the costs of making public money available. These costs should then be market reflective and should take into account

risk differentiation, etc. They add that it is far from sure that any governmental institution is equipped well enough to assume this difficult task, and thus whether such an institution could do so in a more efficient manner than an insurance company or mutual insurance scheme.

89. The discussion here focuses on the U.S. pooling system. However, some features of the German system are also worth noting. While the U.S. system is based on a statutory obligation or duty of the individual operator to contribute, the German system is formed by a voluntarily concluded contract under civil law among the four leading German energy producing companies, the "Solidarvereinbarung" (a "Solidarity Agreement").

Since 2002, when German nuclear operators had to respond to their Government's decision to increase the amount of financial security to be provided by the private sector to up to €2.5 billion. The nuclear insurance industry still provides a portion of coverage which currently rests at up to €255.6 million. However the remainder of the security, up to €2.24 billion, is to be provided by an operator pooling system which enables the operators of those plants to provide the financial security required of them by the German Atomic Energy Act. Under this new arrangement, each partner accepts liability, vis-à-vis the others, to contribute a percentage of the total amount of coverage to be provided by the liable operator, and the total of all partners' commitments in this regard is agreed to be 100 percent of that amount.

The size of each partner's guarantee is determined on the basis of the number of shares it holds in each and every nuclear power plant. Where a nuclear incident occurs, the guarantee must be paid to the liable operator as long as neither it nor its parent company can provide the required financial security; proof of that inability must be proved by a public accountant's certificate. As under the American system, no money is required to be paid in advance of a nuclear accident; and where the partners do pay following an accident, they will have a right of recourse against the operator. In the meantime, they will offer claims handling support to the liable operator through deployment of their infrastructure and expertise. The German regulatory authorities accept this arrangement in satisfaction of obligations under both domestic legislation and international conventions, on condition that the partners annually submit a public accountant's certificate attesting to their financial capacity to meet their obligations under the scheme. See, for example, Schwartz, "Alternative Financial Security," p. 18; and Pelzer, "International Pooling," pp. 43-45.

90. Pelzer, "International Pooling," in particular p. 46.

91. Faure, and Vanden Borre, *Economic Analysis of the Externalities*, p. 32.

92. See "Legislative updates," *NEA News*, Vol. 23, No. 2, 2005, p. 32.

93. On the basis of $300 million of the first tier + [(95.8 + 5%) x 104 = 10,461] of the second tier. This implies the following if a nuclear accident occurs in the United States causing $7 billion (roughly €5.2 billion) of damage. In a first layer, the liability insurer will have to pay $300 million. This leaves a remainder of $6.7 billion to be covered in the second tier of the U.S. compensation scheme. This will be financed collectively by all the 104 nuclear operators in the United States. This means that every U.S. nuclear operator will have to pay, in the second layer, a total amount of $64.4 million ($6.7 billion/104 nuclear power plants) per power plant.

94. The second tier payment, in the scenario described above, will be collected through retrospective premiums which are currently limited to $15 million per year and per reactor. The result is that the second layer ($6.7 billion) will be financed by the operators of all 104 nuclear power plants in a period of 5 years, whereby each will pay $15 million during the next 4 years and $4.4 million in the fifth year.

95. Faure and Vanden Borre, "Economic Analysis of the Externalities," p. 32.

96. *Ibid.*

97. *Ibid.*, p. 33. To address these concerns, Faure and Vanden Borre note the following. The model is to be applied on a limited international, e.g., European basis — at least at the start of setting up such a model. The reason for this is quite simple: such a model can only work if the (operational) safety of the participating nuclear power plants is similar or at least comparable. Those opera-

tors wanting to participate in the model will have an incentive to enhance the safety of their power plants. Lastly, the model will only work if major regulatory issues have been resolved. In their view, by far the most important one is the creation of a European independent regulatory body (a kind of European Nuclear Regulatory Agency); this body will deliver permits to nuclear installations falling under the international nuclear liability regime and will determine the way the operators will insure their liability.

98. Pelzer, "International Pooling," pp. 50-52.

99. *Ibid.*, p. 50. Note that this conclusion does not, however, exclude state measures designed to support respective efforts of operators if states deem them useful.

100. The parties to the Agreement are: Energie Baden-Württemberg AG (EnBW), E.ON Energie AG, Hamburgische Electricitäts-Werke AG (now: Vattenfall Europe AG), and RWE AG.

101. The approximate percentages read as follows: E.ON, 42 percent, RWE, 25.9 percent; EnBW, 23.9 percent; and Vattenfall, 8.2 percent.

102. Under the current German pooling arrangement, there is a guaranteed amount of compensation available of €2.5 billion (approximately $.32 billion). But the liability of the German operator is not affected, the liability remains unlimited, and in the event that the damage caused exceeds the financing of the pool the other assets of the operator are available to add to compensation.

103. Pelzer, "International Pooling," p.51. It is conceivable that operators develop formal mechanisms in order to enable the partners to decide on the eligibility of an installation—these might include direct monitoring, inspection, and assessments by or on behalf of the pool.

104. Illustrative Nuclear Programme.

105. Pelzer, "International Pooling," p. 55.

106. EESC 2007: Opinion of the European Economic and Social Committee on the Communication from the Commission to the Council and the European Parliament— Illustrative Nuclear Program, Presented Under Article 40 of the Euratom Treaty for the Opinion of the European Economic and Social Committee, *Communication from the Commission to the Council and the European Parliament*, COM(2006) 844 Final, European Commission, July 12, 2007.

107. Andris Piebalgs, "The Euratom Treaty and development of the Nuclear Industry," Keynote speech at the International Nuclear Law Association Congress, Brussels, Belgium, October 3, 2007.

# APPENDIX 1

## CONVENTION LIMITATION AMOUNTS[1]

| Convention | Operator Liability | Installation State[2] | Combined States Party | TOTAL |
|---|---|---|---|---|
| Paris Convention, 1960 | *At least* Special Drawing Rights (SDR) 5 million and up to a maximum of 15 million SDRs. *(a) (b)*<br><br>(*At least* € ±6 million and up to €17.83 million) | - | - | *At least* SDR 5 million and up to a maximum of 15 million SDRs<br><br>(*At least* € ±6 million and up to €17.83 million) |
| Brussels Suppl. Convention, 1963 | *At least* SDR 5 million. *(c)*<br><br>(*At least* €+6 million) | The difference between the operator liability amount and SDR 175 million<br><br>(€202.13 million) | 125 million SDRs *(d)*<br><br>(€148.62 million) | SDR 300 million<br><br>(€356.7 million) |
| Paris Convention, 2004 | *At least* €700 million *(e) (f)* | - | - | *At least* €700 million |
| Brussels Suppl. Convention, 2004 | *At least* €700 million | The difference between the operator liability amount and €1200 million | €300 million *(g)* | €1500 million |
| Vienna Convention, 1963 | $ 5 million gold<br><br>(€±50 million) | - | - | $ 5 million gold<br><br>(€±50 million) |

# CONVENTION LIMITATION AMOUNTS[1] (Cont.)

| Convention | Operator Liability | Installation State[2] | Combined States Party | TOTAL |
|---|---|---|---|---|
| Vienna Convention, 1997 | *At least* SDR 150 million *(h)*<br><br>(€178.25 million) | The difference between the operator liability and SDR 300 million *(i)*<br><br>(€356.7 million) | - | SDRs 300 million<br><br>(€356.7 million) |
| Convention on Supplementary Compensation for Nuclear Damage, 1997 | Not specified. *(j) (k)* | At least SDRs 300 million<br><br>(At least € 356.7 million) | If damage exceeds 300 million SDR, calculated separately for each individual state party. *(l)* | At least SDRs ±600 million. *(m)(n)*<br><br>(At least €±713.4 million) |

## NOTES:

**Paris Convention, 1960**

*(a)* Switzerland introduced a system of unlimited liability which it considered incompatible with the Paris Convention system and therefore it elected not to become a party to the Paris Convention. However, in practice, some Paris Convention parties did not implement this provision too strictly as to the maximum amount of liability. Some imposed a higher amount of liability (e.g. in Belgium an amount of €300 million was set) or even by introducing a system of unlimited liability (Germany).

*(b)* The Steering Committee of the NEA recommended Contracting Parties to set a maximum liability of not less than 150 million SDRs (€178.35 million or $217.13 million). Recommendation of the Steering Committee of April 20, 1990, NE/M(90)1, *Paris Convention: Decisions, Recommendations, Interpretations*, Paris, OECD/NEA, 1990, p. 13.

**Brussels**    *(c)*    Figures given are for the Paris Convention as amended by the
**Supplementary**        1982 Protocol which entered into force on August 1, 1991.
**Convention,**        Prior to the entry into force of the 1982 Protocol, the amounts
**1963**        were SDRs 70 million (€±59.5 million) for the Installation
State, SDRs 50 million (€±83 million) for the combined state
contribution, to a total of SDRs 120 million (€±142.7 million).

    *(d)*    Only half (50 percent) of the fund comes from contributions
from those states party who have nuclear power plants. The
other 50 percent comes from all states party, independent of
whether or not they have nuclear power plants.

The formula for contributions is:
a.    as to 50 percent, on the basis of the ratio between
the gross national product at current prices of each
Contracting Party and the total of the gross national
products at current prices of all Contracting Parties
as shown by the official statistics published by
the Organization for Economic Cooperation and
Development for the year preceding the year in
which the nuclear incident occurs;
b.    as to 50 percent, on the basis of the ratio between
the thermal power of the reactors situated in
the territory of each Contracting Party and the
total thermal power of the reactors situated in
the territories of all the Contracting Parties. This
calculation shall be made on the basis of the
thermal power of the reactors shown at the date
of the nuclear incident in the list referred to in
Article 2(a)(i): provided that a reactor shall only be
taken into consideration for the purposes of this
calculation as from the date when it first reaches
criticality.

**Paris**    *(e)*    The Protocol revising the Paris Convention now explicitly
**Convention,**        provides for the possibility of unlimited operator liability.
**2004**

    *(f)*    States adhering after January 1, 1999, may limit an operator's
liability to €350 million for a period of 5 years starting from
February 12, 2004.

**Brussels** *(g)* Most of the fund (65 percent) comes from contributions
**Supplementary** from states party with nuclear power plants. The
**Convention,** remaining 35 percent comes from all states party,
**2004** independent of whether or not they have nuclear power
plants.

The formula for contributions is:
a. as to 35 percent, on the basis of the ratio
between the gross domestic product at current
prices of each Contracting Party and the total of
the gross domestic products at current prices of
all Contracting Parties as shown by the official
statistics published by the Organization for
Economic Cooperation and Development for
the year preceding the year in which the nuclear
incident occurs;

b. as to 65 percent, on the basis of the ratio
between the thermal power of the reactors
situated in the territory of each Contracting
Party and the total thermal power of the reactors
situated in the territories of all the Contracting
Parties.

**Vienna** *(h)* For a transitional period of 15 years from the date of
**Convention,** opening up for signature of the Protocol (September 12,
**1997** 1997) a lesser amount of 100 million SDRs or less might be
stipulated. If it is less than 100 million SDRs, the state must
make available the difference up to 100 million SDRs, during
the transitional period.

*(i)* For a transitional period of 15 years from the date of opening
up for signature of the Protocol (September 12, 1997) a
lesser amount of 100 million SDRs might be stipulated.

**Convention on** *(j)* According to Art. III.1.a of the Convention on Supplementary
**Supplementary** Compensation (CSC), the Installation State shall ensure the
**Compensation,** availability of at least 300 million SDRs ( 356.7 million €).
**1997** This provision provides for an obligation of the Installation
State to ensure that 300 million SDRs are available; the
Installation State is free to choose how this amount is funded
(private insurance, regional agreement, . . .). A State meets
its obligation under Art. III.1.a of the CSC when it imposes a
nuclear liability on the operator for the entire amount.

**Convention on**  *(k)*  For a transitional period of 10 years (from September
**Supplementary**  12, 1997) a lesser amount (150 million SDRs) might be
**Compensation,**  stipulated.
**1997**

*(l)*  Most, but not all, of the contributions to the international fund
will come from States with nuclear power plants. Specifically,
90 percent of the contributions to the international fund
will be based on the installed nuclear capacity in a member
country and thus will come from only those member
countries where reactors are located. The remaining 10% of
the contributions will be based on the UN rate of assessment
of a member country. Given that many nuclear power
generating States have a large UN rate of assessments, it
is likely that, as a group, non-nuclear-generating States will
provide no more than 2 or 3% of the contributions to the
international fund.
The contributions will be made according to the following
formula:
- the amount which shall be the product of the installed
  nuclear capacity of that Contracting Party multiplied by
  300 SDRs per unit of installed capacity;
and
- the amount determined by applying the ratio between the
  United Nations rate of assessment for that Contracting
  Party as assessed for the year preceding the year in
  which the nuclear incident occurs, and the total of such
  rates for all Contracting Parties to 10 percent of the sum
  of the amounts calculated for all Contracting Parties.

*(m)*  One-half of the international fund is reserved exclusively
for transboundary damages (that is, damages outside the
Installation State).

*(n)*  This requirement is set out in Art. XI of the CSC, which
states that the funds of the second tier shall be distributed
as follows: 50 percent of the funds shall be available to
compensate claims for nuclear damage suffered in or outside
the Installation State; 50 percent of the funds shall be
available to compensate claims for nuclear damage suffered
outside the territory of the Installation State to the extent that
such claims are uncompensated from the former amount.

# ENDNOTES - APPENDIX 1

1 . The exact value of the SDR is determined by the International Monetary Fund (IMF) and is published on its website. For this Table, we used the exchange rate of March 20, 2006: €1.189 $1.44757 USD.

2 . That is, the State party to the Convention in which the nuclear installation is operated.

# APPENDIX 2

# OPERATOR LIABILITY AMOUNTS AND FINANCIAL SECURITY LIMITS IN EU COUNTRIES

**Operator Liability Amounts And Financial Security Limits in EU Countries (as of October 2006, OECD Unofficial)**

| State | Paris/Brussels Convention (PC/BC) or Vienna Convention (VC) | Liability Amount in National Currency or Special Drawing Rights with USD Equivalent [*] | Financial Security Limit if Different from Liability Amount with USD Equivalent |
|---|---|---|---|
| Austria | -[1] | Unlimited | €407 million (USD = 498 M) |
| Belgium | PC/BC | SDR 300 million (USD = 438 M) (12 billion BEF) | |
| Bulgaria | VC | Approximately €49 million. (BGL 96 million) | |
| Cyprus | - | - | |
| Czech Republic | VC | CZK 6 billion (USD = 252,8 M) | CZK 1.5 billion (USD = 63 M) |
| Denmark | PC/BC | SDR 60 million (USD = 87,6 M) | |
| Estonia | VC[2] | No specific legislation | |
| Finland | PC/BC | SDR 175 million (USD = 255,5 M) [1] | €700 million under new legislation (not yet EIF) |
| France | PC/BC | SDR 76 million (USD = 111,5 M) [2] | €700 million under new legislation (not yet EIF) |

**Operator Liability Amounts And Financial Security Limits in EU Countries (*cont.*)**
**(as of October 2006, OECD Unofficial)**

| State | Paris/Brussels Convention (PC/BC) or Vienna Convention (VC) | Liability Amount in National Currency or Special Drawing Rights with USD Equivalent [*] | Financial Security Limit if Different from Liability Amount with USD Equivalent |
|---|---|---|---|
| Latvia | Revised VC[3] | Approximately €122 million. (LVL 80 million) | |
| Lithuania | VC[4] | €50 million[5] | |
| Luxembourg | -[6] | No specific legislation | |
| Malta | - | - | |
| Netherlands | PC/BC | SDR 285 million (USD = 416 M) | |
| Poland | VC | SDR 150 million (USD = 219 M) | |
| Portugal | PC (not BC) | No specific legislation | |
| Romania | Revised VC and CSC[7] | SDR 300 million[8] (USD 438 M) | |
| Slovakia | VC | Approximately €75 million (2 billion SKK) | |
| Slovenia | PC/BC | SDR 150 million (USD = 219 M) | |
| Spain | PC/BC | ESP 25 billion (approx SDRs 150 million) | |

**Operator Liability Amounts And Financial Security Limits in EU Countries (*cont.*)**
**(as of October 2006, OECD Unofficial)**

| State | Paris/Brussels Convention (PC/ BC) or Vienna Convention (VC) | Liability Amount in National Currency or Special Drawing Rights with USD Equivalent [*] | Financial Security Limit if Different from Liability Amount with USD Equivalent |
|---|---|---|---|
| Sweden | PC/BC | SDR 300 million (USD = 438 M) | |
| | | New proposal is unlimited. | New proposal is for state guaranteed reinsurance to complement private insurance, together this should cover SDR 1200 million.[9] |
| United Kingdom | PC/BC | SDR 150 million (USD = 219 M) | |

Notes:

[1] New Nuclear Liability Act (not yet EIF) provides for unlimited liability where BSC coverage exhausted and damage remaining

[2] New liability provisions (not yet EIF) provide for 700 million EUR

[*] As of 19 September 2005, 1 SDR = 1.46 USD

## ENDNOTES - APPENDIX 2

1. Austria signed the 1960 Paris Convention and the 1963 Brussels Supplementary Conventions upon their adoption, but has not ratified these instruments.

2. With reservation that reservation that Estonia would not be liable for damage resulting from nuclear installations or nuclear material located on its territory if the operator is of foreign nationality.

3. Latvia ratified the 1997 Protocol to amend the Vienna Convention (it ratified on December 5, 2001, and the revised Convention entered into force on October 4, 2003).

4. Lithuania has signed the 1997 Convention on Supplementary Compensation.

5. Minimum amount under 1963 VC.

6. Luxembourg signed the 1960 Paris Convention and the 1963 Brussels Supplementary Conventions upon their adoption, but has not ratified these instruments.

7. Romania ratified the 1997 Protocol to amend the Vienna Convention (it ratified on December 29, 1998, and the revised Convention entered into force on October 4, 2003) and the Convention on Supplementary Compensation (March 2, 1999).

8. Less than SDR 300 million but at least SDR 150 million, provided that the amount of SDRs 300 million is made available from public funds. For a 10-year transitional period of 10 years (from December 3, 2001) it may be limited to less than 150 million SDRs, but not less than SDRs 75 million, provided that the difference up to SDRs 150 million SDRs shall be made available from public funds.

9. For Swedish operators, private nuclear insurance will not be available to fully cover the €700 million of liability to be imposed upon a nuclear operator under the 2004 Protocol to Amend the Paris Convention. Under the new proposals, the Government (should) be authorised by the Swedish Parliament to provide alternative financial security to supplement the amount of (currently) available insurance, subject to charges that are calculated on the basis of standard commercial terms and that conform to European Union regulations regarding restrictions against competition, within the framework of a state guarantee. This self-financed commitment should preferably take the form of a rein-

surance commitment so that financial coverage of the operator's liability may be available for up to 1200 million euros, the amount required to be paid by operators and by their governments under the first two tiers of the Brussels Supplementary Convention as amended by the 2004 Protocol to Amend that Convention.

# CHAPTER 14

## MARKET-BASED NUCLEAR NONPROLIFERATION[1]

### Henry Sokolski

A world full of nuclear weapons-ready states is not inevitable. Nor does avoiding this fate necessarily require massive new government spending programs; development of new, advanced technology; negotiation of new international treaties; or any heroic military maneuvers. It will, however, at a minimum, require that the United States and other states with nuclear power programs to do two things they should have done long ago but have yet to tackle seriously — identify the full costs of nuclear power as compared to its nonnuclear alternatives, and make nuclear power operators obtain as much private financing and insurance as possible to pay for these expenses.

This requirement may seem unusual, but given the increasing political imperative to make the right choices to avoid global warming in the cheapest quickest manner, the United States, the European Union (EU), and many other countries already have good reason to begin to take such steps. In fact, identifying the full cost of nuclear power as compared to its alternatives will be difficult to avoid as we move toward a carbon-constrained world with serious carbon taxes. Certainly, if we fail to identify these costs — including all the direct and indirect subsidies, and the security and environmental costs that have yet to be internalized — then imposing such taxes will simply propel nuclear power much further both here and abroad than would otherwise be the case.

On the other hand, identifying all costs of nuclear power and doing the same for its alternatives would go a long way to assure that any energy choices that are made would be reached on the basis of sound economic comparisons rather than political whim. Given the potential for using peaceful nuclear programs for military purposes, a state that chooses nuclear power over much cheaper, emission-compliant alternatives should set off both economic and security alarms.

To secure the full benefits of this approach ultimately requires taking a second step — getting private banks and insurers to bear much more of nuclear power's full costs. To a great extent, we already do this for most non-nuclear forms of electricity. Yet, governments both here and abroad have held off doing this out of concerns that the nuclear industry, after nearly a half-century of government funding and support, is not quite yet mature enough to be subjected to such market demands. In some respects, this has actually kept the nuclear industry from doing its best. Certainly, if nuclear power had to cover all of its insurance costs against accidents and security, the industry would literally place a much higher premium on building and operating only the most modern and safest plants and do even more on their own (rather than wait for government regulation) to physically secure their plants.

More important, if nuclear operators had to cover most or all their costs, the most dangerous and economically uncompetitive forms of nuclear energy would have far greater difficulty proceeding as far as they have to date. Certainly, nuclear fuel making, which can bring a state within days or weeks of acquiring nuclear weapons, and large nuclear reactor projects in the energy-rich and unstable regions of the world, such as the Middle East, would be much harder to sell to private investors and insurers than almost any non-nuclear alternative.

Few, in or out of the nuclear industry, dispute these points. It would be useful to exploit this consensus to promote some level of nuclear restraint. This is a particularly important as more and more countries use the Nuclear Nonproliferation Treaty (NPT), the example of the United States, and the nuclear power practices of other states as justifications to engage in the most uneconomical and dangerous nuclear activities themselves. What will be required to discipline such dangerous enthusiasm? Public recognition and emphasis of the following points:

1. *Nuclear energy is not just another way to boil water.* Spreading nuclear power reactors worldwide with nuclear cooperation agreements, generous government-backed export loans, and guaranteed financing is a surefire way to increase the number of nuclear weapons ready nations. Unfortunately, even "proliferation resistant" light water reactors require tons of low enriched fresh fuel to be kept at the site, and they can also produce scores of bomb's worth of weapons-usable plutonium that is contained in the reactor's spent fuel. Research commissioned by my center, which was subsequently authenticated by experts at our national laboratories and the U.S. State Department, details just how little is required to take these materials and convert them into weapons fuel.

Under one scenario, a state could build a small, covert reprocessing line, divert spent fuel without tipping off International Atomic Energy Agency (IAEA) inspectors, produce its first bomb's worth of material in less than 2 weeks, and continue to make a bomb's worth of material a day.[2] There is no technical fix in sight for this problem for decades or, perhaps, ever. Even the Global Nuclear Energy Partnership (GNEP), which originally claimed it could develop nearly "proliferation proof" fuel-cycles, no longer makes this claim and it warns against spreading its "proliferation resistant" UREX system to nonweapons states for fear it too might be diverted to make bombs.[3] What this means is that large nuclear reactors and even light water reactors ought not to be for everyone; they should only go to those states that we can be confident are out of the bomb making business and that can make a compelling case for the economic profitability of these activities.

2. *Adam Smith's "Invisible Hand" is trying to help us* since the most dangerous nuclear activities — fuel making and large reactors in energy-rich regions of the Middle East — are also the most economically uncompetitive when compared with their alternatives. Rather than fight this natural and helpful selection of the financially and economically fittest by pushing government-guaranteed financing for nuclear exports and government-funded nuclear commercialization projects, states interested in pursuing nuclear programs should rely far more on private firms to finance and insure nuclear and non-nuclear power projects and allow these firms to determine which of these projects is the most cost effective.

3. *Pushing government-backed nuclear sales and subsidized fuel assurances can be self-defeating both for nonproliferation and nuclear power's own long-term health.* Backing the construction of large nuclear reactors in Libya, Jordan, Egypt, the United Arab Emirates (UAE), and Turkey (as the United States is currently doing) and the construction of similar plants in Saudi Arabia and Yemen is not only uneconomic in the near and mid-

term when compared with developing fossil-fuel-fired alternatives, but could prompt a not-so-peaceful nuclear competition in one of the world's most war-torn regions. The nuclear industry may initially benefit from the construction of a few additional reactors, but the security fallout from any war in this region could eliminate these gains.[4] As for extending fuel assurances to nations that do not currently make their own fuel, these offers, if not properly conditioned, could increase the pace of proliferation. This is particularly so if they are designed to deal less with narrowly defined "market disruptions" caused by natural disasters, breach of contract, and terrorism than to make fuel "affordable." In fact, some nuclear fuel market observers believe that nuclear ore and fuel products are about to come into much more demand even if the world's current fleet of nuclear reactors does not expand. Their projections focus on how relatively cheap Russian blend-down uranium, and U.S. surplus uranium supply fuel contracts, and older, lower cost fuel contracts associated with terminated reactor projects are about to run out over the next few years. Meanwhile, the licensed operating-lives of many reactors are being extended by 20 or more years. As a result, uranium prices jumped significantly in 2007 and 2008. Such price spikes, nuclear fuel market experts argue, could easily reoccur in the future.[5]

Fuel assurances or fuel banks ought not to be designed to address such market trends. Certainly, if they emphasize the need to assure affordable fuel and financial incentives, they will act on nuclear proliferation much as throwing kerosene on a smoldering fire might—as an accelerant rather than as a moderator. Much like a loss leader in a department store, the effect of such subsidized assurances will be to get more nations interested in acquiring reactors that might not have otherwise.

With the reactors will come all the nuclear training, which will not stop at just lessons on running nuclear power plants. Indeed, even as the IAEA develops its own fuel bank proposals to reduce the need for nations to make their own nuclear fuel, the Agency is adamant that no nation should give up what it currently believes is their natural right to do—make nuclear fuel. This means that any nation that might take advantage of fuel assurances could, at any time, change its mind and proceed to make nuclear fuel.

Finally, even narrowly defined assurances once offered are likely to prompt demands for more generous subsidized assurances later. For these reasons, it is important that any effort to back the further development of fuel assurances stay clear of any effort to make nuclear fuel more affordable or to encourage the development of financial incentives to get nations to avail themselves of such assurances. Draft legislation, which Senators Dick Lugar and Evan Bayh developed, is careful to avoid any encouragement of any financial subsidies, and furthermore helps the IAEA meet its safeguarding mission as well. Neither does it rush to fund any specific fuel assurance option as there are several still under development. These desirable features make sense.[6]

4. *Nuclear operators should pay the full costs of engaging in dangerous nuclear activities.* Fortunately, the nuclear activities that are most dangerous—making nuclear fuel and making nuclear power in regions where there is ready access to natural gas and oil—are also the most difficult to economically justify as compared to their nonnuclear alternatives. Internalizing as many of the external security costs associated with

operating such plants would help to keep this so. Since civilian fuel making is virtually indistinguishable from bomb fuel making, it would make sense to demand that physical security requirements for such plants be equivalent to that of nuclear weapons facilities. These additional costs should be borne by the owners of these facilities. Even the IAEA's own safeguards reviewers admit that nuclear fuel making cannot be inspected to detect diversions in a timely fashion,[7] it would be reasonable to insist on monitoring them more extensively. Such increased monitoring — which the owners of these facilities, again, should pay for — is unlikely ever to provide for timely detection of diversions but would, at least, make detection of diversions more likely. Also, ultimately the full cost of insuring nuclear plants against attacks and accidents should be borne by their owners. The Price-Anderson Nuclear Industries Indemnity Act, which capped the amount of insurance coverage for nuclear accidents, was originally intended to last only for 10 years. That was a half century ago. All of these costs should be identified and internalized into the real costs of nuclear power. The less economic sense that paying the full costs of a civilian nuclear project makes as compared to paying the full costs of non-nuclear alternatives and the more that a government chooses nonetheless to subsidize such nuclear activities, the more international security alarms should be set off.

5. *Identifying and charging for the full costs of civilian projects should help us return to a more sane reading of the nuclear rules.* Currently, many governments (including our own) have mistakenly read the NPT as entitling nations to a per se right to any nuclear activity, no matter how uneconomic or unsafeguardable it is. This has bedeviled our dealings with nations such as

Iran. In fact, a proper understanding of the negotiating history, law, and technology of safeguards makes it clear that there is no per se right to engage in unbeneficial (read, money-losing) activities that can bring one within days or weeks of acquiring nuclear weapons. We already understand that sharing the potential benefits of peaceful nuclear explosives under the NPT has been a nonstarter because there are no economic benefits to using nuclear explosives for excavation and because a peaceful nuclear explosion is impossible to distinguish from a nuclear weapons test. The same economic and security discipline needs to be applied to the sharing of the benefits of the applications of peaceful nuclear energy.[8] So far, members of the NPT have not been so disciplined because they see the potential security benefits of acquiring a near nuclear-weapons option through development of peaceful nuclear power. If we are serious about preventing the spread of nuclear weapons, though, we should be much more active in smoking out this motive by being much stricter about economic rationales.

6. *We have always spoken about the need to meet certain economic criteria before developing large nuclear energy programs;* we need to do this more. The French, the United States, and the IAEA have all quietly noted that nuclear power programs only make sense for nations that have a large electrical grid, a major nuclear regulatory and science infrastructure, and proper financing. The British government, after an extensive analysis, concluded last year that if carbon emissions are properly priced (or taxed), then British nuclear power operators should be able to cover nearly all of their own costs without government support.[9] In the first Bush administration, U.S. officials rightly noted the absurdly negative economics for Iran of building

the Bushehr reactor, as well as the nuclear fuel making plant at Natanz, as compared to exploiting natural gas. Critics did the same to reverse U.S. policy in backing the building of large nuclear power plants in North Korea. U.S. bank analysts, meanwhile, are still divided over whether to invest heavily in U.S. nuclear power construction. They and the nuclear industry would feel more comfortable building new reactors if they were able to secure more government guarantees and subsidies for this work. Economic judgments and criteria, in short, have long been used by several key governments, private firms, and institutions in judging the merits of proposed nuclear projects. More can be done to more honestly cost these projects and to compare them against non-nuclear alternatives. Here, internationally, two places to start would be to back the principles contained in the Charter Energy Treaty and the Global Energy Charter for Sustainable Development. In concert, these international agreements encourage countries to open their energy sectors to fair competition and to state the full price of any energy option.[10]   In addition, it is not too early to consider what might be developed as a follow-on to the Kyoto Protocol after 2012. Whatever is finally agreed to here would be improved if it fostered the principles of full costing and international open-market competitions. This is also a set of principles that the G-20 ought to adopt.

7. *Promoting market-based nonproliferation is worthwhile, but it will not solve all problems.* Would a market-fortified NPT regime of this sort eliminate the problems already posed by a nuclear-ready Iran or a nuclear-armed North Korea? Unfortunately, the answer is no. Those problems can now only be dealt with by military, economic, and diplomatic efforts to

squeeze Iran and North Korea to transition to less hostile rule — such as those used against the Soviet Union during the Cold War. But the market-fortified system suggested would help prevent Iran's and North Korea's patently uneconomic ploys from becoming an international model of nuclear behavior for countries now professing an earnest desire to back peaceful nuclear power development. These countries include Indonesia, Libya, Saudi Arabia, South Korea, Nigeria, Egypt, Turkey, Morocco, Jordan, the UAE, and Yemen (all of which are bizarrely receiving active U.S. or IAEA nuclear cooperative technical assistance to complete their first large power stations). Also, unlike the situation under today's interpretation of the NPT, which ignores suspicious "civilian" nuclear undertakings even when they obviously lack any economic rationale, the market-fortified system described above would help to flag worrisome nuclear activities much sooner — well before a nation came anywhere near to making bombs. Such an approach, in short, would encourage an NPT-centered world worthy of the name, a world in which the NPT would clearly restrain the further spread of nuclear weapons-related technology rather than foster it.

## ENDNOTES - CHAPTER 14

1. This essay is drawn from testimony given before a hearing of The House Committee on Foreign Affairs, "Every State a Superpower? Stopping the Spread of Nuclear Weapons in the 21st Century," held on May 10, 2007, Rayburn House Office Building, Room 2172, Washington, DC.

2. See Victor Gilinsky, Harmon Hubbard, and Marvin Miller, *A Fresh Examination of the Proliferation Dangers of Light Water Reactors*, Washington, DC: The Nonproliferation Policy Education Center, October 22, 2004, available from *www.npec-web.org/files/20041022-GilinskyEtAl-LWR.pdf*.

3. See Office of Fuel Cycle Management, *Global Nuclear Energy Partnership Strategic Plan*, Washington, DC: U.S. Department of Energy, GNEP-167312, Rev.0, January 2007, p. 5, where the DoE notes that

> there is no technology 'silver bullet' that can be built into an enrichment plant or reprocessing plant that can prevent a country from diverting these commercial fuel cycle facilities to non-peaceful use. From the standpoint of resistance to rogue-state proliferation there are limits to the nonproliferation benefits offered by any of the advanced chemical separations technologies, which generally can be modified to produce plutonium. . . .

4. In 2006, 13 new, additional nations announced their intention to construct and operate large power reactors on their soil. To get some idea of how large a jump this is, one need only consider that that number constitutes a 42 percent increase in the number of nations (31) currently operating large reactors within their borders. The nations in question were Turkey, Egypt, Saudi Arabia, Libya, Yemen, Jordan, Vietnam, Australia, Bangladesh, Morocco, Tunisia, Indonesia, and Nigeria.

5. See e.g., Jeff Combs, Ux Consulting Company, "Price Expectations and Price Formation," presentation to the Nuclear Energy Institute International, Uranium Fuel seminar, October 2006; Tom Neff (MIT), "Uranium and Enrichment: Enough Fuel for the Nuclear Renaissance?" December 2006. See presentation to the North West Power Council, "New Nuclear Reactors," February 2007, available from *www.nwcouncil.org/news/2007/02/p1.pdf*.

6. See S1138, "The Nuclear Safeguards and Supply Act of 2007" introduced April 18, 2007 by Senators Lugar and Bayh, soon to be available from *thomas.loc.gov/cgi-bin/query/z?c110:S.1138*.

7. See, e.g., the comments of the chairman of the IAEA's Standing Advisory Group on International Safeguards, John Carlson, Australian Safeguards and Non-Proliferation Office, "Addressing Proliferation Challenges from the Spread of Uranium Enrichment Capability," Paper for the Annual Meeting of the Institute for Nuclear Materials Management, Tucson, July 8-12, 2007 (available

from NPEC upon request). Also see Paul Leventhal, "Safeguards Shortcomings: A Critique," Washington, DC: Policy Education Center, October 22, 2004, p. 38, available from *www.npec-web.org/ files/20041022-GilinskyEtAl-LWR.pdf*; Marvin Miller, "Are IAEA Safeguards in Plutonium Bulk-Handling Facilities Effective?" Washington, DC: NCI, August 1990; Brian G. Chow and Kenneth A. Solomon, *Limiting the Spread of Weapons-Usable Fissile Materials*, Santa Monica, CA: RAND, 1993, pp. 1-4; and Marvin Miller, "The Gas Centrifuge and Nuclear Proliferation," in Victor Gilinsky *et al., A Fresh Examination of the Proliferation Dangers of Light Water Reactors*, Washington, DC: The Nonproliferation Policy Education Center, 2004.

8. On the proper reading of the NPT, see Eldon V.C. Greenberg, "NPT and Plutonium: Application of NPT Prohibitions to 'Civilian' Nuclear Equipment, Technology and Materials Associated with Reprocessing and Plutonium Use," Washington, DC: Nuclear Control Institute, 1984 (Revised May 1993); Paul Lettow, "Fatal Flaw? The NPT and the Problem of Enrichment and Reprocessing," unpublished essay, April 27, 2005; Henry D. Sokolski, "Clarifying and Enforcing the Nuclear Rules," prepared testimony before Weapons of Mass Destruction: Current Nuclear Proliferation Challenges, a hearing before the Committee on Government Reform's Subcommittee on National Security, Emerging Threats, and International Relations, U.S. House of Representatives, September 6, 2006 p. 3, fn. 2, available from *www.npec-web. org/files/20060921-FINAL-Sokolski-TestimonyHouseSubcommittee. pdf*; and Robert Zarate, "The NPT, IAEA Safeguards and Peaceful Nuclear Energy: An Inalienable Right But Precisely to What?" an essay presented to *Assessing the Ability of the IAEA to Safeguard Peaceful Nuclear Energy*, a conference held in Paris, France, and hosted by the French Ministry of Foreign Affairs, the *Fondation pour la Recherche Stratégique*, and NPEC, November 11-12, 2006, available from *www.npec-web.org/files/20070301-Zarate-NPT-IAEA -PeacefulNuclear.pdf*.

9. See *The Energy Challenge: Energy Review Report 2006*, London, UK: British Department of Trade and Industry, July 11, 2006, available from *www.dti.gov.uk/energy/review/* .

10. More on the current membership and investment and trade principles of the Energy Charter Treaty is available from

*www.encharter.org/.* The second principle of the Global Energy Charter for Sustainable Development calls for:

> The establishment of guidelines and internationally standardized methods of evaluation for determining the external effects and total lifecycle costs and risks for all energy systems, taking into account the environmental, health and other damage caused by energy-related activities.

See *The Global Energy Charter for Sustainable Development,* available from *www.cmdc.net/echarter.html.*

# ABOUT THE CONTRIBUTORS

JAMES ACTON is an associate in the Nonproliferation Program at the Carnegie Endowment for International Peace. Prior to joining the Carnegie Endowment in October 2008, Acton was a lecturer at the Centre for Science and Security Studies in the Department of War Studies at King's College London, where he conducted the research for this project. He has also worked as the science and technology researcher at the Verification Research, Training and Information Centre (VERTIC), where he was a participant in the UK–Norway dialogue on verifying the dismantling of warheads. He is currently the joint UK member of the International Panel on Fissile Materials. Dr. Acton co-authored the Adelphi Paper, *Abolishing Nuclear Weapons*, and co-edited the follow-up book, *Abolishing Nuclear Weapons: A Debate* (both with George Perkovich). He has published widely on topics related to nonproliferation and disarmament in such journals as the *Bulletin of the Atomic Scientists*, *Nonproliferation Review*, and *Survival*. Dr. Acton holds a Ph.D. in theoretical physics from the University of Cambridge.

WYN BOWEN is Professor of Non-Proliferation & International Security, and Director of the Centre for Science & Security Studies, in the Department of War Studies at King's College London. In 1994, he spent 5 months as a Center Associate of the Matthew B. Ridgway Center for International Studies, Graduate School of Public and International Affairs, University of Pittsburgh. From September 2005 until August 2007, he was Professor of International Security in the Defence Studies Department at King's College. In 1997-

98 he served as a weapons inspector on several missile teams in Iraq with the UN Special Commission and has also worked as a consultant to the International Atomic Energy Agency. He served as a Specialist Advisor to the House of Commons' Foreign Affairs Committee for inquiries into "The Decision to go to War with Iraq" (2003) and "Weapons of Mass Destruction" (2000). Dr. Bowen holds a BA (Hons) from the University of Hull, and an MA and Ph.D. from the University of Birmingham.

PETER BRADFORD is a former member of the United States Nuclear Regulatory Commission and former chair of the New York and Maine Utility Regulatory Commissions. He has taught at the Yale School of Forestry and Environmental Studies and currently is an Adjunct Professor at Vermont Law School teaching "Nuclear Power and Public Policy." A member of the China Sustainable Energy Policy Council, he served on a recent panel evaluating the reliability of the Vermont Yankee nuclear power plant; on the European Bank for Reconstruction and Development Panel advising how best to replace the remaining Chernobyl nuclear plants in Ukraine; a panel on the opening of the Mochovce nuclear power plant in Slovakia; and the Keystone Center collaborative on nuclear power and climate change; and is Vice Chair of the Board of The Union of Concerned Scientists. Professor Bradford is the author of "Fragile Structures: A Story of Oil Refineries, National Securities and the Coast of Maine" and many other articles. He is a graduate of Yale University and the Yale Law School.

SIMON CARROLL is an independent environmental consultant, with a specialist focus on matters related to nuclear energy, including nuclear liability and compensation. As a senior advisor with Greenpeace International, he participated in the negotiations of the International Atomic Energy Agency (IAEA) Standing Committee on Nuclear Liability on the 1997 Protocol to amend the Vienna Convention on Civil Liability for Nuclear Damage and the 1997 Convention on Supplementary Compensation. In March 2009, he was appointed to the United Kingdom's newly-created Nuclear Liabilities Financing Assurance Board (NLFAB), a body which provides independent advice on financial arrangements for the decommissioning of nuclear reactors and radioactive waste management in the UK.

ANTONY FROGGATT has worked as a freelance consultant on energy and nuclear issues in the European Union (EU) and neighbouring states since 1997. He has worked at length on EU energy policy issues for European Governments, the European Commission and Parliament, and commercial bodies. He has also worked extensively with environmental groups and public bodies in Central Europe and neighboring states, particularly in the run up to enlargement, assisting in the development of policies, initiatives and capacity building. He is also a regular speaker at conferences, universities, and training programs across Europe and is a Senior Research Fellow at Chatham House. Prior to working freelance, Mr. Froggatt was employed for 9 years as a nuclear campaigner and coordinator for Greenpeace International.

STEVE KIDD is Director of Strategy & Research at the World Nuclear Association, the international association for nuclear energy based in London. After a brief period as an economics tutor at Sheffield University, he followed a career as an industrial economist with leading UK companies. These specialized in the raw materials and engineering sectors and included Rio Tinto and Rover Cars. He practiced as an independent consultant from 1990 onwards and then joined the former Uranium Institute as Senior Research Officer in 1995. He assumed his present position when the Institute became the World Nuclear Association in 2001. He organizes and teaches training courses for nuclear professionals in developing nuclear countries on behalf of the World Nuclear University, and is a frequent speaker at conferences and meetings around the world. Mr. Kidd is the author of the recent book, *Core Issues — Dissecting Nuclear Power Today*, and authored many articles on the commercial aspects of nuclear power. Mr. Kidd holds a bachelor's and a master's degree in economics from Queens' College and the University of Cambridge, respectively, and was the winner of the Adam Smith Prize.

DOUG KOPLOW is the founder of Earth Track in Cambridge, MA, which focuses on making the scope and cost of environmentally harmful subsidies more visible and identifying reform strategies. His work on natural resource subsidies spans nearly 20 years and has been widely cited across the political spectrum. Clients have included the Organization for Economic Cooperation and Development, the United Nations Environment Program, Greenpeace, the Alliance to Save Energy, the International Institute for Sustainable Development, the U.S. National Commission on

Energy Policy, and the U.S. Environmental Protection Agency. Mr. Koplow's most recent work has been on U.S. subsidies to biofuels and the nuclear fuel cycle. His analysis documents the many ways these sectors are socializing their investment risks and quantifies the high subsidy cost per unit energy produced or ton of greenhouse gas avoided. Mr. Koplow holds an MBA from the Harvard Graduate School of Business Administration and a BA in economics from Wesleyan University.

AMORY LOVINS, a MacArthur Fellow and consultant physicist, is among the world's leading innovators in energy and its links with resources, security, development, and environment. He has advised the energy and other industries for more than 3 decades as well as the U.S. Departments of Energy and Defense. His work in 50+ countries has been recognized by the "Alternative Nobel," Blue Planet, Volvo, Onassis, Nissan, Shingo, Goff Smith, and Mitchell Prizes, the Benjamin Franklin and Happold Medals, 10 honorary doctorates, honorary membership of the American Institute of Architects, Foreign Membership of the Royal Swedish Academy of Engineering Sciences, honorary Senior Fellowship of the Design Futures Council, and as a recipient of the Heinz, Lindbergh, Jean Meyer, Time Hero for the Planet, Time International Hero of the Environment, Popular Mechanics Breakthrough Leadership, and World Technology Awards. A former Oxford don, he advises major firms and governments worldwide and has briefed 19 heads of state. Mr. Lovins cofounded and is Chairman and Chief Scientist of Rocky Mountain Institute, an independent, market-oriented, entrepreneurial, nonprofit, nonpartisan think-and-do tank that creates abundance by design.

Much of its pathfinding work on advanced resource productivity (typically with expanding returns to investment) and innovative business strategies is synthesized in *Natural Capitalism* (1999, with Paul Hawken and L. H. Lovins). This intellectual capital provides most of RMI's revenue through private-sector consultancy that has served or been invited by more than 80 Fortune 500 firms, lately redesigning more than $30 billion worth of facilities in 29 sectors. In 1992, RMI spun off E SOURCE, and in 1999, Fiberforge Corporation, a composites engineering firm that he chaired until 2007; its technology, when matured and scaled, will permit cost-effective manufacturing of the ultra-light-hybrid Hypercar® vehicles he invented in 1991. Mr. Lovins is the author of 29 books, including *Small Is Profitable: The Hidden Economic Benefits of Making Electrical Resources the Right Size* (2002, an Economist book of the year blending financial economics with electrical engineering) and the Pentagon-cosponsored *Winning the Oil Endgame* (2004, a roadmap for eliminating U.S. oil use by the 2040s, led by business for profit). His most recent visiting academic chair was in spring 2007 as MAP/Ming Professor in Stanford's School of Engineering, offering the University's first course on advanced energy efficiency.

MYCLE SCHNEIDER works as independent international energy and nuclear policy consultant. He is a member of the International Panel on Fissile Materials (IPFM), based at Princeton University, and teaches within the International MSc in Project Management for Environmental and Energy Engineering at the Ecole des Mines in Nantes. Since 2000, he is an advisor to the German Ministry of the Environment. Between 1998 and 2003 he was an advisor to the French

Environment Minister's Office and to the Belgian Minister for Energy and Sustainable Development. Between 1983 and April 2003, he was executive director of the energy information service WISE-Paris. He has given evidence and held briefings at Parliaments in Australia, Belgium, France, Germany, Japan, South Korea, Switzerland, United Kingdom (UK), and at the European Parliament. He has lectured extensively including at universities and engineering schools in various countries. Media representatives from around the world have inquired for his information, advice, or complete features, including many TV and radio stations, electronic, and print media. His numerous publications cover the analysis of nuclear proliferation, security, and safety, as well as environmental and energy planning issues. In 1997, he was honored with the Right Livelihood Award ("Alternative Nobel Prize") together with Jinzaburo Takagi for their work on plutonium issues. Mr. Schneider is the co-editor of *International Perspectives on Energy Policy and the Role of Nuclear Power* (Multi-Science Publishing, 2009), and lead author of *The World Nuclear Industry Status Report 2009*, commissioned by the German Environment Ministry (August 2009).

HENRY SOKOLSKI is the Executive Director of the Nonproliferation Policy Education Center (NPEC), a Washington, DC-based nonprofit organization founded in 1994 to promote a better understanding of strategic weapons proliferation issues among policymakers, scholars, and the media. He currently serves as an adjunct professor at the Institute of World Politics in Washington, DC, and as a member of the Congressional Commission on the Prevention of Weapons of Mass Destruction Proliferation and Terrorism. He previously served as Deputy for Nonproliferation Policy in the Department of Defense, for which he received a medal for outstanding public service from Secretary of Defense Dick Cheney. He also worked in the Office of the Secretary of Defense's Office of Net Assessment, as a consultant to the National Intelligence Council, and as a member of the Central Intelligence Agency's Senior Advisory Group. In the U.S. Senate, he served as a special assistant on nuclear energy matters to Senator Gordon Humphrey (R-NH), and as a legislative military aide to Dan Qualye (R-IN). Mr. Sokolski has authored and edited a number of works on proliferation, including *Best of Intentions: America's Campaign Against Strategic Weapons Proliferation* (Westport, CT: Praeger, 2001), *Nuclear Heuristics: Selected Writings of Albert and Roberta Wohlstetter* (Strategic Studies Institute, U.S. Army War College, 2009), *Falling Behind: International Scrutiny of the Peaceful Atom* (Strategic Studies Institute, U.S. Army War College, 2008); *Pakistan's Nuclear Future: Worries Beyond War* (Strategic Studies Institute, U.S. Army War College, 2008); *Gauging U.S.-Indian Strategic Cooperation* (Strategic Studies Institute, U.S. Army War College, 2007); *Getting Ready for a Nuclear-Ready Iran* (Strategic Studies Institute, U.S. Army War College, 2005); and *Getting MAD: Nuclear Mutual Assured Destruction, Its Origins and Practice* (Strategic Studies Institute, U.S. Army War College, 2004).

SHARON SQUASSONI serves as director and senior fellow of the Proliferation Prevention Program at the Center for Strategic and International Studies (CSIS). Prior to joining CSIS, she was a senior associate in the Nuclear Nonproliferation Program at the Carnegie Endowment for International Peace. From 2002-07, she advised Congress as a senior specialist in weapons of mass destruction at the Congressional Research Service, Library of Congress. Before joining CRS, she worked briefly as a reporter in the Washington bureau of Newsweek magazine. Ms. Squassoni also served in the executive branch of government from 1992 to 2001. Her last position was as Director of Policy Coordination for the Nonproliferation Bureau at the State Department. She also served as a policy planner for the Political-Military Bureau at State. She began her career in the government as a nuclear safeguards expert in the Arms Control and Disarmament Agency. She is the recipient of various service awards and has published widely. She is a frequent commentator for U.S. and international media outlets. Ms. Squassoni holds a BA in political science from the State University of New York at Albany, a master's in public management from the University of Maryland, and a master's in national security strategy from the National War College.

JOHN STEPHENSON is a Project Manager at Dalberg Global Development Advisors, a strategy consulting firm focused on international development. He has consulted to the senior management teams of leading international financial institutions, multilateral development organizations, foundations, and multinational corporations on strategy, organizational effectiveness, stakeholder and change management, and development policy. He has experience in several development sectors, including energy and the environment, access to finance, health, private sector development, post-conflict reconstruction, and governance and public sector reform. Some of his most recent engagements include: (1) evaluating fund manager proposals and conducting due diligence as part of a $500 million global call for Renewable Energy funds in emerging markets; (2) serving as a strategic advisor on an innovative $50 million fund for post-conflict countries; (3) working with the United Nations Foundation and Vodafone Group Foundation on their public-private partnership in mobile health and emergency response; and (4) assisting the East African Community to formulate an energy access scale-up strategy to support attainment of the Millennium Development goals with a focus on alternative energy sources. Prior to joining Dalberg, he worked at the World Bank where he participated in the formulation of the Bank's Country Assistance Strategy for the Democratic Republic of Congo. Mr. Stephenson holds a bachelor's degree magna cum laude in government and East Asian studies from Harvard University, and a master's degree from Georgetown University's School of Foreign Service.

STEPHEN THOMAS is Professor of Energy Policy at the Public Services International Research Unit in the University of Greenwich, London, where he leads the energy research. He has worked as an independent energy policy researcher for more than 20 years. From 1979-2000, he was a member of the Energy Policy Programme at SPRU, University of Sussex, and in 2001, he spent 10 months as a visiting researcher in the Energy Planning Programme at the Federal University of Rio de Janeiro, Brazil. He was a member of the team appointed by the European Bank for Reconstruction and Development to carry out the official economic due diligence study for the project to replace the Chernobyl nuclear power plant (1997). He was a member of an international panel appointed by the South African Department of Minerals and Energy to carry out a study of the technical and economic viability of a new design of nuclear power plant, the Pebble Bed Modular Reactor (2001-02). He was part of an independent team appointed by Eletronuclear (Brazil) to carry out an assessment of the economics of completing the Angra dos Reis 3 nuclear power plant (2002). Mr. Thomas has a BSc in chemistry from Bristol.

PETER TYNAN is a Partner at Dalberg Global Development Advisors, a boutique global strategy and policy advisory firm focused on global issues and emerging markets. He leads the Global Access to Finance Practice Group and works in Dalberg's Washington, DC, office. He has advised, governments, corporations and development institutions in strategy and policy, organizational reform, and finance and energy issues. He has advised the Overseas Private Investment Corporation in placing $500m in renewable energy investments, and the Asian Development Bank in placing $100m in renewable energy investments in Asia. He is also advising a number of clients on regional renewable energy policy and investment. Prior to joining Dalberg, he advised the Minister of Finance in the Democratic Republic of the Congo, and the Minister of Finance in the Republic of Egypt. He has worked with multiple U.S. Government agencies, including for the Administrator and CFO of the General Services Administration (GSA) in strategy, strategic planning, and the reorganization of the GSA. He previously worked in private equity, where he sourced and evaluated middle market private equity investments. Mr. Tynan is the co-author of *India's Integration into the Global Economy: Lessons and Opportunities for Latin America and the Caribbean* (Inter-American Development Bank, 2009); "Will the U.S.-India Civil

Nuclear Cooperation Light India?" in *Gauging US-Indian Strategic Cooperation* (Strategic Studies Institute, U.S. Army War College, 2006); and *Imagining Australia: Ideas For Our Future* (Allen & Unwin, 2004). He regularly speaks on access to finance, renewable energy, and SME financing issues. Mr. Tynan holds a Bachelor in Business with First Class Honours degree and the University Medal from the University of Technology in Sydney, Australia; a master's in public policy from the Kennedy School of Government at Harvard University, and an MBA from Harvard Business School.

FRANK VON HIPPEL, a theoretical physicist, is a Professor of Public and International Affairs and co-principal investigator with Harold Feiveson of Princeton's research program on Science and Global Security. From September 1993 through 1994, he was on leave as Assistant Director for National Security in the White House Office of Science and Technology Policy, and played a major role in developing U.S.-Russian cooperative programs to increase the security of Russian nuclear-weapon materials. He is an ex-chair of the American Physical Society's Panel on Physics and Public Affairs. He chairs the editorial board of *Science & Global Security*. During the following 10 years, while his research focus was in theoretical elementary-particle physics, he held research positions at the University of Chicago, Cornell University, and Argonne National Laboratory, and served on the physics faculty of Stanford University. In 1974, Dr. von Hippel's interests shifted to "public-policy physics." After spending a year as a Resident Fellow at the National Academy of Science, during which time he organized the American Physical Society's Study on Light Water Reactor Safety, he was invited to join the research and in 1984 the teaching faculty of Princeton University. During the late 1970s, his research focused on technical questions relating to the containment and mitigation of nuclear-reactor accidents, alternatives to recycling plutonium in nuclear-reactor fuel, and the potential for major improvements in automobile fuel economy. Since the early 1980s, his research has focused on developing the analytical basis for deep cuts in the U.S. and Soviet/Russian nuclear stockpiles, and removal of their nuclear missiles off launch on warning alert; verifying nuclear warhead elimination, a universal cutoff of the production of weapon-usable fissile materials and the phasing out of their use in nuclear reactor fuel; and a comprehensive nuclear-warhead test ban. Dr. von Hippel has served on advisory panels to the Congressional Office of Technology Assessment, U.S. Department of Energy, National Science Foundation, and U.S. Nuclear Regulatory Commission, and on the boards of directors of the American Association for the Advancement of Science and the *Bulletin of the Atomic Scientists*. For many years, he was the elected chairman of the Federation of American Scientists. In 1977, Dr. von Hippel shared with Joel Primack the American Physical Society's 1977 Forum Award for Promoting the Understanding of the Relationship of Physics and Society for their book, *Advice and Dissent: Scientists in the Political Arena*. In 1989, he was awarded the Federation of American Scientists' Public Service Award for serving as a "role model for the public interest scientist." In 1991, the American Institute of Physics published a volume of von Hippel's selected works under the title *Citizen Scientist*, as one of the first three books in its "Masters of Physics" series. In 1993, he was awarded a 5-year MacArthur Prize fellowship. In 1994, he received the American Association for the

Advancement of Sciences' Hilliard Roderick Prize for Excellence in Science, Arms Control and International Security. Dr. von Hippel holds a BS in physics from MIT in 1959 and D.Phil. in theoretical physics from Oxford, where he was a Rhodes Scholar.